Arsenic Pollution Control in Nonferrous Metallurgy

Li-Yuan Chai
Editor

# Arsenic Pollution Control in Nonferrous Metallurgy

Springer

*Editor*
Li-Yuan Chai
School of Metallurgy and Environment
Central South University
Changsha, Hunan, China

Chinese National Engineering Research
Center for Control and Treatment of Heavy
Metal Pollution (CNERC-CTHMP)
Changsha, Hunan, China

ISBN 978-981-13-6720-5 ISBN 978-981-13-6721-2 (eBook)
https://doi.org/10.1007/978-981-13-6721-2

Library of Congress Control Number: 2019931868



This Springer imprint is published by the registered company Springer Nature Singapore Pte Ltd.
The registered company address is: 152 Beach Road, #21-01/04 Gateway East, Singapore 189721, Singapore

# Preface

Arsenic is an abundant element in the Earth's crust and occurs in more than 300 minerals, generally in combination with sulfur and metals with the formula MAsS and $MAs_2$ (M=Fe, Ni, Co) to form arsenopyrite. Geogenic and anthropogenic activities can liberate arsenical compounds that range in toxicity, mobility, and bioavailability from arsenic-bearing minerals to the environment. Arsenic is highly toxic and adversely affects human health. It has been identified as the number one environmental chemical of concern according to the World Health Organization.

Arsenic-containing minerals that occur in base metal ores and/or concentrates enter metallurgical processing circuits accordingly. Arsenic is found in the flue gas, dust, slag, residue, and sludge discharged from the copper, gold, and lead smelting process. Due to human activities, mainly through mining and smelting, arsenic can now be found in many more places than where they never existed before. Therefore, arsenic pollution has become a worldwide environmental challenge especially in the metal industry. Hence, arsenic in the process streams must be treated and/or immobilized properly prior to the discard of waste.

This book helps readers to gain an understanding of the following aspects of the science and technology of arsenic pollution control, especially for the metallurgical industry. It is composed of five chapters that concern the principles and technologies of arsenic pollution control.

Chapter 1 describes a basic overview of the mineralogy, distribution, pollution source, and characteristics of arsenic pollution.

Chapter 2 provides the behavior, pollution source distribution, and environmental influence of arsenic in typical lead smelters.

Chapter 3 deals with arsenic behaviors and pollution control technologies in aqueous solution. Redox behavior and chemical species of arsenic in acidic aqueous system, photochemical oxidation of trivalent arsenic, and the molecular reaction mechanism, formation mechanism, and characteristics of tooeleite are stated. Cascade sulfide precipitation and separation of copper and arsenic from high-arsenic acid wastewater are well detailed. Finally, the chapter introduces a new process to remove arsenic efficiently by $Fe_3O_4$ hierarchical particles via adsorption in aqueous solution.

Chapter 4 gives a state of arsenic behaviors and pollution control technologies for solid waste. This chapter particularly expatiates on the stabilization, solidification, and vitrification technologies; most of the technologies have been commercialized by at least one nonferrous metal smelter.

Chapter 5 mainly presents the processes of recovery of valuable resources from arsenic-bearing anode slime from lead smelting, separation of copper and arsenic in arsenic-bearing materials; the results demonstrate the efficiency of the proposed technologies, which are the fundamental of the clean utilization of arsenic-bearing materials.

This book contains the potential research and the industrial application. It provides an overview to researchers, graduate, and undergraduate students, as well as academicians who get interested in arsenic control in metallurgical industry. This book could be attractive to scientists working in the field of arsenic pollution control.

The book is written based on our working experience in the field of arsenic pollution control for more than 15 years. In addition, much of this book would not have been possible without many graduate students (Jin-Qin Yang, Meng-Qing Yue, Guo-Min Jiang, Ting Wang, Ting-Ting Song, Suo Dai, Jie Lei, Mi Wang, Zong-Wen Zhao, Jing-Jing Ma, Yuan-Cheng Li, De-Gang Liu, Chen Shen, et al.) who contributed technically and colleagues with whom we have collaborated over the years. I would like to thank all the contributors who participate in the above-mentioned chapters.

These researches were supported by National Key R&D Program of China (2017YFC0210400, 2018YFC1900301, and 2018YFC1903301), Special Program on Environmental Protection for Public Welfare (201509050), Key project of National Natural Science Foundation of China (51634010), and Natural Science Foundation of China (51304251, 51474247, 51774338, and 51704337).

Changsha, China Li-Yuan Chai

# Contents

# Contributors

**Li-Yuan Chai** School of Metallurgy and Environment, Central South University, Changsha, Hunan, China;
Chinese National Engineering Research Center for Control and Treatment of Heavy Metal Pollution (CNERC-CTHMP), Changsha, Hunan, China

**Yong-Ming Chen** School of Metallurgy and Environment, Central South University, Changsha, Hunan, China

**Yong Ke** School of Metallurgy and Environment, Central South University, Changsha, Hunan, China

**Qing-Zhu Li** School of Metallurgy and Environment, Central South University, Changsha, Hunan, China

**Yan-Jie Liang** School of Metallurgy and Environment, Central South University, Changsha, Hunan, China

**Xiao-Bo Min** School of Metallurgy and Environment, Central South University, Changsha, Hunan, China

**Mei-Qing Shi** Chinese National Engineering Research Center for Control and Treatment of Heavy Metal Pollution (CNERC-CTHMP), Changsha, Hunan, China

**Hai-Ying Wang** School of Metallurgy and Environment, Central South University, Changsha, Hunan, China

**Qing-Wei Wang** School of Metallurgy and Environment, Central South University, Changsha, Hunan, China

**Yun-Yan Wang** School of Metallurgy and Environment, Central South University, Changsha, Hunan, China

**Wei-Chun Yang** School of Metallurgy and Environment, Central South University, Changsha, Hunan, China

**Wei-Zhi Zeng** School of Metallurgy and Environment, Central South University, Changsha, Hunan, China

# Chapter 1
# Arsenic Distribution and Pollution Characteristics

**Yun-Yan Wang, Li-Yuan Chai and Wei-Chun Yang**

Arsenic is a chemical element symboled "As", and ranked 33rd on the periodic table. It is abundant in the Earth's crust and has been found in more than 300 minerals in nature. Arsenic present in the minerals is usually mobilized through geogenic and anthropogenic activities.

Arsenic is viewed as being synonymous with toxicity and listed as the most hazardous substance according to the USEPA agency for toxic substance and disease registry. Chronic exposure to arsenic causes adverse health effects such as skin lesions, hyperkeratosis, and cancer. Anthropogenic sources of arsenic pollution originate in several industries, such as refining or smelting of metal ores, microelectronics, wood preservation, and battery manufacturing. At present, arsenic contamination is a growing global concern. As one of the most prevalent toxic elements in the environment, the toxicity, mobility, and fate of arsenic in the environment are determined by a series of complicated processes including mineralogy, chemical speciation, and biological processes [1].

Y.-Y. Wang (✉) · L.-Y. Chai · W.-C. Yang
School of Metallurgy and Environment, Central South University,
Changsha, Hunan, China
e-mail: wyy@csu.edu.cn

L.-Y. Chai
e-mail: lychai@csu.edu.cn

W.-C. Yang
e-mail: yang220222000@yahoo.com.cn

L.-Y. Chai
Chinese National Engineering Research Center for Control and Treatment
of Heavy Metal Pollution (CNERC-CTHMP), Changsha, Hunan, China



L.-Y. Chai (ed.), *Arsenic Pollution Control in Nonferrous Metallurgy*,
https://doi.org/10.1007/978-981-13-6721-2_1

## 1.1 Mineralogy and Distribution of Arsenic

### *1.1.1 Arsenic-Containing Minerals*

Arsenic comprises about 2–5 mg $kg^{-1}$ of the Earth's crust. As a compound with other elements such as oxygen, chlorine, and sulfur, arsenic is widely distributed in minerals and ores that contain copper or lead.

There are hundreds of arsenic minerals known in nature. Among them, approximately 60% are arsenates, approximately 20% are sulfides and sulfosalts, 10% are oxides, and the rest are arsenites, arsenides, native elements, and metal alloys [2].

In the natural environment, arsenic is present as a major constituent of many sulfide and oxide minerals such as realgar ($As_4S_4$), orpiment ($As_2S_3$), arsenopyrite (FeAsS), arsenolite ($As_2O_3$), and scorodite ($FeAsO_4 \cdot 2H_2O$) [3].

High-arsenic concentrations have been found in many oxide minerals and hydrous metal oxides, either as part of their periodic structure or as the sorbed and occluded species. Iron oxides are particularly well known to accumulate arsenic up to contents of several weight percentages. Arsenic (as As(III) or As(V)) can substitute for P(V), Si(IV), Al(III), Fe(III), and Ti(IV) in various mineral structures and is therefore present in many rock-forming minerals, albeit at much lower concentrations. The element is primarily concentrated in sulfide minerals where it can occur as an arsenide or sulfarsenide anion bound to transition metals, e.g., lollingite ($FeAs_2$), arsenopyrite (FeAsS), or more rarely in minerals where arsenic forms nominally a cation, e.g., realgar ($As_4S_4$). Despite a large number of arsenic minerals, the largest reservoir of arsenic in crustal rocks is probably pyrite which contains trace to minor contents of this element [1].

The most important primary arsenic-bearing minerals are those where the arsenic occurs as the anion (arsenide) or dianion (diarsenide), or as the sulfarsenide anion (s); these anions are bonded to metals such as Fe (Lollingite, arsenopyrite), Co (cobaltite), and Ni (gesdorffite).

Although arsenic is not easily substituted into the main diagenetic mineral structure, it is easily to be a minor component in the abundant iron sulfide mineral pyrite. When the primary minerals are exposed to the atmosphere and surface or groundwaters, alteration reactions cause the formation of secondary arsenic minerals, such as simple arsenic oxides or more complex phased with arsenic, oxygen, and various metals. The secondary minerals comprise arsenite and arsenate minerals that are formed by linking As(III)-oxo-anion groups or As(V)-oxo-anion groups, respectively, to a variety of mono-, di-, and trivalent metal cations. Secondary arsenite minerals are rare in natural environments, usually occurring as the products of hydrothermal alterations under mildly reducing conditions. Conversely, secondary arsenate compounds comprise a large class of minerals that have been found in many oxidized environments.

A large number of secondary arsenic minerals have been identified and inferred in both natural high-arsenic geochemical anomalies and old mining and industrial sites contaminated by arsenic. The genesis of these secondary arsenic minerals is dependent on the original ore mineralogy, and the processes that have occurred during the formation of the mineral (particularly redox and pH changes).

Under acidic conditions of the Fe(III)–As(V) system, scorodite is by far the most common secondary arsenic mineral, usually originating from arsenopyrite or arsenic-bearing pyrite oxidation. Scorodite is the main secondary arsenic mineral in many environments, e.g., natural weathered mineralized rocks [4], various types of naturally contaminated soils in the different climatic. Tooeleite has similar characteristics to scorodite and is thus known as the trivalent arsenic form of scorodite. Otherwise, Tooeleite is the first arsenite sulfate mineral [5], which was found on waste dumps of the US Mine at Gold Hill, Tooele County, Utah [6].

### 1.1.2 Distribution of Arsenic Minerals

Arsenic, the twentieth most abundant element, has become an anathema over the world, particularly for Asian countries, viz., Bangladesh, India. China, and Japan. Arsenic scarcely exists as a native element in nature because it exhibits a high affinity to form chemical bonds with other elements and species. Arsenic, arsenide, and arsenosulfides are typically found in anoxic hydrothermal ore deposits and metamorphic and intrusive igneous rocks. The arsenic mineral resources in Chile, the United States, Canada, Mexico, and Philippines account for about half of the world's arsenical reserves, and other countries with rich arsenic rescurces include France, Sweden, Namibia, Peru, and so on. Table 1.1 displays the main arsenic minerals and their distribution.

In China, arsenic-bearing minerals were mainly distributed in the central-southern and western regions, especially Guangxi province, Yunnan province, and Hunan province, accounting for 61.6% of the total proven reserves, these arsenic-containing minerals coexisted with other metallic sulfide ores.

**Table 1.1** The main arsenic minerals and their distribution

| Arsenic minerals | Distribution |
|---|---|
| Orpiment | Cili County, Hunan, China; Nanhua County, Yunnan, China; Romania; Free State of Saxony, Germany, etc. |
| Realgar | Shimen county, Hunan, China, etc. |
| Arsenopyrite | Freiberg, Germany; Cornwall, England; Cobalt, Canada; Hunan, Xiangxi, Yunnan, China, etc. |
| Arsenic-bearing copper ore | USA; Canada; Mexico; Chile; Sweden |
| Liroconite | Cornwall, England; Huanaco, Chile; Southern Bolivia; Arizona, USA, etc. |
| Enargite | Peru; Chile; Argentina; Luzon; USA, etc. |

http://baike.asianmetal.cn/metal/as/resources&production.shtml. Accessed 27 Nov. 2018

Arsenic minerals are mainly arsenopyrite, realgar, and orpiment, in some oxidation zones of sulfide deposits is arsenate mineral. China has the world's unique realgar ($As_4S_4$) resources, widely distributed in Hunan, Guizhou, Sichuan, Yunnan, and other provinces, and Hunan Shimen realgar mine is the largest realgar mine in the world.

In 2014, China was the top producer of white arsenic with almost 70% world share, followed by Morocco, Russia, and Belgium, according to the British Geological Survey and the United States Geological Survey. Most of the arsenic refinement operations in the US and Europe have closed over environmental concerns. Arsenic is found of the smelter dust from copper, gold, and lead smelters, and is recovered primarily from copper refinement dust.

### *1.1.3 Transformation Behavior of Arsenic Mineralogical Phase*

Arsenic appears in the Earth's crust at an average concentration of 2–5 mg $kg^{-1}$, with low levels commonly found in the air, water, and soil [7]. The most common form of arsenic is arsenite and arsenate compounds. Arsenic is usually very toxic, especially in the cases of inorganic arsenic compounds such as arsenic trichloride and arsenous acid. The toxicity of arsenic follows the order of inorganic As (III) > organic As(III) > inorganic As(V) > organic As(V) > elemental Arsenic. Arsenite is 60 times more toxic than arsenate.

Arsenic is unique among the trace metalloids and oxyanion-forming trace metals (e.g., As, Se, Sb, Mo, V, Cr, U, and Re). It is sensitive to mobilization at pH values typical of natural waters (pH 6.5–8.5) under both oxidizing and reducing conditions. Arsenic can occur in the environment in several oxidation states (−3, 0, +3, and +5), but in natural waters, it is mostly found in inorganic forms as oxyanions of trivalent arsenite or pentavalent arsenate. Organic forms of arsenic are produced by biological activity, mostly in surface waters, but are rarely quantitatively important. Organic arsenic compounds may, however, occur where waters are significantly impacted by industrial pollution [8].

Arsenic may be solubilized by various processes. When pH is high, arsenic may be released from surface binding sites that lose their positive charge. When water level drops and sulfide minerals are exposed to air, arsenic trapped in sulfide minerals can be released into the water. When organic carbon is present in water, bacteria are fed by directly reducing As(V) to As(III) or by reducing the element at the binding site, releasing inorganic arsenic [9].

The aquatic transformations of arsenic are affected by pH, reduction–oxidation potential, organic matter concentration, and the concentrations and forms of other elements, especially iron and manganese. The main factors are pH and the redox potential. Generally, the main forms of arsenic under oxic conditions are $H_3AsO_4$,

$H_2AsO_4^-$, $HAsO_4^{2-}$, and $AsO_4^{3-}$ at pH 2, 2–7, 7–11, and 11, respectively. Under reducing conditions, $H_3AsO_4$ is predominant at pH 2–9.

Oxidation and reduction affect the migration of arsenic in subsurface environments. Arsenite is the most stable soluble form of arsenic in reducing environments and arsenate, which is less mobile than arsenite, is dominant in oxidizing environments at neutral pH. Therefore, arsenic may be more mobile under reducing conditions. The reducing environment is also rich in organic matter, which may enhance the solubility of arsenic compounds. As a result, the adsorption of arsenic is reduced and dissolved arsenic accumulates in groundwater. That is why the arsenic content is higher in reducing environments than in oxidizing environments [10].

The presence of sulfur is another factor that affects the transformation of arsenic in natural water. Arsenic can precipitate when metal sulfides form. In this way, arsenic is removed from the water and its mobility decreases. When oxygen is present, bacteria oxidize reduced sulfur to generate energy, potentially releasing bound arsenic.

Redox reactions involving iron also appear to be essential factors in the fate of arsenic in aquatic systems. The reduction of iron oxyhydroxides plays a key role in the release of arsenic to water. So, arsenic can be enriched in water with elevated iron concentrations [11]. Under oxidizing conditions, arsenic can be mobilized from pyrite or iron oxides especially at elevated pH value. Under reducing conditions, arsenic can be mobilized by reductive desorption or dissolution when associated with iron oxides. The reductive desorption occurs under two circumstances. One is when arsenate is reduced to arsenite which adsorbs to iron oxides less strongly. The other results from a change in the charge on the mineral surface which leads to the desorption of bound arsenic [12].

Some species of bacteria catalyze redox transformations of arsenic. Dissimilatory arsenate-respiring prokaryotes (DARP) speed up the reduction of As(V) to As(III). DARP use As(V) as the electron acceptor of anaerobic respiration and obtain energy to survive. Other organic and inorganic substances can be oxidized in this process. Chemoautotrophic arsenite oxidizers (CAO) and heterotrophic arsenite oxidizers (HAO) convert As(III) into As(V). CAO combine the oxidation of As(III) with the reduction of oxygen or nitrate. They use obtained energy to fix produce organic carbon from $CO_2$. HAO cannot obtain energy from As(III) oxidation. This process may be an arsenic detoxification mechanism for the bacteria [13].

Fe, Al, and Mn oxides are known to be the primary minerals controlling the arsenic concentration due to their chemistry and abundance [8, 14]. Among them, Fe oxide minerals such as hydrous ferric oxide (HFO), goethite, and hematite are considered to be the most important sinks for arsenic. However, under reduced conditions, sulfide minerals such as pyrite ($FeS_2$) could play an important role in regulating the dissolved metal concentrations [15].

The environmental stability of the minerals is then dependent on the processes that have occurred during the post-formational development of the environment (e.g., changes in solution chemistry and pH due to consumption of primary sulfides, changes in redox due to variations of groundwater level, and reduction/increase). Scorodite, ferric arsenate, and arsenical ferrihydrite are important arsenic carriers occurring in a wide range of environments and are also common precipitates used by metallurgical industries to control arsenic in effluents [16].

## 1.2 Pollution Sources of Arsenic

The arsenic presented in environment derives from both natural and anthropogenic sources. Anthropogenic sources of arsenic contamination include industrial process, such as mining, smelting and coal-fired power plants, agricultural pesticides and chemicals for timber preservation.

### *1.2.1 Natural Pollution Sources*

In addition to human influence, many environmental arsenic problems are the result of mobilization under natural conditions such as biological activity, volcanic emissions, and so on. The main natural pollution sources of arsenic include minerals, rocks, sediments, soils, and the atmosphere [9, 17, 18].

**(1) Minerals**

There are more than 300 minerals that have been found, in which arsenic is a major constituent, such as realgar, orpiment, and so on. The existing forms of arsenic include elemental arsenic, arsenides, sulfides, oxides, etc. Among the minerals, arsenopyrite (FeAsS) is the most abundant arsenic mineral. Arsenopyrite derives from hydrothermal solution and formed under the temperature of 100 °C or higher, which is almost the earliest-formed mineral. Then, it generates rarer native arsenic and arsenian pyrite. And then, there are realgar and orpiment that generally generate. However, arsenopyrite is often present in ore, but it has relatively less abundant compared to arsenite pyrite. Therefore, the latter may be the most important source of arsenic in minerals. Arsenic-bearing mineral is the primary natural pollution sources of arsenic [19].

**(2) Rocks**

The rocks containing arsenic mainly include igneous, metamorphic, and sedimentary rocks. Arsenic concentrations in igneous rocks are usually low. Different igneous rock types usually have different arsenic concentrations, but this difference is relatively slight. Averages of these igneous rocks are less than 5 mg $kg^{-1}$ but higher than 1.5 mg $kg^{-1}$. The rocks that is higher than 5 mg $kg^{-1}$ are volcanic glasses, whose arsenic concentrations are 5.9 mg $kg^{-1}$. Arsenic concentrations in metamorphic rocks are similar to those of igneous rocks, which are about 5 mg $kg^{-1}$. Among metamorphic rocks, pelitic rocks have the highest concentrations with an average of 18 mg kg$^{-1}$. Sedimentary rocks have relatively higher concentrations of arsenic than the above two kinds of rocks, which are in the range of 5–10 mg kg$^{-1}$. For example, arsenic concentrations (about 13 mg kg$^{-1}$) in argillaceous rocks are often high. It is higher than the arsenic concentrations of sandstones (4.1 mg kg$^{-1}$). Coals and the bituminous also have high-arsenic concentrations, some of which can be up to 35,000 mg kg$^{-1}$.

**(3) Sediments**

There is no significant difference between the arsenic concentrations of the unconsolidated sediments and the concentration in the consolidated sediments. Among all types of sediments, concentrations of arsenic in muds and clays are usually higher than that in sands and carbonates, whose values are about 3–10 mg kg$^{-1}$. When pyrite or iron oxide is present, the values would be elevated. In mineralized areas, these phenomena are also common. For instance, if there are abundant sulfide minerals, placer deposits in streams will have high concentrations of arsenic.

**(4) Soils**

The main source of arsenic in soils is geological and depends to some extent on the concentration of the source rock. The baseline concentrations of arsenic in soils are usually between 5 and 10 mg kg$^{-1}$. Concentrations of arsenic in peats and bog soils are about 13 mg kg$^{-1}$, which are higher than the baseline concentrations, primarily due to increased prevalence of sulfide mineral phases under the reduced conditions. The oxidation of pyrite can form acid sulfate soils in sulfide-rich geography, which can also accumulate arsenic.

**(5) Atmosphere**

Arsenic can occur in the open atmosphere and the air of soils, sediments, and buildings. About 89–98.6% of the atmospheric arsenic is attached onto particles rather than existing as vapors. The natural pollution source of arsenic in the atmosphere comes mainly from volcanic eruptions. And, burning vegetation also releases a large amount of arsenic, especially forest fires. In addition, microbial activity in soils is also a natural source of atmospheric arsenic. Arsenic concentrations of atmosphere are usually low before anthropogenic influences, such as smelting and other industrial activities, which are around $10^{-5}$ to $10^{-3}$ μg m$^{-3}$. Finally, atmospheric arsenic can be moved to the water, soils, and other Earth's surface by wet and dry deposition.

### *1.2.2 Anthropogenic Pollution Sources*

Anthropogenic sources of arsenic pollution include mining and smelting operations, burning of fossil fuels, use of arsenical fungicides, herbicides and insecticides in agriculture, wood preservatives, and disposal of arsenic-containing wastes. Mining and smelting operations are an important source of arsenic pollution. Arsenic has long been a problem associated with the extraction of nonferrous metals. During mineral processing, some 60–90% of arsenic minerals entered the tailings in the processes of ore dressing, 0.05% mineral processing wastewater, and 15–40% the concentrate [20].

The extractive process of nonferrous metal is one of the most important sources of arsenic contamination [21–23]. In the smelting of copper or lead–zinc ore, larger amounts of acidic wastewater are discharged when the gaseous emission containing $SO_2$, As, Cu, Pb, Zn, and other impurities are washed with the diluted acid [24]. Moreover, wasted electrolyte with high concentration of arsenic is produced abundantly in copper and lead-zinc electrolyte processes [25, 26].

## 1.3 Arsenic Pollution Characteristics and Status in Environment

### *1.3.1 Arsenic Pollution in Groundwater*

Arsenic in groundwater exists primarily as oxyanions representing two oxidation states: arsenic As(III) (arsenite) and arsenic As(V) (arsenate). Both As (III) and As (V) exist within the pH range of 6–9. The predominant As(III) species are uncharged $H_3AsO_3$ while the primary arsenate species are monovalent $H_2AsO_4^-$ and divalent $HAsO_4^{2-}$. Geology and groundwater environment make one form, either As(III) or As(V) dominant. Although As(V) is thermodynamically favored in oxic waters and As(III) in anoxic waters, they can coexist in both types of waters. The variability of the arsenic concentration in groundwater is ascribed to the arsenic content of the aquifer and the varying dissolution/desorption processes releasing the arsenic from the solid phase into the liquid phase. Reductive dissolution of iron oxides is considered as the principal cause of arsenic release from aquifer sediments [27].

Arsenic contamination of groundwater in different parts of the world is an outcome of natural and/or anthropogenic sources, leading to adverse effects on human health and ecosystem. Millions of people from different countries are heavily dependent on groundwater containing an elevated level of arsenic for drinking purposes. The maximum contaminant level for arsenic in drinking water is set by the United States Environmental Protection Agency (USEPA) and the World Health Organization (WHO) is 0.01 mg $L^{-1}$ [28]. Arsenic contamination in groundwater poses a significant health risk to humans. About 140 million people worldwide drink groundwater containing unsafe levels of arsenic [29]. Many countries, including Argentina, Chile, China, West Bengal, Bangladesh, and Vietnam [30, 31] are the well-known high-arsenic groundwater areas.

Both Argentina and Chile are arid oxidizing environments. The surface water and groundwater from Administrative Region II of northern Chile have high concentrations of arsenic, which caused a lot of health problems, including skin-pigmentation changes, squamous-cell carcinoma (skin cancer). The main region of high-arsenic groundwater of Argentina is the Chaco-Pampean Plain. The primary reason for the high concentration of arsenic in this area is the arid conditions, under which silicate and carbonate weathering reactions are pronounced and the groundwater often has high pH values. In this case, a large amount of arsenic has been dissolved by the metal oxides in the sediments.

China, West Bengal, Bangladesh, and Vietnam all belong to reducing environments. Bangladesh and West Bengal rely heavily on groundwater to meet public drinking water needs. It is estimated that up to 30 million people in Bangladesh and 6 million in West Bengal are exposed to high-concentration arsenic in drinking water. In these contaminated areas, the formation of aquifer sediments and the presence of solid organic matter have contributed to the development of reducing conditions, in which arsenic is more likely to be mobilized. In China, high-arsenic concentrations groundwater has been found in Xinjiang [32], Shanxi [33], Ningxia, Jilin, Inner Mongolia [34], Taiwan, and so on. The groundwater in Hexigten Banner, Hetao basin, and Tumed basin of Inner Mongolia has high concentrations of arsenic that are more than 50 μg $L^{-1}$. The affected areas reach 3000 $km^2$, and more than 1 million people are under threat. The high-arsenic groundwater in the Hexigten Banner is mainly caused by the mining of arsenopyrite ore, while the reason for Hetao basin and Tumed basin is mainly the geological origin. China's high-arsenic groundwater is mainly found in the north, but it is also found in the Yangtze River delta [35] and Jianghan Plain [36].

### *1.3.2 Arsenic Pollution in Atmosphere*

**(1) Pathway for arsenic to atmosphere**

Generally, atmospheric arsenic inputs can be divided into the following parts: (1) terrestrial volcanic eruptions; (2) anthropogenic releases of arsenic; (3) volatilization of arsenic in soils on the land surface and wind erosion of soil; (4) volatility from ocean (marine droplets). Volcanic eruptions and weathering of arsenic-bearing minerals are important natural sources of arsenic in the atmosphere. Anthropogenic emissions of arsenic, such as nonferrous metal smelting, fossil fuel combustion, and the use of arsenic-containing insecticides, herbicides, and wood preservatives, are more than three times larger than that of natural emissions [37–39].

Arsenic enters the water or soil through precipitation and subsidence and then into the atmosphere. Its life cycle in the atmospheric environment is about 7–10 days. Only about 800–1740 tons of arsenic is present in the atmosphere. Among them, anthropogenic arsenic released the largest amount of arsenic, and the total anthropogenic arsenic released into the atmosphere reached 28,070 tons each year, accounting for more than 70% of the total. The distribution of arsenic in the air in the northern and southern hemispheres is not balanced, with the amount of arsenic remaining in the air of the northern hemisphere 1480 tons, which is much larger than that of the southern hemisphere (260 tons). The difference is related to the large land area in the northern hemisphere and to the emissions of the more densely populated and highly industrialized countries.

### Natural sources

Arsenic in remote areas mainly comes from natural sources, such as volcanic eruptions released into the atmosphere. The release amount is about 17,150 t $a^{-1}$, naturally occurred from forest fires, oil, and wood combustion emissions the amount is about 125–3345 t $a^{-1}$. The ocean accounts for 27 t $a^{-1}$ and the low-temperature activity of soil microbes accounts for 160–26,200 t $a^{-1}$ [40]. Weathered erosion release of arsenic minerals is another important natural source of atmospheric arsenic. The arsenic minerals in the environment generally coexist with sulfur, such as $As_2S_3$, FeAsS, AsS, CoAsS, $Cu_3AsS_4$, and NiAs and arsenic tetrahedrite ($Cu_{12}As_4S_{13}$), which are weathered, eroded, and released approximately 1980 t $a^{-1}$ arsenic into the atmosphere. In nature, some inorganic arsenic is released into the atmosphere through the methylation of microorganisms, such as methylated arsenic and dimethyl arsenic, or further biotransformed into inorganic arsenic or oxidized to arsine ($AsH_3$, $CH_3AsH_2$, $(CH_3)_2AsH$, and $(CH_3)_3As$). It volatiles and eventually enters the atmosphere in a gaseous form [41–43].

### Metal mining and smelting

Smelting of Cu, Ni, Pb, and Zn ores/concentrates is one of the most important arsenic sources. Approximately, 6200 t $a^{-1}$ arsenic is emitted globally, 80% of which comes from copper smelting industry [44].

### Fossil fuel combustion

Another important anthropogenic source of arsenic is the combustion of fossil fuels. Arsenic is an element which has a strong affinity to organic or inorganic components in coal. In the raw coal and lignite, the world average arsenic contents are (9.0 ± 0.8) and (7.4 ± 1.4) mg $kg^{-1}$, respectively. During the combustion of coal, arsenic is oxidized to $As_2O_3$ under high temperature, and volatilizes into flue gas. And, the higher the temperature benefits, the higher arsenic gasification rate. It is calculated that about 3.9% of arsenic in coal enters the flue gas.

### Use of arsenic-containing chemicals

Since 1890, inorganic arsenic compounds such as sodium arsenate have been widely used as herbicides before the alternatives emerge. Approximately 8000 t $a^{-1}$ arsenic is used, and about 3440 t $a^{-1}$ arsenic is volatilized into the atmosphere due to the volatility of arsenic [45]. About 8–95% of the arsenic is volatilized into the atmosphere during the heat treatment of arsenic-preservative-treated wood. About 150 t $a^{-1}$ arsenic passes into the atmosphere via the use of preservatives.

**(2) Speciation and transform of arsenic in the atmosphere**

The migration of arsenic occurs mainly through anthropogenic and natural activities such as weathering, biological activities, and volcanic eruptions. Atmospheric arsenic conversion is specifically controlled by reduction, oxidation reactions, methylation, and so on [46].

At present, most of the arsenic is enriched in PM2.5 for atmospheric transport. Therefore, studies on atmospheric arsenic transport and transformation mainly focus on atmospheric particulates (TSP, $PM_{10}$, and $PM_{2.5}$, etc.). Speciation of arsenic in particulate matter mainly in the following forms: (1) forming insoluble arsenate (such as calcium arsenate, aluminum arsenate, iron arsenate, etc.), which deposited in atmospheric particulate matter, (2) being adsorbed by atmospheric particle matters which acts as a buffer against arsenic, (3) being adsorbed on the surface of atmospheric particles and other metal-poor precipitation salts, (4) existing in the crystal structure of atmospheric particles, and (5) dissolving in the atmospheric particulate solution into a water-soluble arsenic form [47, 48].

Recent studies have found that arsenic in particulate matter is mainly present in the form of inorganic arsenic As(III) and As(V), with As(V) accounting for more than 70% due to the oxidation effects, or a very small amount of organic arsenic present as methylarsonic acid and dimethylarsinic acid [49–51]; Arsenic oxides or water-soluble arsenic are hydrolyzed and precipitated into arsenic (arsine, monomethylarsine, dimethylarsine, and trimethylarsine) which gradually adhere to the surface of the particulates and are gradually converted to the organic state through the action of microorganisms in the atmosphere. However, it is still hard to define whether the organic arsenic in the airborne particles detected by the current method is derived from this process. Therefore, due to the detection methods, the source of organic arsenic carried by atmospheric particles is not very clear, which may be derived from the application of methyl arsenic pesticides, microbial methylation of arsenic in atmospheric migration, or the release of volatile organic arsenic produced during the microbial methylation process.

## *1.3.3 Arsenic Pollution in Solid Phase*

According to an estimation from the WHO, contextual levels of arsenic in soil range from 1 to 40 mg $kg^{-1}$ [52]. Soils that are highly contaminated with arsenic can pose a significant risk to groundwater quality, food safety, and eventually human health. High concentrations of arsenic in the soils may be derived from both natural enrichment and anthropogenic pollution such as mining, smelting, the use of arsenic-containing pesticides and animal manures, and irrigation with arsenic-contaminated groundwater [53–55]. Soil contamination with arsenic is an increasing worldwide concern and exists in quite a number of places, especially in the areas where arsenic-bearing minerals were mined, smelted, and processed and where soils have extra high-arsenic concentrations [56–59]. Total arsenic content in non-contaminated soils is usually below 10 mg $kg^{-1}$ [60]. However, it can be increased up to three or four magnitudes in serious arsenic-contaminated soils [61]. The soils in such “hot-spot-areas” not only exhibit high-arsenic concentration locally but may also cause environmental concern. In those soils, possible mobilization and subsequent leaching into ground or surface water or entry into the food chain should always be considered as a serious hazard [62]. For example, Shimen realgar mine, located in Shimen County, Hunan Province in central south China, is famous in the world for its plenty of realgar ore ($As_4S_4$) deposit. It is the largest realgar deposit in Asia with more than 1500 years mining histories. In the past several decades, intensive mining, arsenic trioxide smelting, sulfuric acid, and phosphate fertilizer producing were under way until 2011 when it was out of production. A large amount of waste such as mining waste, arsenic residue, and arsenic ash was dumped on the soil surface and subjected to erosion and weathering processes, causing long-term and severe impact on soil and water. Therefore the water and soil in the mine area were serious and widely contaminated with arsenic and have posed a significant risk to human health [63]. It has been reported that average arsenic concentrations in local farmland soils can be as high as 99.51 mg $kg^{-1}$, some ten times as high as the concentrations in local unpolluted soils, and arsenic concentrations in rivers in this area are as high as 14.5 mg $L^{-1}$ [64]. Additionally, thousands of people in this mine area and its vicinity area suffered from arsenic poisoning [65].

## References


1. Bowell, R.J., Alpers, C.N., Jamieson, H.E., et al.: The environmental geochemistry of arsenic—an overview. Rev. Mineral. Geochem. **79**(1), 1–16 (2014)
2. Drahota, P., Filippi, M.: Secondary arsenic minerals in the environment: a review. Environ. Int. **35**(8), 1243–1255 (2009)
3. Walker, S.R., Parsons, M.B., Jamieson, H.E., et al.: Arsenic mineralogy of near-surface tailings and soils: influences on arsenic mobility and bioaccessibility in the Nova Scotia gold mining districts. Can. Miner. **47**(3), 533–556 (2009)
4. Utsunomiya, S., Peters, S.C., Blum, J.D., et al.: Nanoscale mineralogy of arsenic in a region of New Hampshire with elevated As-concentrations in the groundwater. Am. Miner. **88**(11–12), 1844–1852 (2003)
5. Morin, G., Rousse, G., Elkaim, E.: Crystal structure of tooeleite, $Fe_6(AsO_3)_4SO_4(OH)_4 \cdot 4H_2O$, a new iron arsenite oxyhydroxysulfate mineral relevant to acid mine drainage. Am. Miner. **92**(1), 193–197 (2007)
6. Nishimura, T., Robins, R.G.: Confirmation that tooeleite is a ferric arsenite sulfate hydrate, and is relevant to arsenic stabilization. Miner. Eng. **21**(4), 246–251 (2008)
7. Akter, A., Ali, M.H.: Arsenic contamination in groundwater and its proposed remedial measures. Int. J. Environ. Sci. Technol. **8**(2), 433–443 (2011)
8. Smedley, P.L., Kinniburgh, D.G.: A review of the source, behavior and distribution of arsenic in natural waters. Appl. Geochem. **17**(5), 517–568 (2002)
9. How Does Arsenic Get into the Groundwater. Civil and Environmental Engineering. University of Maine. https://umaine.edu/arsenic/how-does-arsenic-get-into-the-groundwater/. Accessed 27 Nov 2018

10. Zeng, Z.H., Zhang, Z.L.: The formation of As element in groundwater and the controlling factor. Shanghai Geol. **87**(3), 11–15 (2002)
11. Zheng, Y., Stute, M., Geen, A.V., et al.: Redox control of arsenic mobilization in Bangladesh groundwater. Appl. Geochem. **19**(2), 201–214 (2004)
12. Thomas, M.A.: The association of arsenic with redox conditions, depth, and ground-water age in the glacial aquifer system of the northern United States. USGS U.S. Geological Survey, Virginia, pp. 1–18(2007)
13. Hong, B.: Influence of microbes on biogeochemistry of arsenic mechanism of arsenic-mechanism of arsenic mobilization in groundwater. Adv. Earth Sci. **21**(1), 77–82 (2006)
14. Su, C.: Environmental implications and applications of engineered nanoscale magnetite and its hybrid nanocomposites: a review of recent literature. J. Hazard. Mater. **322**(Part A), 48–84 (2017)
15. Kim, E.J., Batchelor, B.: Macroscopic and X-ray photoelectron spectroscopic investigation of interactions of arsenic with synthesized pyrite. Environ. Sci. Technol. **43**(8), 2899–2904 (2009)
16. Paktunc, D., Dutrizac, J., Gertsman, V.: Synthesis and phase transformations involving scorodite, ferric arsenate and arsenical ferrihydrite: implications for arsenic mobility. Geochim. Cosmochim. Acta **72**(11), 2649–2672 (2008)
17. Henke, K.: Arsenic: Environmental Chemistry, Health Threats and Waste Treatment, vol. 20, no. 3, pp. 199–201. Wiley (2009)
18. Pollutants, M.A.B.E.: Arsenic, vol. vii, p. 332. National Academy of Sciences, Washington, D.C. (1977)
19. Singh, R., Singh, S., Parihara, P., et al.: Arsenic contamination, consequences and remediation techniques: a review. Ecotoxicol. Environ. Saf. **112**, 247–270 (2015)
20. Yang, B., Zhang, G.L., Deng, W., et al.: Review of arsenic pollution and treatment progress in nonferrous metallurgy industry. Adv. Mater. Res. **634–638**(1), 3239–3243 (2013)
21. Luo, T., Cui, J.L., Hu, S., et al.: Arsenic removal and recovery from copper smelting wastewater using $TiO_2$. Environ. Sci. Technol. **44**(23), 9094–9098 (2010)
22. Morales, A., Cruells, M., Roca, A., et al.: Treatment of copper flash smelter flue dusts for copper and zinc extraction and arsenic stabilization. Hydrometallurgy **105**(1–2), 148–154 (2010)
23. Li, Y.C., Min, X.B., Chai, L.Y., et al.: Co-treatment of gypsum sludge and Pb/Zn smelting slag for the solidification of sludge containing arsenic and heavy metals. J. Environ. Manag. **181**, 756–761 (2016)
24. Ke, Y., Shen, C., Min, X.B., et al.: Separation of Cu and As in Cu-As-containing filter cakes by $Cu^{2+}$-assisted acid leaching. Hydrometallurgy **172**, 45–50 (2017)
25. Peng, Y.L., Zheng, Y.J., Chen, W.M., et al.: The oxidation of arsenic from As(III) to As(V) during copper electrorefining. Hydrometallurgy **129–130**, 156–160 (2012)
26. Zheng, Y.J., Peng, Y.L., Lang, K.E., et al.: Separation and recovery of Cu and As from copper electrolyte through electrowinning and $SO_2$ reduction. Trans. Nonferrous Met. Soc. China **23** (7), 2166–2173 (2013)
27. Shankar, S., Shikha, U.: Arsenic contamination of groundwater: a review of sources, prevalence, health risks, and strategies for mitigation. Sci. World J. **2014**(1), 179–186 (2014)
28. Pena-Pereira, F., Villar-Blanco, L., Lavilla, I., et al.: Test for arsenic speciation in waters based on a paper-based analytical device with scanometric detection. Anal. Chim. Acta **1011**, 1–10 (2018)
29. Michael, H.A.: An arsenic forecast for China. Science **341**(6148), 852–853 (2013)
30. Mukherjee, A., Sengupta, M.K., Hossain, M.A., et al.: Arsenic contamination in groundwater: a global perspective with emphasis on the Asian scenario. J. Health Popul. Nutr. **24**(2), 142–163 (2006)
31. Nordstrom, D.K.: Worldwide occurrences of arsenic in ground water. Science **296**(5576), 2143–2145 (2002)

32. He, Y., Luo, Y.: Status of arsenic pollution in Xinjiang and research progress in prevention and treatment of arsenism. People's Military Surg. **(06)**, 616–618 (2017)
33. Zhang, L., Xie, X., Li, J., et al.: Hydrochemical and geochemical investigations on high arsenic groundwater from Datong Basin, Northern China. Asian J. Ecotoxicol. **02**, 213–221 (2013)
34. Deng, Y.: Geochemical Processes of High Arsenic Groundwater System at Western Hetao Basin. China University of Geosciences, Wuhan (2008)
35. Zhao, S., Liu, G., Yang, B., et al.: Screening report on endemic arsenism and high content of arsenic in Xiantao City, Hubei Province. Chin. J. Endemiol. **28**(1), 71–74 (2009)
36. Deng, Y., Wang, Y., Li, H., et al.: Seasonal variation of arsenic speciation in shallow groundwater from endemic arsenicosis area in Jianghan Plain. Editor. Comm. Earth Sci. J. China Univ. Geosci. **40**(11), 1876–1886 (2015)
37. Matschullat, J.: Arsenic in the geosphere—a review. Sci. Total Environ. **249**(1–3), 297–312 (2000)
38. Pacyna, J.M., Pacyna, E.G.: An assessment of global and regional emissions of trace metals to the atmosphere from anthropogenic sources worldwide. Environ. Rev. **9**(4), 269–298 (2001)
39. Sánchez-Rodas, D., Campa, A.M.A.S.D.L., Rosa, J.D.D.L., et al.: Arsenic speciation of atmospheric particulate matter (PM10) in an industrialised urban site in southwestern Spain. Chemosphere **66**(8), 1485–1493 (2007)
40. Chilvers, D., Peterson, P.: Global cycling of arsenic. Lead, mercury, cadmium and arsenic in the environment. 279–301 (1987)
41. Goldberg, S.: Geochemistry, groundwater and pollution. Vadose Zone J. **5**(1), 510–510 (2006)
42. Turner, A.W.: Bacterial oxidation of arsenite. Nature **164**(4158), 76–77 (1949)
43. Francesconi, K.A., Kuehnelt, D.: Arsenic compounds in the environment. Environ. Chem. Arsen. (Boca Raton) 51–94 (2002)
44. Wang, P., Wang, S.L., Liu, S.Q., et al.: Occurrence, speciation, source and geochemical cycle of arsenic. Environ. Sci. Technol. **33**(7), 96–103 (2010)
45. Garelick, H., Jones, H., Dybowska, A., et al.: Arsenic pollution sources. Rev. Environ. Contam. Toxicol. **197**, 17–60 (2008)
46. Aurilio, A.C., Mason, R.P., Hemond, H.F.: Speciation and fate of arsenic in three lakes of the aberjona watershed. Environ. Sci. Technol. **28**(4), 577–585 (1994)
47. Fu, Q.Y., Zhuang, G.S., Li, J., et al.: Source, long-range transport, and characteristics of a heavy dust pollution event in Shanghai. J. Geophys. Res. **115**, 1–12 (2010)
48. Pey, J., Alastuey, A., Querol, X., et al.: A simplified approach to the indirect evaluation of the chemical composition of atmospheric aerosols from PM mass concentrations. Atmos. Environ. **44**(39), 5112–5121 (2010)
49. Song, C.H.: Analysis of arsenic speciation in atmospheric particulate matters. J. Wuhan Univ. Technol. **32**(13), 45–47 (2010)
50. He, T.T.: Analysis of Total Arsenic and Speciation in Atmospheric Particles in Shijing, Beijing. Nanhua University, Hengyang
51. He, T.T., Li, B., Xu, D.D., et al.: Ultrasonic extraction of arsenic speciation in atmospheric particles with phosphoric acid. Chin. J. Anal. Chem. **39**(4), 491–495 (2011)
52. Gupta, D.K., Chatterjee, S.: Arsenic Contamination in the Environment. The Issues and Solutions. Springer, Cham, Springer International Publishing (2017). ISBN: 9783319543543
53. Mandal, B.K., Suzuki, K.T.: Arsenic round the world: a review. Talanta **58**(1), 201–235 (2002)
54. García-Sánchez, A., Alonso-Rojo, P., Santos-Francés, F.: Distribution and mobility of arsenic in soils of a mining area (Western Spain). Sci. Total Environ. **408**(19), 4194–4201 (2010)
55. Kien, C.N., Noi, V.N., Bang, N.D., et al.: Arsenic and heavy metal concentrations in agricultural soils around tin and tungsten mines in the Dai Tu district, N. Vietnam. Water Air Soil Pollut. **197**(1–4), 75–89 (2009)
56. Krysiak, A., Karczewska, A.: Arsenic extractability in soils in the areas of former arsenic mining and smelting, SW Poland. Sci. Total Environ. **379**(2–3), 190–200 (2007)

57. Niazi, N.K., Singh, B., Shah, P.: Arsenic speciation and phytoavailability in contaminated soils using a sequential extraction procedure and XANES spectroscopy. Environ. Sci. Technol. **45**(17), 7135–7142 (2011)
58. Otones, V., Álvarez-Ayuso, E., García-Sánchez, A., et al.: Arsenic distribution in soils and plants of an arsenic impacted former mining area. Environ. Pollut. **159**(10), 2637–2647 (2011)
59. Mikutta, C., Mandaliev, P.N., Mahler, N., et al.: Bioaccessibility of arsenic in mining-impacted circumneutral river floodplain soils. Environ. Sci. Technol. **48**(22), 13468–13477 (2014)
60. Adriano, D.C.: Trace Elements in Terrestrial Environments: Biogeochemistry, Bioavailability, and Risks of Metals, 2nd edn. Springer, New York (2001)
61. Kim, E.J., Yoo, J.C., Baek, K.: Arsenic speciation and bioaccessibility in arsenic-contaminated soils: sequential extraction and mineralogical investigation. Environ. Pollut. **186**, 29–35 (2014)
62. Kabata-Pendias, A.: Trace Elements in Soils and Plants, 4th edn. CRC Press, Boca Raton (2010)
63. Tang, J.W., Liao, Y.P., Yang, Z.H., et al.: Characterization of arsenic serious-contaminated soils from Shimen realgar mine area, the Asian largest realgar deposit in China. J. Soils Sediments **16**(5), 1519–1528 (2016)
64. Zhu, X.Y., Wang, R.C., Lu, X.C., et al.: Secondary minerals of weathered orpiment-realgar-bearing tailings in Shimen carbonate-type realgar mine, Changde, Central China. Mineral. Petrol. **109**(1), 1–15 (2015)
65. Li, H., Ben, B.: Arsenic pollution sows despair in Chinese cancer village. Reuters. HESHAN China Mon Jun 23, 2014 6:42am EDT. www.360doc.com/content/14/0625/08/26286_389523917.shtml. Accessed 27 Nov 2018

# Chapter 2
# Pollution Source Distribution of Arsenic in the Typical Smelter

**Mei-Qing Shi, Li-Yuan Chai and Yan-Jie Liang**

Arsenic has long been a problem associated with the extraction of nonferrous metals. During the last decade, a combination of lower grade ores and environmental concerns has resulted in accelerated technological developments to meet arsenic challenges [1].

Arsenic present in the minerals is usually mobilized through hydrometallurgical and pyrometallurgical processes. Arsenic-containing minerals that occur in base metal ores and concentrates enter metallurgical processing systems accordingly, and subsequently release into the environment. Therefore, arsenic contamination has become a worldwide environmental challenge especially in the metal smelting industries. Most of the arsenic is volatilized and oxidized to $As_2O_3$ and $As_4O_6$ during the pyrometallurgical operations of nonferrous metals, including roasting, converting, and smelting.

## 2.1 Behavior and Distribution of Arsenic in a Typical Lead Smelter

Lead smelting industry is a significant anthropogenic source of arsenic contamination [2]. China has been the world's largest lead producer and consumer for several years [3]. SKS (Shuikoushan) lead smelting process is one of the main lead

M.-Q. Shi (✉) · L.-Y. Chai
Chinese National Engineering Research Center for Control and Treatment of Heavy Metal Pollution (CNERC-CTHMP), Changsha, Hunan, China
e-mail: shimeiqing0925@126.com

L.-Y. Chai
e-mail: lychai@csu.edu.cn

L.-Y. Chai · Y.-J. Liang
School of Metallurgy and Environment, Central South University, Changsha, Hunan, China
e-mail: liangyanjie2015@csu.edu.cn



L.-Y. Chai (ed.), *Arsenic Pollution Control in Nonferrous Metallurgy*,
https://doi.org/10.1007/978-981-13-6721-2_2

smelting technologies in China due to its advantages such as reduced energy consumption, higher metal recovery, and longer furnace life [4]. Despite the advantages of SKS lead smelting method, the arsenic emission from the pyrometallurgical process cannot be neglected. Gaining insights into the behavior, fate, transport, and effects on the environment of arsenic in typical SKS lead smelter would be very essential to reduce ecological risk and potential negative impacts of arsenic on the environment.

The flow scheme of SKS lead smelting is illustrated in Fig. 2.1. Lead sulfide concentrates, flux, pulverized coal, and dust from dust collection systems of bottom blowing furnace and blast furnace are sent to bottom blowing furnace after mixing

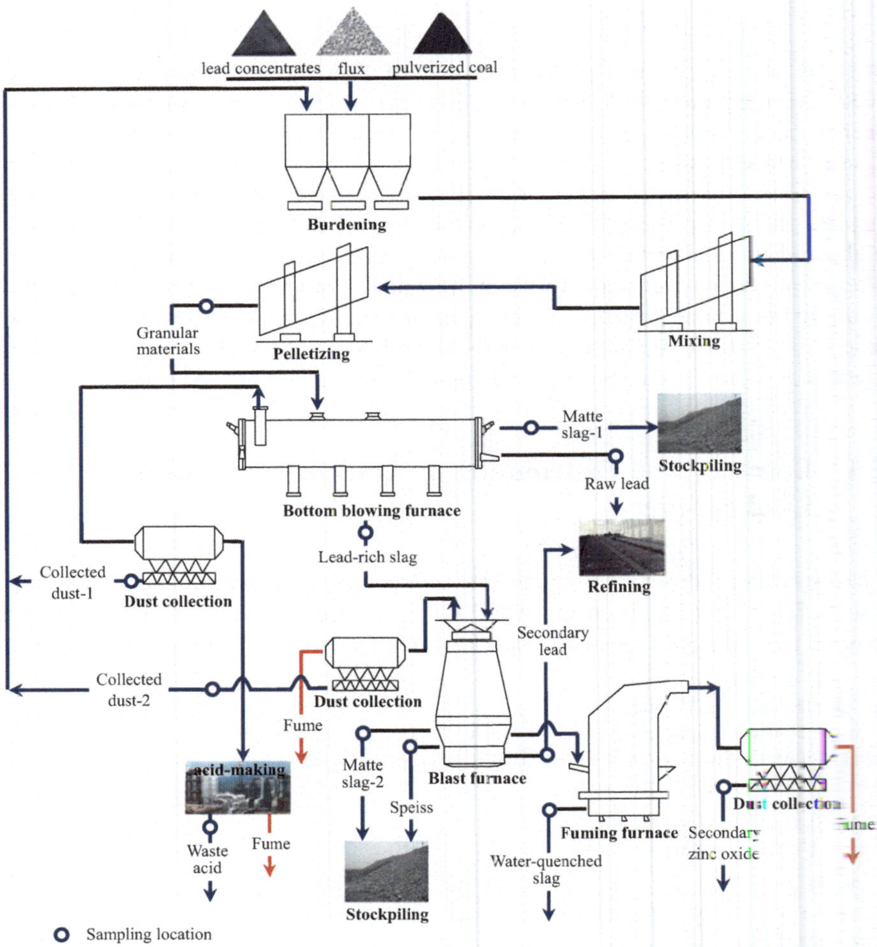


**Fig. 2.1** Flow scheme of the investigated SKS lead smelting and sampling locations

**Table 2.1** Samples classification

| Classification | Samples |
|---|---|
| Raw materials | Granular materials (GM) |
| Intermediate materials | Raw lead (RL), lead-rich slag (LS), collected dust-1 (CD1), secondary lead (SL), collected dust-2 (CD2) |
| Final materials | Waste acid (WA), matte slag-1 (MS1), speiss (S), matte slag-2 (MS2), water-quenched slag (WS), secondary zinc oxide (SZO), post dust (PD) |

and pelletizing. The process is used to produce lead metal and lead-rich slag. Then the lead-rich slag is charged to a blast furnace for reduction, where blast furnace slag, lead bullion, and speisses are generated. And the slag is directly transferred into a fuming furnace to recover lead and zinc in the slag. Lead bullion generated in bottom blowing furnace and blast furnace is electrolyzed after fire-refining.

All the samples are divided into three categories, namely, raw materials, intermediate materials, and final materials as indicated in Table 2.1.

### *2.1.1 Arsenic Behavior in SKS Lead Smelting Process*

Arsenic contents in all the samples are analyzed to investigate the partitioning and transformation behavior of arsenic in the SKS lead smelting process and the calculated arsenic percentages in liquid and solid phases along the entire smelting process are illustrated in Fig. 2.2. The mass flows of arsenic during individual processing stages are presented in Fig. 2.3.

**(1) Arsenic contents**

Arsenic is found at quantifiable contents in all samples from input to final output samples (Table 2.2). Of all the solid samples, CD2 shows the highest arsenic content, followed by SZO, MS2, and LS. The contents of arsenic in GM are observed at 0.99 ± 0.02%. The high arsenic contents in WA (4064.03 ± 476.60 mg $L^{-1}$), CD2 (6.60 ± 2.10%), and SZO (4.33 ± 0.75%) imply a large scale of arsenic volatilizing in the high-temperature smelting process. And there is more arsenic volatilized in blast furnace and fuming furnace than in bottom blowing furnace, for arsenic contents in CD2 and SZO generated from blast furnace and fuming furnace, respectively, are much higher than that in CD1 (0.20 ± 0.01%) from bottom blowing furnace. Besides, the arsenic content in LS (1.62 ± 0.07%) is found to increase by about 63.60% compared to GM. While the arsenic contents decrease by 85.80% and 73.46%, respectively, in WS (0.23 ± 0.17%) and S (0.43 ± 0.01%) compared to arsenic content of LS which is the source of arsenic for blast furnace and fuming furnace. This can be explained by the following two considerations. First, the volatilized arsenic decreases with oxygen partial pressure increasing. In oxidizing atmosphere, arsenic can be oxidized to As(V) which is less volatile than As(III).

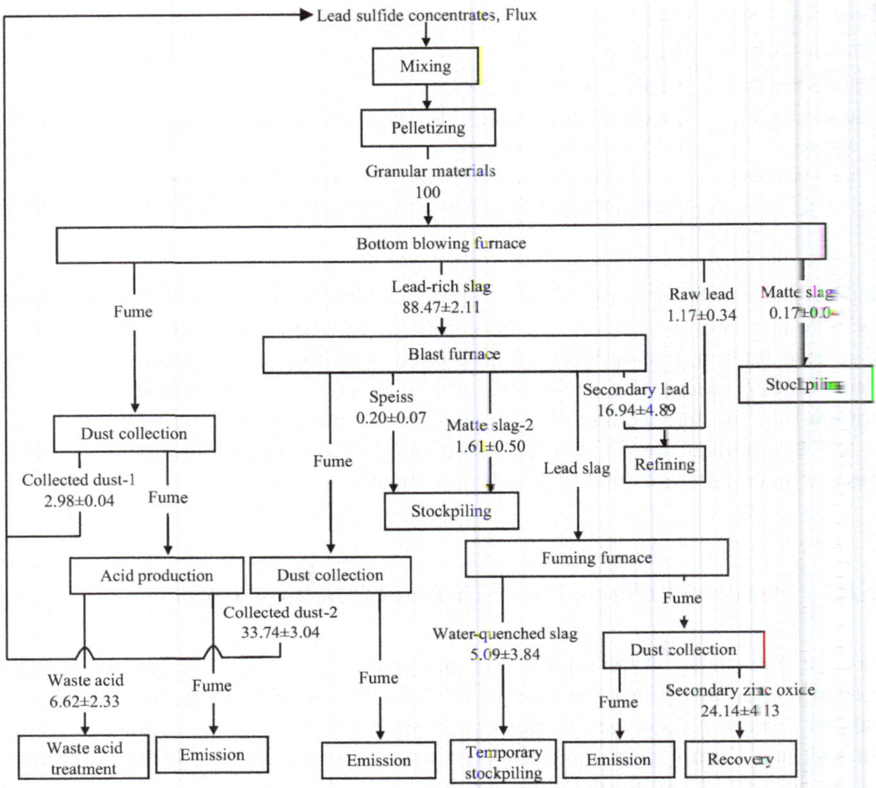


**Fig. 2.2** Total material flow (%) of arsenic in the entire SKS lead smelting process

Second, the temperature in the blast furnace and the fuming furnace is higher than that in the bottom blowing furnace. But it is not fit for matte slag of these two stages, for MS1 shows an arsenic decrease to 0.14 ± 0.01% and arsenic content in MS2 increases to 2.87 ± 0.89%. It suggests that the ratio of arsenic distribution in the matte slag increases with oxygen partial pressure decreasing. Arsenic content in the raw lead (0.11 ± 0.04%) is lower than that in the secondary lead (0.70 ± 0.24%), implying that arsenic is much easier to enter the molten lead under reducing atmosphere during the smelting process.

### (2) Material flow analysis of arsenic

The final flow direction of arsenic can be divided into three categories: intermediate materials, stockpiled materials, and unorganized emissions. Intermediate materials are the major media for arsenic loading, accounting for about 78.97% of the total arsenic, which subsequently pass to next smelting process or other production process. These include arsenic in RL and SL which accounts for 1.17 ± 0.34% and 16.94 ± 4.89%, respectively, getting into lead electrolysis workshop, as well as

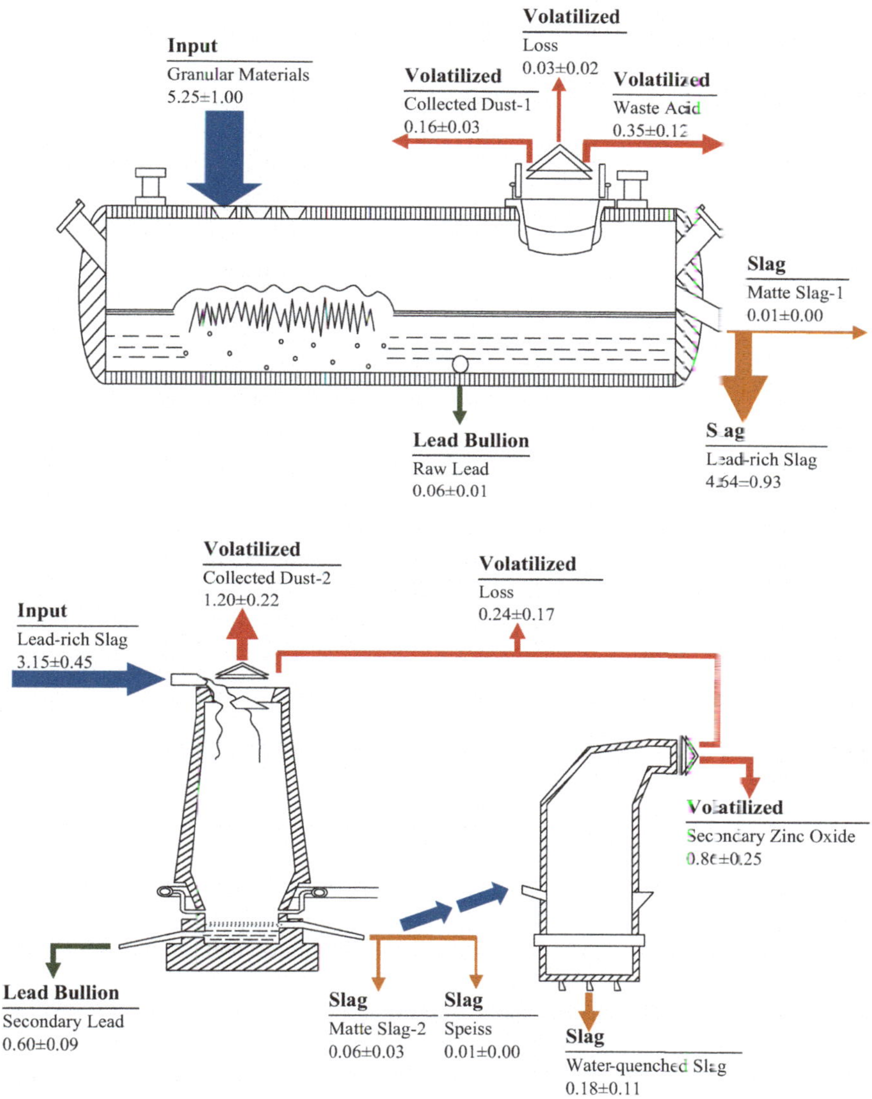


**Fig. 2.3** Mass flows (t $d^{-1}$) of arsenic in individual processing stages of SKS lead smelting process

arsenic in CD1 (2.98 ± 0.04%) and CD2 (33.74 ± 3.04%) which are turned back to bottom blowing furnace and SZO (24.14 ± 4.13%) used for zinc recovery. The ratio of arsenic in stockpiled materials is nearly 13.69%, which consists of MS1 (0.17 ± 0.04%), S (0.20 ± 0.07%), MS2 (1.61 ± 0.50%), WS (5.09 ± 3.84%), and WA (6.62 ± 2.33%) (precipitated into sludge after waste acid treatment).

**Table 2.2** Arsenic contents of the input and output materials (% except for mg $L^{-1}$ for waste acid) at each smelting stage[a]

| Bottom blowing furnace | | Blast furnace and fuming furnace | |
|---|---|---|---|
| GM | 0.99 ± 0.02 | WS | 0.23 ± 0.07 |
| LS | 1.62 ± 0.07 | SOZ | 4.33 ± 0.75 |
| MS1 | 0.14 ± 0.01 | SL | 0.70 ± 0.24 |
| RL | 0.11 ± 0.04 | MS2 | 2.87 ± 0.89 |
| WA | 4064.03 ± 476.60 | S | 0.43 ± 0.04 |
| CD1 | 0.20 ± 0.01 | CD2 | 6.60 ± 2.20 |

[a]Values shown are mean concentrations ± SD (n = 3) of the three sampling days

This part of arsenic could be leached into underground water or dispersed into atmosphere. There is around 7.31% of total arsenic lost during the pyrometallurgical process through unorganized emissions, which includes 0.59 ± 0.37% of total arsenic lost in bottom blowing furnace and 6.71 ± 3.45% lost in blast furnace and fuming furnace. The condition of workplace can be deteriorated because of unorganized emission of arsenic, and it poses a great threat to the environment. Therefore, arsenic in stockpiled materials and unorganized emissions is the main source of arsenic contamination from lead smelting process. Besides, arsenic in dust cannot be neglected due to its dispersion during the collection and transportation process.

**(3) Mass balance analysis of arsenic**

Mass balance is calculated by multiplying contents of arsenic by daily input/output mass values.

$$\sum C_{iinput} \cdot M_{iinput} = \sum C_{ioutput} \cdot M_{ioutput} + W_{lost} \quad (2.1)$$

where $C_{iinput}$ and $C_{ioutput}$ represent the arsenic contents of input and output samples (t $t^{-1}$), respectively; $M_{iinput}$ and $M_{ioutput}$ are daily input and output values (t), respectively; and $W_{lost}$ represents the weight of lost arsenic (t) which includes arsenic loss in final flue gas discharge and unorganized emission.

To assess distributions of arsenic in the whole smelting process, the mass percentage of arsenic in individual processing stages is first calculated using $W_{ioutput}/W_{tinput} \times 100\%$, where $W_{ioutput}$ and $W_{tinput}$ represent the mass of arsenic in a certain output material and total input of individual processing stage, respectively. Then arsenic percentages in the second stage are calculated by multiplying the mass percentage of arsenic of the individual processing stage by that of the lead-rich slag.

We take arsenic mass loading in GM and LS as the arsenic input of bottom blowing furnace and blast furnace-fuming furnace, respectively. The output consists of arsenic in lead bullion and slag and volatilized arsenic. The volatilized arsenic includes arsenic in dust, waste acid, and unorganized emissions.

In GM, the mass flow of arsenic is 5.25 ± 1.00 t $d^{-1}$, which is the arsenic input of the bottom blowing furnace as well as the input of the entire smelting system every day. In the first stage, there is only 0.06 ± 0.01 t $d^{-1}$ of arsenic distributed to the RL. Major arsenic, which is accounted for about 88.57% of the total amount of arsenic, is distributed to slag, which includes 4.64 ± 0.93 t $d^{-1}$ arsenic in LS, 0.01 ± 0.00 t $d^{-1}$ arsenic in MS1. The volatilized output is 0.35 ± 0.12 t $d^{-1}$ arsenic in WA, 0.03 ± 0.02 t $d^{-1}$ of lost arsenic and 0.16 ± 0.03 t $d^{-1}$ in CD1.

The mass flow of arsenic entered blast furnace is 3.15 ± 0.45 t $d^{-1}$ by treating the LS. It is about 1.49 t $d^{-1}$ of arsenic less than that generates in bottom blowing furnace process because not all of the LS is charged to the blast furnace during the sampling campaigns. Only about 0.25 t $d^{-1}$ of arsenic is found in the slag, namely, 0.06 ± 0.03 t $d^{-1}$ in MS2, 0.01 ± 0.00 t $d^{-1}$ in S and 0.18 ± 0.11 t $d^{-1}$ in WS, much less than the fraction in the first stage. While it is noteworthy that the relatively high mass fractions of arsenic are present in CD2 (1.20 ± 0.22 t $d^{-1}$), SZO (0.86 ± 0.25 t $d^{-1}$), and lost arsenic (0.24 ± 0.17 t $d^{-1}$). Volatilized arsenic in the second stage, which is accounted for a total of about 73.02% of the input arsenic of the second stage, is much higher than that in the first stage (10.29% of the initial arsenic). In the view of arsenic flow direction, there is a total of about 0.26 t $d^{-1}$, 0.27 t $d^{-1}$ of arsenic distributed to stockpiled materials and unorganized emissions, respectively, and 1.36 t $d^{-1}$ of arsenic in collected dust cycles within the two stages. Stockpiled materials, unorganized emissions as well as all kinds of dust are the main objects that would be a threat to the environment due to the direct exposure to the environment. Therefore, further investigation of these materials is necessary.

### *2.1.2 Characterization of Dust and Slag Samples*

XRD patterns for selected slag samples and dust samples from different furnaces contain numerous peaks corresponding to several crystallized species (Fig. 2.4). Arsenic in dust samples mainly existed as arsenate combined with Pb, Na, Ca, K or Zn. Arsenic combined with Zn and K is more easily leached out which implies a higher leaching rate of arsenic for CD1 and SZO. In addition, cadmium sulfate ($CdSO_4$) and anglesite ($PbSO_4$) are identified in PD and CD1, respectively. CD2 and SZO have similar XRD patterns and zincite (ZnO) is the major phase in these two samples. The only difference is that SZO contained more ZnO and part of the Pb is replaced by K in the arsenic compound. This may be due to the similar atmosphere of the blast furnace and fuming furnace.

For slag samples, galena (PbS) and arsenic zinc ($Zn_3As_2$) are clearly observed in MS1. This may result from incomplete oxidation of raw materials in the first stage. Hardystonite ($Ca_2ZnSi_2O_7$) is found in both speiss and MS2, and major arsenic species in these two samples include lead iron arsenate hydroxide ($PbFe(AsO_4)(AsO_3(OH))$) and sodium aluminum arsenate ($NaAl_{1.5}As_{4.5}O_{14}$), respectively. Magnesium aluminum iron oxide ($MgFeAlO_4$) and sodium iron arsenate

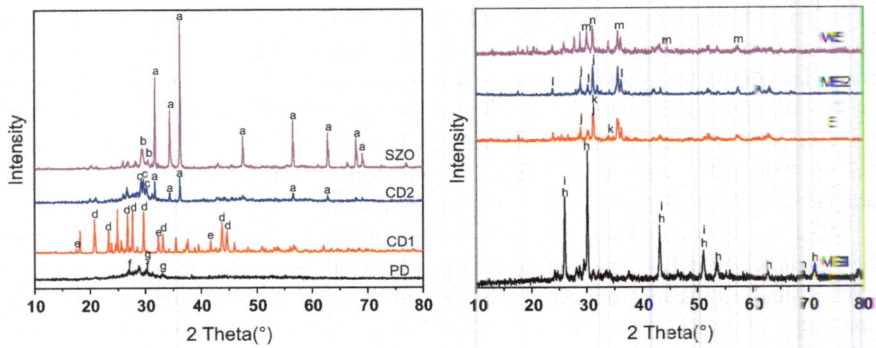


**Fig. 2.4** XRD patterns of selected samples (a) zincite, (b) potassium lead arsenate, (c) mimetite, (d) anglesite, (e) zincroselite, (f) cadmium sulfate, (g) calcium arsenate, (h) galena, (i) arsenic zinc, (j) hardystonite, (k) lead iron arsenate hydroxide, (l) sodium aluminum arsenate, (m) magnesium aluminum iron oxide, and (n) sodium iron arsenate

($Na_7Fe_3(As_2O_7)_4$) are mainly presented in WS, which also implies higher leaching rates of arsenic in MS2 and WS.

## 2.2 Environment Risk Assessment

The final materials as well as dust could enter into the environment and probably pose a potential risk to the environment.

**(1) Leachable arsenic concentration**

The Toxicity Characteristic Leaching Procedure (TCLP) developed by the US Environmental Protection Agency (USEPA) is used to simulate the leaching of contaminated materials by organic acid. The overall purpose of this analysis is to classify waste according to its ultimate hazard and to predict the long-term behavior of contaminants. Leachable arsenic concentrations of selected samples are shown in Table 2.3. MS2 (19.55 ± 0.54 mg $L^{-1}$), CD1 (34.18 ± 3.00 mg $L^{-1}$), and SZO (225.62 ± 22.00 mg $L^{-1}$) are hazardous because the arsenic leaching concentrations

**Table 2.3** Leachable arsenic concentrations based on TCLP test

| Slag samples | TCLP (mg $L^{-1}$) | Leaching rate (%) | Dust samples | TCLP (mg $L^{-1}$) | Leaching rate (%) |
|---|---|---|---|---|---|
| MS1 | 0.30 ± 0.18 | 0.43 ± 0.16 | CD1 | 34.18 ± 3.00 | 34.18 ± 3[illegible]4 |
| S | 0.70 ± 0.04 | 0.33 ± 0.02 | CD2 | 0.27 ± 0.03 | 0.01 |
| MS2 | 19.55 ± 0.54 | 1.36 ± 0.13 | SZO | 225.62 ± 22.00 | 10.42 ± 1[illegible]9 |
| WS | 0.96 ± 0.14 | 0.84 ± 0.06 | PD | 0.05 ± 0.01 | 0.01 |

are higher than the acceptable value of 5 mg $L^{-1}$. While the MS1 (0.30 ± 0.18 mg $L^{-1}$), S (0.70 ± 0.04 mg $L^{-1}$), WS (0.96 ± 0.14 mg $L^{-1}$), CD2 (0.27 ± 0.03 mg $L^{-1}$), and PD (0.05 ± 0.06 mg $L^{-1}$) are much safer to the environment.

The leaching rate (R) can be calculated by the following equation:

$$R(\%) = \frac{L_{Asi}}{T_{Asi}} \times 100\% \tag{2.2}$$

where $i$ is the certain sample, $L_{As}$ is leaching mass of arsenic (mg), and $T_{As}$ is the total mass of arsenic (mg).

The calculated results of leaching rate show that arsenic leaching rates of MS2 (1.36 ± 0.13%) and WS (0.84 ± 0.06%) are higher than those of MS1 (0.43 ± 0.16%) and S (0.33 ± 0.02%) among the slag samples, and for the dust samples, the leaching rates of CD1 and SZO are 34.18 ± 3.04% and 10.42 ± 1.29%, respectively, higher than the other two dust samples. This could be explained that arsenic combined with Na, K, and Zn would be more easily extracted into the acid solution.

**(2) Environment risk assessment**

Measurement of the total concentration of metals provides inadequate information to assess the bioavailability or toxicity of metals. The toxicity of metals depends especially on their chemical forms. The three-stage Community Bureau of Reference (BCR) sequential extraction procedure [5] is applied to analyze the effective combination forms of arsenic in the solid samples. The acid soluble/exchange fraction (F1) is a loosely bound phase and liable to change with environmental conditions such as pH changes. The reducible fraction (F2) is thermodynamically unstable and available under anoxic conditions [6], therefore, these two fractions (F1 and F2) are classified as unstable fractions which are more harmful to the environment. The oxidizable fraction (F3) containing metals binding to organic matter and metals in the sulfide combination state is identified as a potential effect fraction. The residual fraction (F4) is a stable fraction in which the metals are not expected to be released in solution under environmental conditions. The speciation of As is shown in Fig. 2.5. For the dust samples, arsenic distribution in unstable fractions (F1 + F2) ranked in the order: CD1 (90.06%) > SZO (10.19%) > PD (2.49%) > CD2 (0.01%), while for the slag samples, the order is S (11.02%) > MS2 (10.89%) > WS (4.25%) > MS1 (1.26%).

A risk assess code (RAC) [7] is also applied to estimate the environmental risk associated with arsenic pollution in obtained samples. The RAC assesses the availability of arsenic by applying a scale to the percentage of arsenic present in acid soluble/exchange fraction (F1). The classification of risk has been categorized in terms of the RAC (Table 2.4).

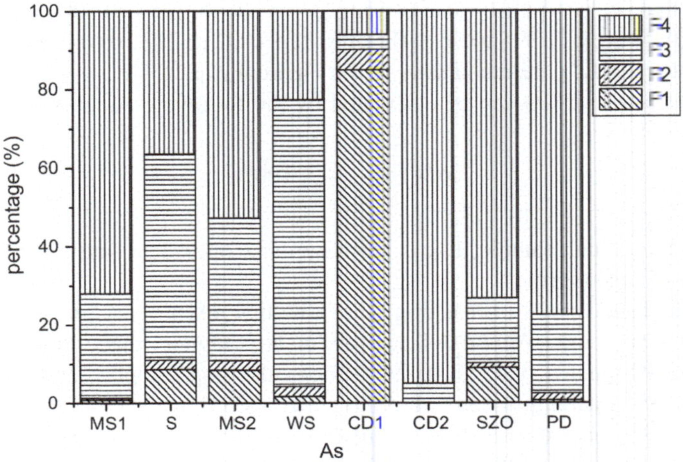


**Fig. 2.5** Speciation of arsenic by BCR

**Table 2.4** Risk assess code (RAC)

| Category | Risk | Metal in acid soluble/exchange fraction (F1) (%) |
|---|---|---|
| I | No risk (NR) | <1 |
| II | Low risk (LR) | 1–10 |
| III | Medium risk (MR) | 11–30 |
| IV | High risk (HR) | 31–50 |
| V | Very high risk (VHR) | >50 |

The F1 fraction of arsenic in CD1 accounts for 84.81% of the total arsenic content, which is classified in very high-risk category based on RAC, while F1 in CD2 and PD is totally environmentally insignificant (0.006% and 0.66% of total arsenic content, respectively). Therefore, CD2 and PD are categorized as having no risk to the environment. The percentage of arsenic in SZO distributed in F1 is 8.99%, and it is classified in the low-risk category. Sample S, MS2, and WS with arsenic in F1 of 8.57%, 8.42%, and 1.70%, respectively, are categorized as low risk to the environment and MS1 with 0.84% of arsenic in F1 as no risk. There is little difference between S and MS2 on the arsenic distribution in the F1 and F2.

In conclusion, arsenic volatilizes more in blast furnace and fuming furnace than that in bottom blowing furnace. The final flow direction of arsenic includes 78.[illegible]7% in intermediate materials, 13.69% in stockpiled materials, and 7.31% in unorganized emissions, and the latter two as well as all kinds of dust are the main sources of arsenic contamination from lead smelting process.

## References

1. Reddy, R.G., Ramachandran, V.S.: Arsenic Metallurgy. The Minerals, Metals & Materials Society, Wiley (2005). ISBN:Y9780873395854
2. Basha, C.A., Selvi, S.J., Ramasamy, E., et al.: Removal of arsenic and sulfate from the copper smelting industrial effluent. Chem. Eng. J. **141**(1), 89–98 (2008)
3. Gao, W., Wang, C.Y., Yin, F., et al.: Situation and technology progress of lead smelting in China. Adv. Mater. Res. **581**, 904–911 (2012)
4. Dai, X., Huang, G.H.: The Major Lead Smelting Process and Development Proposals of China. Hunan Nonferrous Metals, China (2010)
5. Davidson, C.M., Duncan, A.L., Littlejohn, D., et al.: A critical evaluation of the three-stage BCR sequential extraction procedure to assess the potential mobility and toxicity of heavy metals in industrially-contaminated land. Anal. Chim. Acta **363**(1), 45–55 (1998)
6. Xie, X.D., Min, X.B., Chai, L.Y., et al.: Quantitative evaluation of environmental risks of flotation tailings from hydrothermal sulfidation–flotation process. Environ. Sci. Pollut. Res. **20**, 6050–6058 (2013)
7. Sundaray, S.K., Nayak, B.B., Lin, S., et al.: Geochemical speciation and risk assessment of heavy metals in the river estuarine sediments—a case study: Mahanadi basin, India. J. Hazard. Mater. **186**(2), 1837–1846 (2011)

# Chapter 3
# Arsenic Behaviors and Pollution Control Technologies in Aqueous Solution

**Li-Yuan Chai, Qing-Zhu Li, Qing-Wei Wang, Yun-Yan Wang, Wei-Chun Yang and Hai-Ying Wang**

The removal of arsenic from solutions has been investigated for decades and continues to be a topic of intense research studies. The most commonly used arsenic removal methods are oxidation, precipitation, ion exchange, adsorption, lime treatment, biological removal processes, and reverse osmosis. A number of arsenic treatment techniques have been recommended or they are being used so far, but still paucity exists where practically feasible, less expensive, and eco-friendly technique for remediation of arsenic is urgently required.

L.-Y. Chai (✉) · Q.-Z. Li · Q.-W. Wang · Y.-Y. Wang · W.-C. Yang · H.-Y. Wang
School of Metallurgy and Environment, Central South University, Changsha, Hunan, China
e-mail: lychai@csu.edu.cn

Q.-Z. Li
e-mail: qingzhuli@csu.edu.cn

Q.-W. Wang
e-mail: qw_wang@csu.edu.cn

Y.-Y. Wang
e-mail: wyy@csu.edu.cn

W.-C. Yang
e-mail: yang220222000@yahoo.com.cn

H.-Y. Wang
e-mail: haiyw25@163.com

L.-Y. Chai
Chinese National Engineering Research Center for Control and Treatment of Heavy Metal Pollution (CNERC-CTHMP), Changsha, Hunan, China

 
L.-Y. Chai (ed.), *Arsenic Pollution Control in Nonferrous Metallurgy*,
https://doi.org/10.1007/978-981-13-6721-2_3

## 3.1 Introduction

The over standard arsenic concentration has been detected in ground and surface waters, which arises from the discharge of arsenic-containing wastewaters [1, 2]. Arsenic is often found in association with nonferrous metal ore [3]. In nonferrous metal smelting process, acidic wastewater containing high arsenic is generated from the wet scrubber process of smelting fume [4] (acidic wastewater samples in Table 3.1 from a typical copper smelter and lead-zinc smelter).

Arsenic prevalently exists as arsenite (As(III)) and arsenate (As(V)) in aquatic environment. As(III) is predominant inorganic arsenic species in smelting and mining effluents [1, 5], and constitutes over 70% of dissolved arsenic in groundwater samples [6]. Including precipitation [7], coagulation [8, 9], ion exchange [10, 11], adsorption [12, 13], and so forth, many techniques are available for arsenic removal but inclined to remove As(V). Consequently, pre-oxidation is an essential anchor for As(III)-contained wastewater treatment. Various methods have been studied for As (III) oxidation (e.g., ozone [6], manganese and chlorine compounds [14, 15], ferrate [16], and hydrogen peroxide [17]). In comparison, the catalyzed oxidation systems are more attractive due to rapid reaction, such as, Fenton ($Fe(II)/H_2O_2$) [18], Fenton-like ($Fe(III)/H_2O_2$) [19], photo-Fenton ($UV/H_2O_2$) [20] $TiO_2$ photocatalytic system ($UV/TiO_2$) [21], etc. In these systems, the most important reactive species devoted to As(III) oxidation is determined as hydroxyl radical ($HO^{\cdot}$) and its reaction rate toward As(III) is very high ($8.5 \times 10^9\ M^{-1}\ s^{-1}$) at acidic conditions [22]. Similar to $HO^{\cdot}$, sulfate radical ($SO_4^{\cdot-}$) is also a strong oxidizing agent. It has been reported that $SO_4^{\cdot-}$ is more advantageous than $HO^{\cdot}$ due to a longer half-life, and thus capable to escape from the solvent cage to oxidize As(III) [23, 24]. Moreover, $SO_4^{\cdot-}$ is a very strong electron acceptor enabling reactions which are impossible for $HO^{\cdot}$ [25]. For example, perfluorinated carboxylic acids are inert toward $HO^{\cdot}$ but can be degraded by $SO_4^{\cdot-}$ [26, 27]. A promising oxidation technique based on $SO_4^{\cdot-}$ has emerged over the past years to mineralize organic pollutants and oxidize As(III) [28–31]. The generation of $SO_4^{\cdot-}$ from peroxydisulfate (PDS, $S_2O_8^{2-}$) resembles that of

**Table 3.1** The composition of acidic wastewaters from a certain lead-zinc smelter and copper smelter (mg $L^{-1}$)

| Components | #1 | #2 | Components | #1 | #2 | Components | #1 | #2 |
|---|---|---|---|---|---|---|---|---|
| S | 8182 | 26,435 | Zn | 2.6 | 369 | Al | 16.65 | 27.8 |
| As(total) | 2335.5 | 9010 | Bi | – | 256.1 | Pb | 4.15 | 4.2 |
| As(III) | 1557 | 7660 | Cu | 0.85 | 184.1 | Co | 0.25 | 2.4 |
| Na | 14371 | 7038 | Fe | 4896.5 | 124.9 | P | 27.7 | 1.8 |
| Ni | 0.55 | 735.6 | Mg | 2.5 | 108.9 | Mn | 5.55 | 1.3 |
| Si | 373.7 | 674.8 | K | 15.4 | 72.5 | Hg | 0.45 | 0.05 |
| Ca | 23.05 | 537.3 | Cd | 12.25 | 42.8 | | | |

*Note* #1, Acidic wastewater from lead-zinc smelter; #2, Acidic wastewater from copper smelter
Trivalent arsenic As(III) was determined by HG-AFS and other pollutants were determined by ICP-AES

$HO^{\cdot}$ from $H_2O_2$. The peroxide bond in radical precursors PDS and $H_2O_2$ can be broken by heat, ultraviolet, ultrasound, radiolysis, etc. [28] Numerous activated methods (UV light [32], acoustic cavitation [31], and Fe(II) activation [29]) to generate $SO_4^{\cdot-}$ have thus been studied for As(III) oxidation. For comparison with the $HO^{\cdot}$-based oxidation, As(III) oxidation based on $SO_4^{\cdot-}$ should be discussed. The high intensive xenon (Xe) light can simulate natural solar light, which is effective to activate PDS and then stimulate As(III) oxidation at room temperature. In addition, we noticed that As(III) oxidation was rapid in the ferrous-activated PDS system, i.e., the initial few-minute reaction dominated in a 60-min oxidation process [29]. As for oxidation mechanism, both $SO_4^{\cdot-}$ and $HO^{\cdot}$ were reported to play roles in As(III) oxidation via electron transfer reaction [29]. Tetravalent As(IV) as an intermediate has been postulated in the metal (Fe(II), Fe(III) and Cu(II)) catalyzed As(III)-PDS systems [33, 34]. In addition to the resultant $As(OH)_3^{\cdot+}$ of electron transfer reaction, $As(OH)_4^{\cdot}$, the product of $HO^{\cdot}$-based addition reaction, was also reported previously [35, 36]. Unfortunately, there is no suitable detection method for various arsenic intermediates and few studies on molecular mechanism of As(III) oxidation.

The behavior of arsenic in the environment strongly depends on its chemical forms. Most authors investigated adsorption and precipitation techniques for treatment of arsenic-containing wastewater and focused on the distribution of arsenic species such as $H_3AsO_4$, $H_2AsO_4^-$, $HAsO_4^{2-}$, $AsO_4^{3-}$, $H_3AsO_3$, $H_2AsO_3^-$, $HAsO_3^{2-}$, and $AsO_3^{3-}$ [37]. Arsenic is removed from wastewaters mostly based on iron materials through precipitation and adsorption. The arsenic removal efficiency will be determined by the complexation between arsenic and iron to some extent. Arsenic and iron generally occur in water system as arsenite As(III), arsenate As (V), ferrous Fe(II), and ferric Fe(III). Many solid compounds of ferrous arsenite and ferrous arsenate have been reported [38] but no evidence for the existence of ferrous complexes with either arsenite or arsenate has been found. The arsenic disposal procedure currently favored by the industry involves the formation of an insoluble ferric arsenate compound [4]. Well-crystalline iron arsenate (e.g., scorodite) has advantages in lower ferric demand, higher density, and greater stability. Earlier synthesis of scorodite was conducted under autoclave conditions, i.e., under the high temperature and the high-pressure conditions [39]. Recently, atmospheric scorodite synthesis is of great concern due to lower capital investment [40]. Scorodite can be formed at temperatures as low as 40 °C [41], but amorphous precipitates such as ferric arsenate and its assemblages were more easily formed under lower temperatures [42]. Diverse precipitates in Fe–As-acid system are mainly depended on changed species in solutions. For example, we found the transformation of Fe(III)–As(V) complexes into colloid ferric arsenate previously [43]. Therefore, the species distribution and transformation in acidic Fe–As system is meaningful for developing novel phase with high potential in arsenic removal and stabilization by iron salts.

## 3.2 Redox Behavior and Chemical Species of Arsenic in Acidic Aqueous System

### 3.2.1 *Redox Behavior of Arsenic*

The Gibbs free energies of As(III) and As(V) species in HSC 7.0 database are in accordance with most reports [44–49]. An Eh-pH diagram for As (0.1 M)–$H_2O$ system was constructed by using HSC 7.0 software and depicted in Fig. 3.1 [50]. The predominance fields for As(V) species under oxidizing conditions are similar to all published Eh-pH diagrams [51–55]. However, the field boundaries for As(III) species were different because of the inclusion of $H_3AsO_3$ and its conjugate base rather than $HAsO_2$. The $HAsO_2$ stoichiometry was supported by the mass spectrometry because of mass-to-charge-ratios (m/z) of 107 corresponding to $AsO_2^-$ ion in the gas phase [56]. However, $H_3AsO_3$, a moiety comprising one arsenic atom coordinated by three OH ligands was supported by Raman and EXAFS spectroscopy [57–59]. The critical potentials of As(V)/As(III) couples against pH value indicate that the oxidation of As(III) to As(V) is more feasible in alkaline than in acidic solutions. Thus, the redox behavior of As(III) under strong acidic wastewater is challenging.

The redox behavior of As(III) in sulfuric acid (pH = 1.0) was studied by cyclic voltammetry (CV) methods. Figure 3.2 showed the voltammogram of a solution containing 0.1 M As(III) and a blank solution. There are six peaks on the CV curve in 0.1 M As(III) solution (blue line). The peaks B and C correspond to the oxygen evolution and the reduction of the oxide layer on Au electrode. The location and current of peaks A, D, and F are relevant to the concentration of As(III). Peak A (Fig. 3.2) was identified as a multiple electron transfer oxidation of As(III) to As (V), which obviously splits into three small peaks when As(III) concentration is less than 5 mM (Fig. 3.3a). Moreover, the current of the first electron transfer reaction is linearly correlated with As(III) concentration ($R^2$ = 0.9985, Fig. 3.3b) and the potential is ca. 0.9 V over the theoretical potential (ca. 0.3 V derived from

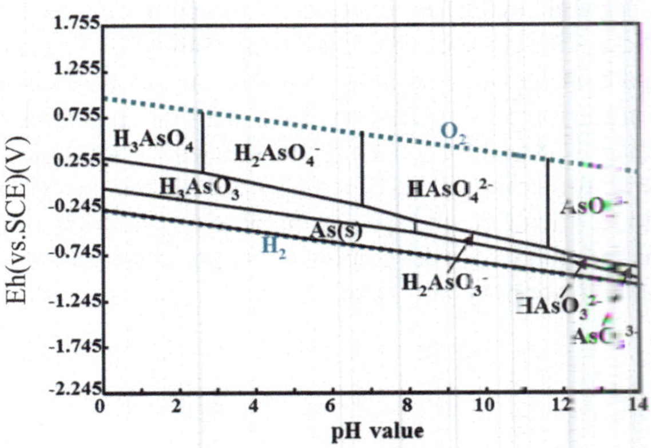


**Fig. 3.1** The Eh (vs. SCE)-pH diagram in As (0.1 M)–$H_2O$ system at 25 °C and 1 bar. Reprinted from Ref. [50] Copyright 2017, with permission from Elsevier

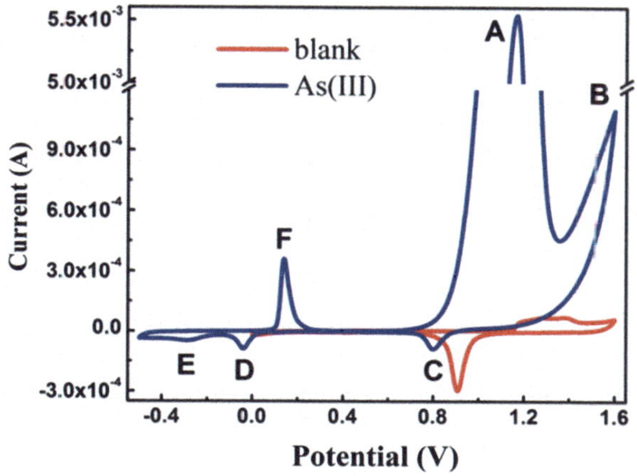


**Fig. 3.2** The voltammogram in 0.1 M As(III) solution (blue line) and blank solution (red line) ($\nu$ = 50 mV $s^{-1}$). Reprinted from Ref. [50] Copyright 2017, with permission from Elsevier

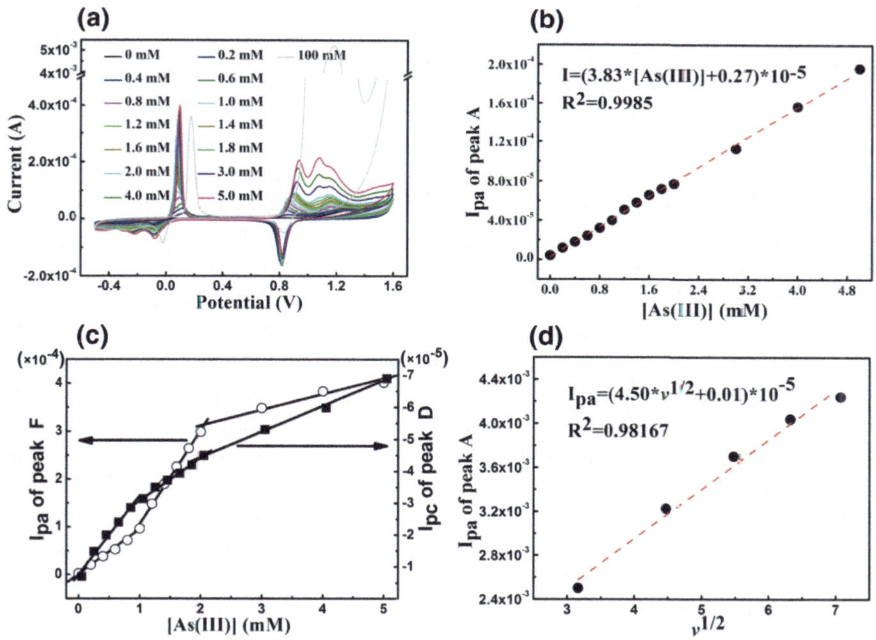


**Fig. 3.3** **a** The voltammogram in As(III) solutions, **b** and **c** linear calibration plots of peak current against As(III) concentration (As(III) = 0–5 mM, $\nu$ = 50 mV $s^{-1}$), and **d** linear calibration plots of peak A current against $\nu^{1/2}$ (As(III) = 0.1 M, $\nu$ = 10, 20, 30, 40, 50 mV $s^{-1}$). Reprinted from Ref. [50] Copyright 2017, with permission from Elsevier

Fig. 3.1). In addition, investigations at different scan rates showed a good linear relationship between the current of peak A and scan rate ($v^{1/2}$) (Fig. 3.3d). Therefore, As(III) oxidation is mainly diffusion-controlled.

Peak D corresponds to the reduction of As(III) and its peak current is linearly correlated to As(III) concentration. However, the different linear calibration plots of peak current against As(III) concentration were found in As(III) range of 0–1, 1–2, and 2–5 mM (Fig. 3.3c), which indicates that multilayer As(s) was electrodeposited from As(III) on Au electrode. The weak peak E can be explained as outmost As(s) electrodeposited from As(III) or from As(V). Although the reduction of As(V) to As(s) is thermodynamic sluggish, the electrogenerated $H_2$ could chemically reduce As(V) to As(s) [60]. Following peak D and E, peak F is no doubt the oxidation of As(s) to As(III) and the linear correlation between peak current and As(III) concentration is in accordance with that of peak D. The consistent trend between peak D and F (Fig. 3.3c) confirmed that multilayer As(s) was electrodeposited on Au electrode.

### 3.2.2 Chemical Species of Arsenic

As(III) and As(V), respectively, exist as $H_nAsO_3^{(n-3)}$ and $H_nAsO_4^{(n-3)}$ ($0 \leq n \leq 3$) in As–$H_2O$ system. The speciation-pH diagram for As(III) and As(V) was calculated at ionic strength 0 based on thermodynamic formation constants reported by Marini et al. [61]. As depicted by thick line in Fig. 3.4a, As(V) predominantly existed as (1) $H_3AsO_4$ and $H_2AsO_4^-$ at pH 0–4.5, (2) $H_2AsO_4^-$ and $HAsO_4^{2-}$ at pH 4.5–9, (3) $HAsO_4^{2-}$ and $AsO_4^{3-}$ at pH 9–14. The major form of As(V) is neutral molecules $H_3AsO_4$ at pH < 2, thus it is hard to be removed by electrostatic adsorption. Figure 3.4b (thick line) showed that As(III) mainly existed as (1) $H_3AsO_3$ at pH 0–7.5, (2) $H_3AsO_3$ and $H_2AsO_3^-$ at pH 7.5–10.5, (3) $H_2AsO_3^-$ and $HAsO_3^{2-}$ at pH 10.5–12, and (4) $H_2AsO_3^-$, $HAsO_3^{2-}$ and $AsO_3^{3-}$ at pH 12–14. Accordingly, $H_3AsO_3$ is the only As(III) species under acidic solutions. It was reported that undissociated $H_3AsO_3$ does not react with $H_2O_2$ [62]. However, rate constant $k$ for the oxidation reaction between As(III) and $H_2O_2$ can be determined by pH, temperature, and IS in pH 7.5–10.3 [63]. This is mainly because the molar fraction of $H_2AsO_3^-$, $HAsO_3^{2-}$, and $AsO_3^{3-}$ species can be affected by pH, temperature, and IS. The speciation-pH diagram at IS = 0.2 M was determined based on the Davies equation. As depicted by thin line in Fig. 3.4, high IS facilitates deprotonation reaction. Accordingly, negative-charged species with higher fraction under high IS solutions can be absorbed on the positive-charged surfaces.

As(V) can complex with metal ion to form aqueous complexes [61]. Fe(III) is usually used as a precipitator for As(V) and Fe(III)–As(V) complexes are meaningful for understanding arsenic species and behaviors in Fe(III)–As(V)–$H_2O$ system. We investigated the complexation between As(V) and Fe(III) by UV-Vis spectroscopy in series of As–Fe–$H_2SO_4$–$H_2O$ acidic solutions (Table 3.2). A new peak at ca. 240–300 nm appeared as As(V) concentration increased and is

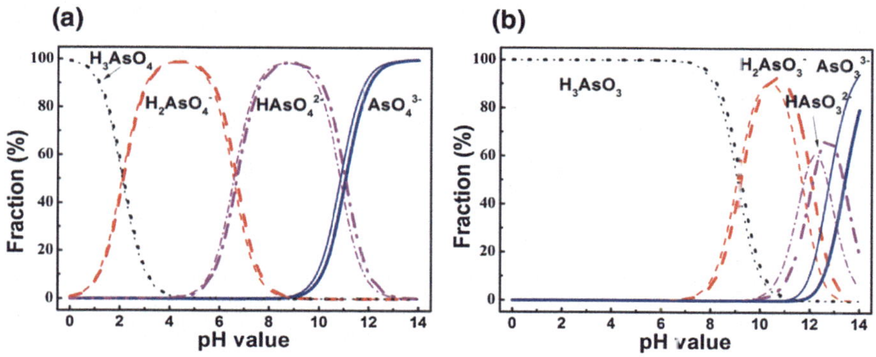


**Fig. 3.4** Speciation-pH diagrams of **a** As(V) species and **b** As(III) species in As–$H_2O$ system at 25 °C. Thick line, IS = 0; thin line, IS = 0.2 M. Reprinted from Ref. [50] Copyright 2017, with permission from Elsevier

**Table 3.2** Solution composition used in UV-Vis experiment and related calculation results

| No. | $[H_2SO_4]_t$ | $[Fe_2(SO_4)_3]_t$ | $[Na_3AsO_4]_t$ | $pH^a$ | $pH^b$ | $IS^b$ | $pH^c$ | $IS^c$ |
|---|---|---|---|---|---|---|---|---|
| | (mM) | | | | | | | |
| 1 | 50 | 0.25 | 0 | 1.35 | 1.293 | 0.0791 | 1.293 | 0.0791 |
| 2 | | | 0.5 | 1.36 | 1.301 | 0.0796 | 1.302 | 0.0796 |
| 3 | | | 1 | 1.41 | 1.310 | 0.0800 | 1.310 | 0.0800 |
| 4 | | | 2 | 1.42 | 1.327 | 0.0810 | 1.328 | 0.0809 |
| 5 | | | 4 | 1.43 | 1.363 | 0.0831 | 1.366 | 0.0829 |
| 6 | | | 6 | 1.44 | 1.400 | 0.0853 | 1.405 | 0.0850 |
| 7 | | | 8 | 1.48 | 1.440 | 0.0878 | 1.448 | 0.0874 |
| 8 | | | 10 | 1.52 | 1.482 | 0.0904 | 1.494 | 0.0899 |
| 9 | | | 15 | 1.63 | 1.597 | 0.0980 | 1.624 | 0.0971 |
| 10 | | | 20 | 1.76 | 1.730 | 0.1071 | 1.786 | 0.1059 |
| 11 | | | 25 | 1.92 | 1.881 | 0.1175 | 1.997 | 0.1162 |
| 12 | | | 30 | 2.10 | 2.056 | 0.1292 | 2.296 | 0.1281 |
| 13 | | | 35 | 2.44 | 2.261 | 0.1415 | 2.714 | 0.1410 |
| 14 | | | 40 | 2.83 | 2.517 | 0.1545 | 3.164 | 0.1543 |
| 15 | | | 45 | 3.75 | 2.893 | 0.1678 | 3.546 | 0.1678 |
| 16 | | | 50 | 6.00 | 4.123 | 0.1818 | 4.730 | 0.1820 |
| 17 | | | 55 | 6.36 | 5.870 | 0.2042 | 5.905 | 0.2048 |

*Note*
[a]Measured results
[b]Calculation results with exclusion of Fe(III)–As(V) complexes
[c]Calculation results with inclusion of Fe(III)–As(V) complexes

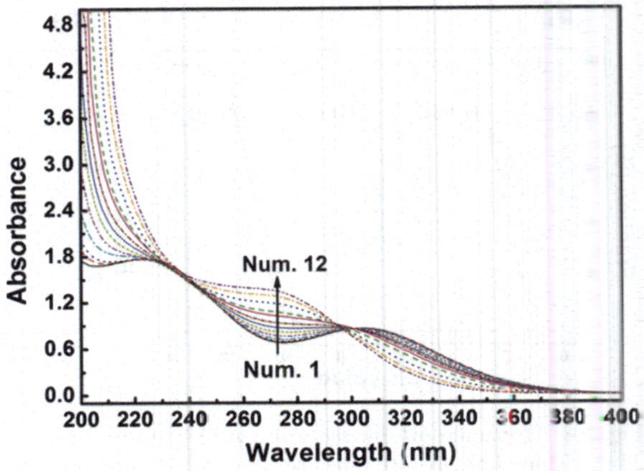


**Fig. 3.5** UV-Vis spectra of $Fe_2(SO_4)_3$–$Na_3AsO_4$–$H_2SO_4$–$H_2O$ solution. Reprinted from Ref. [50] Copyright 2017, with permission from Elsevier

attributed to Fe(III)–As(V) complexes (Fig. 3.5). The lg$K$ of Fe(III)–As(V) complexes were reported by Marini et al. [64] and are in accordance with other reports excluding $FeAsO_4$(a) (Table 3.3). The pH value, IS, and the fraction of various Fe (III) species were thus calculated by MINTEQ 3.1 with the consideration of Fe (III)–As(V) complexes or not. The calculated pH values are slightly higher than measured ones when Fe(III)–As(V) complexes were considered (Table 3.2). Without considering Fe(III)–As(V) complexes, the calculated Fe(III) species predominantly exists as $FeSO_4^+$ and $Fe(OH)_2^+$ when As(V) concentration is less than 45 mM and 45–50 mM, respectively (Fig. 3.6a). The band maxima of $Fe(OH)_2^+$ occur at the wavelengths of 300 nm [65], and thus the new peak at ca. 240–300 nm is not due to $Fe(OH)_2^+$. When Fe(III)–As(V) complexes were considered in species calculations (Fig. 3.6b), $FeH_2AsO_4^{2+}$ and $FeHAsO_4^+$ become major in 15–40 mM As(V) solutions and $FeHAsO_4^+$ is almost the only soluble Fe(III) species when As (V) is 40–50 mM. In addition, when the concentration of As(V) > 30 mM, the

**Table 3.3** The reaction equilibrium constants and Gibbs free energies of Fe–As(V) complexes

| Complexes | Delta G (kJ $mol^{-1}$) | | lg$K(M^n + H_3AsO_4 = MH_mAsO_4^{(n+m-3)} + (3-m)H^+)$ | | | | |
|---|---|---|---|---|---|---|---|
| | Langmuir [66] | Marini [64] | Robin [67] | Whiting [68] | Langmuir [66] | Marini [6[illegible]] | Marini [64] |
| $Fe^{III}H_2AsO_4^{2+}$ | −793.92 | −793.97 | 1.8(1.74) | – | 1.74 | 2.00(1.88) | 2.[illegible]7 |
| $Fe^{III}HAsO_4^+$ | −787.24 | −786.63 | 0.66(0.57) | – | 0.57 | 0.71(0.59) | [illegible]7 |
| $Fe^{III}AsO_4$ | −771.49 | −743.46 | −1.8(−2.19) | – | −2.19 | −6.86(−6.[illegible]7) | −6.[illegible]9 |
| $Fe^{II}H_2AsO_4^+$ | −860.42 | −859.8 | – | 0.44(0.38) | 0.38 | 0.53(0.41) | 0.[illegible]0 |
| $Fe^{II}HAsO_4$ | −825.43 | −823.30 | – | −5.66(−5.75) | −5.75 | −5.87(−5.[illegible]9) | −5.[illegible]1 |
| $Fe^{II}AsO_4^-$ | −778.16 | −780.27 | – | −13.64(−14.03) | −14.03 | −13.41 (−13.53) | −13.[illegible]5 |

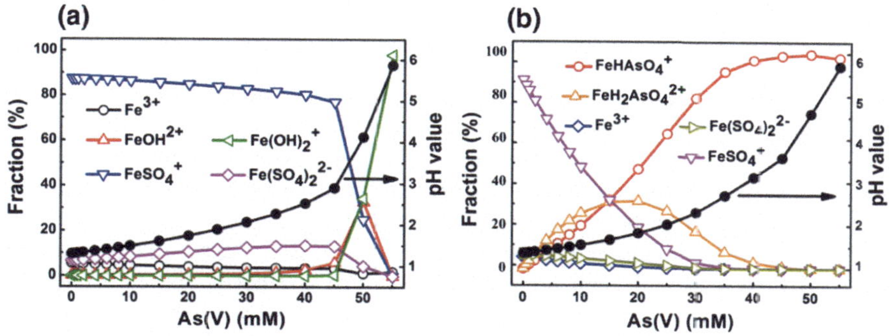


**Fig. 3.6** Fe(III) species analyses in UV-Vis experiment solutions with **a** exclusion and **b** inclusion of Fe(III)–As(V) complexes. Reprinted from Ref. [50] Copyright 2017, with permission from Elsevier

saturation index was higher than 0.577, and thus Hematite, Goethite, Lepidocrocite, $FeAsO_4{\cdot}2H_2O$, etc., are oversaturated, which make experimental solutions turbid. Accordingly, Fe(III)–As(V) complexes are significant to be considered in Fe(III)–As(V)–$H_2O$ acidic solutions. A speciation-pH diagrams with inclusion of Fe(III)–As(V) complexes were calculated for Fe(III)–As(V)–$H_2SO_4$–$H_2O$ system, a representative system of As(V) removal by iron salts from acidic wastewater. As displayed in Fig. 3.7a, As(V) mainly exists as (1) $FeH_2AsO_4^{2+}$ and $H_3AsO_4$ at pH < 1, (2) $FeH_2AsO_4^{2+}$, $FeHAsO_4^{+}$, $H_3AsO_4$, and $H_2AsO_4^-$ at pH = 1–2, and (3) $H_2AsO_4^-$ at pH > 2. As shown in Fig. 3.7b, Fe(III) mainly exists as $FeH_2AsO_4^{2+}$ at pH < 2, and $FeHAsO_4^+$ at pH > 2. However, the total dissolved Fe (III) concentration decreased when pH is higher than 2. Therefore, As(V) mainly exists as $FeH_2AsO_4^{2+}$, $FeHAsO_4^+$, $H_3AsO_4$, and $H_2AsO_4^-$ in acidic wastewater treatment system.

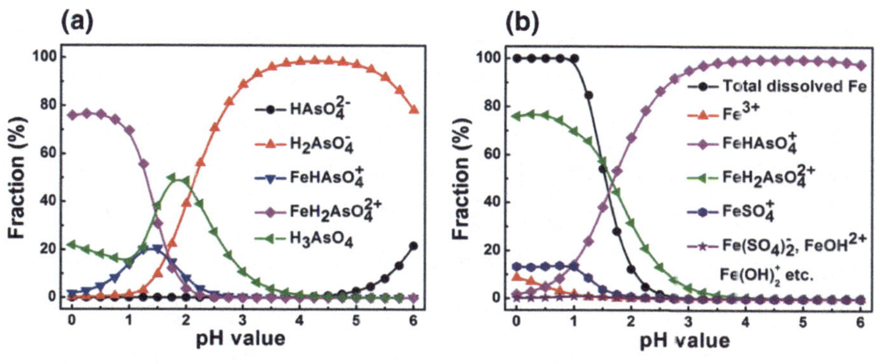


**Fig. 3.7** Speciation-pH diagrams of **a** As(V) and **b** Fe(III) with inclusion of Fe(III)–As(V) complexes. As(V) = Fe(III) = $SO_4$ = 0.1 M. Reprinted from Ref. [50] Copyright 2017, with permission from Elsevier

### 3.2.3 Eh-pH Diagram in Fe–As–$H_2O$ System

Fe(III)–As(V) complexes are significant for the speciation of Fe(III) and As(V) and should also be included in the Eh-pH diagram for As–Fe–$H_2O$ system. Although Fe (III)–As(V), Fe(III)–As(III), Fe(II)–As(V), and Fe(II)–As(III) complexes were all predicted, Fe(III)–As(III) and Fe(II)–As(III) complexes are ruled out because of the possible redox reaction between Fe(III) and As(III) and the oxidation of As(III) in parallel to the dark oxidation of Fe(II) by dissolved $O_2$ [69]. An Eh-pH diagram for Fe–As–$H_2O$ system with inclusion of Fe(III)–As(V) and Fe(II)–As(V) complexes were constructed and compared with the one with exclusion of these complexes. Figure 3.8a with exclusion of Fe–As complexes showed that $FeAsO_4$(s) was predominant in the range of pH 1.2–3.4 under oxidizing conditions. It was reported that amorphous ferric arsenate could transform into scorodite [70]. The kinetics of scorodite formation and its transformation from ferric arsenic is strongly controlled by pH, for example, scorodite precipitated after ~384 h at pH 4.5 but ~13 h at pH 1 [71]. The pH of smelting acidic wastewater is usually less than 3, which is suitable for the production of ferric arsenate and scorodite. However, Fe(III)–As(V) complexes ($FeH_2AsO_4^{2+}$, $FeHAsO_4^+$, and $FeH_2AsO_4^+$) were reported predominant under extremely acidic pH condition [72]. Thus, the Eh-pH diagram with inclusion of Fe–As complexes is more meaningful to understand the arsenic geochemistry in Fe–As–$H_2O$ system. Fe–As complexes were not considered in previous Eh-pH diagram mainly due to the unreliability of thermodynamic data. Accordingly, a new Eh-pH diagram for Fe–As–$H_2O$ system was constructed based on the estimates of equilibrium constants and Gibbs free energies. Figure 3.8b showed that Fe(III)–As (V) complexes shrank the stability field of $H_3AsO_4$ and $FeAsO_4$(s). Moreover, Fe (II)–As(V) complexes occur at pH range of 0.5–3.0. The predominant field of $Fe_3(AsO_4)_2$ was affected by Fe(II)–As(V) complexes. $FeH_2AsO_4^{2+}$, $FeHAsO_4^+$, and $FeH_2AsO_4^+$ were all restricted at acidic pH conditions. Consequently, Fe–As(V)

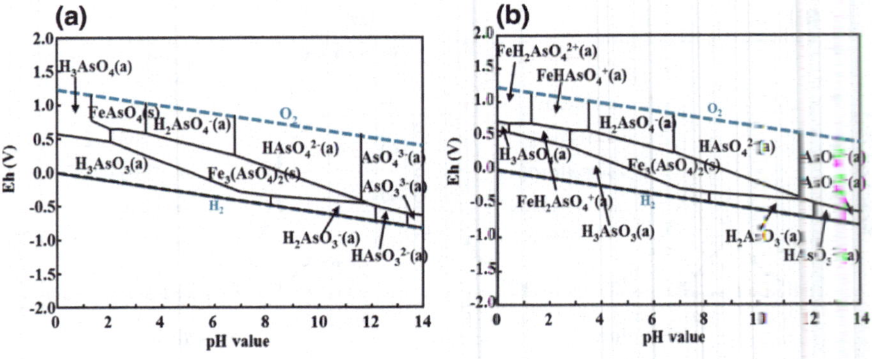


**Fig. 3.8** The Eh-pH diagram in 0.1 M As–0.1 M Fe–$H_2O$ system with the **a** exclusion and **b** inclusion of Fe–As complexes at 25 °C and 1 bar. Eh = E (vs. SCE) + 0.245. Reprinted from Ref. [50] Copyright 2017, with permission from Elsevier

complexes are considerable in the arsenic removal from smelting acidic wastewaters by coprecipitation with iron salts.

### 3.2.4 Fe(III)–As(V) Complexes

In the sample solutions, Fe(III) concentration was 0.5 mmol $L^{-1}$, and As(V) concentration systematically increased from 0 to 30 mmol $L^{-1}$ in both $HClO_4$ and $H_2SO_4$ systems. The pH value of all sample solutions was controlled at lower than 2.1, thus deprotonated As(V) species ($H_2AsO_4^-$, $HAsO_4^{2-}$, and $AsO_4^{3-}$) and Fe(III) hydroxo complexes ($Fe(OH)^{2+}$, $Fe(OH)_2^+$, $Fe(OH)_3^0$, etc.) are ruled out [73, 74]. Fe (III)–As(V) complexes, if any, coexist with protonated As(V) ($H_3AsO_4$), free iron ion ($Fe^{3+}$), and Fe(III)-sulfate complexes in sample solutions. The baseline-corrected UV-Vis spectra were shown in Fig. 3.9[75]. In $HClO_4$ system, perchlorate ($ClO_4^-$) ion does not complex with Fe(III) [76], thus $Fe^{3+}$ ion is the main Fe(III) species in As(V)-free solution. An absorption band on the spectrum of As(V)-free solution was located at ~240 nm (Fig. 3.9a), which is the feature band of $Fe^{3+}$ [77]. When As(V) concentration increase, the spectra curves in the range of 250–340 nm gradually rise. An absorption band was located at around 280 nm, and its absorption strength is positively related to As(V) concentration. The absorption band of $H_3AsO_4$ is not close to 280 nm [43]. Therefore, only Fe(III)–As(V) complexes can explain the band at ~280 nm and the enhanced absorption in range of 250–340 nm. In $H_2SO_4$ system, sulfate can complex with Fe(III) to form Fe(III)-sulfate complexes ($FeSO_4^+$, $Fe(SO_4)_2^-$, $FeHSO_4^{2+}$, etc.) [19]. The absorption bands at ~305 and ~225 nm on the spectrum of As(V)-free solution are the feature bands of $FeSO_4^+$ [19, 78]. When As(V) concentration increases, spectra curves in the range of 240–300 nm gradually rise. A spectral band at ~300 nm is the feature band of $Fe(SO_4)_2^-$ [19]. Another band was located at about 280 nm and its

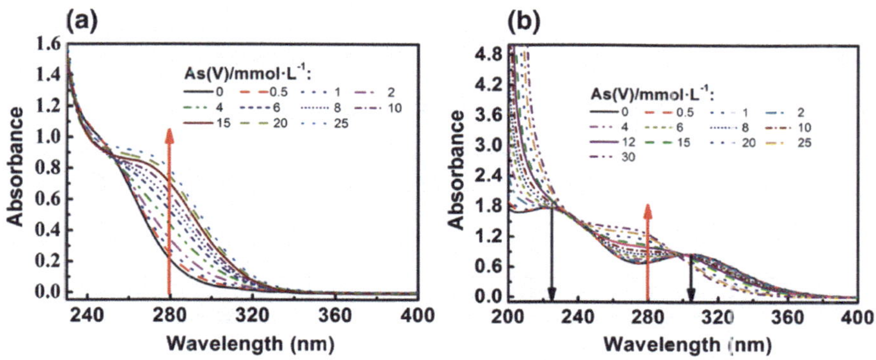


**Fig. 3.9** UV-Vis spectra of sample solutions in **a** Fe(III)–As(V)–$HClO_4$ system **b** Fe(III)–As(V)–$H_2SO_4$ system. Reprinted from Ref. [50] Copyright 2017, with permission from Elsevier

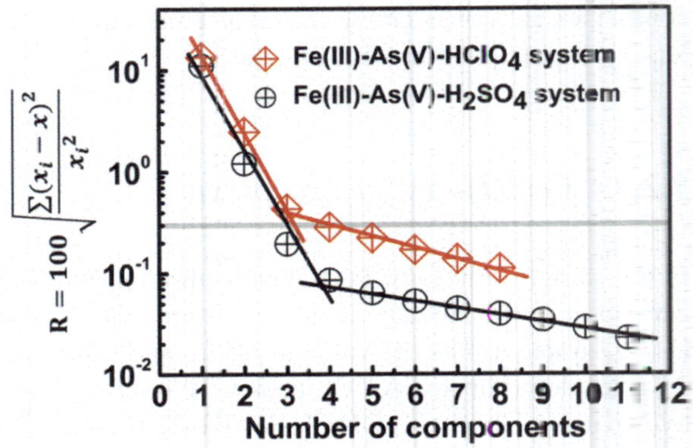


**Fig. 3.10** Principal component analysis (PCA) on spectroscopic data in Fe(III)–As(V)–$HClO_4$ and Fe(III)–As(V)–$H_2SO_4$ system. The grey line indicates the level of analytical uncertainty of 0.3%. Reprinted from Ref. [50] Copyright 2017, with permission from Elsevier

absorption strength is also positively related to As(V) concentration. Accordingly, the ~280 nm band in both the $HClO_4$ and $H_2SO_4$ system is probably owing to Fe(III)–As(V) complexes.

Quantitative analysis of UV-Vis data was performed and principal component analysis (PCA) was applied. PCA of the two series of spectra indicate that at least three and four absorbing species are believable to explain the UV-Vis data obtained in Fe(III)–As(V)–$HClO_4$ system and Fe(III)–As(V)–$H_2SO_4$ system (Fig. 3.10). It is likely that $Fe^{3+}$ and Fe(III)–As(V) complexes are responsible for the spectra obtained in the $HClO_4$ system, while $FeSO_4^+$, $Fe(SO_4)_2^-$, and Fe(III)–As(V) complexes are responsible in the $H_2SO_4$ system (Fig. 3.11). Aqueous Fe(III)–As(V) complexes are mononuclear structure with a formula of $FeH_2AsO_4^{2+}$, $FeHAsO_4^+$ or $FeAsO_4^0$ rather than multinuclear structure [67, 72, 73, 79–81]. Gel-like $FeAsO_4^0$ can make solution turbid and absorb visual light [43]. Hence, $FeH_2AsO_4^{2+}$ and $FeHAsO_4^+$ are proposed in limpid solutions. In addition, given the strong resemblance between $H_3AsO_4$ and $H_3PO_4$, $FeH_2AsO_4^{2+}$ and $FeHAsO_4^+$ are conceivable

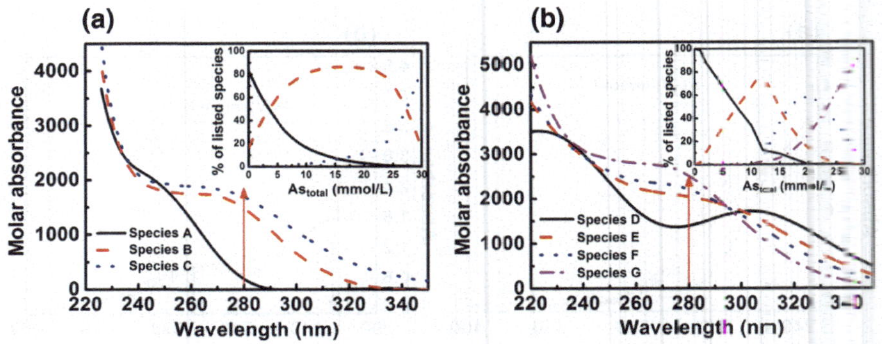


**Fig. 3.11** Molar absorbance spectra from "model-free" analysis in **a** Fe(III)–As(V)–$HClO_4$ system and **b** Fe(III)–As(V)–$H_2SO_4$ system. Reprinted from Ref. [50] Copyright 2017, with permission from Elsevier

because $FeH_2PO_4^{2+}$ and $FeHPO_4^+$ are the main Fe(III)-phosphate complexes [82, 83]. Based on the "model-free" analysis, the molar absorbance spectrum curve of species A resembles the experimental spectrum of As(V)-free solution in $HClO_4$ system, which implies that species A is $Fe^{3+}$. The molar absorbance spectra of the other two species, B and C, have little discrepancy and have an absorption band at ~280 nm. Thus, they are postulated as Fe(III)–As(V) complexes $FeH_2AsO_4^{2+}$ and $FeHAsO_4^+$. Species D has two bands at about 305 and 225 nm, and species E has the feature band at ~300 nm. Thus, species D and E are $FeSO_4^+$ and $Fe(SO_4)_2^-$, respectively. The spectra of species F and G are as same as B and C, which are identified as $FeH_2AsO_4^{2+}$ and $FeHAsO_4^+$ with the same absorption band at ~280 nm. Consequently, two Fe(III)–As(V) complexes, $FeH_2AsO_4^{2+}$, and $FeHAsO_4^+$, were reasonable in both Fe(III)–As(V)–$HClO_4$ and Fe(III)–As(V)–$H_2SO_4$ systems, and they have a feature band at ~280 nm.

X-ray absorption near edge structure (XANES) of Fe K-edge in Fe(III)–As(V)–$HClO_4$ solutions dropped with increased As(V) concentration, but the primary feature was same (Fig. 3.12). The hexahydrate complex ($[Fe(H_2O)_6]^{3+}$) with an octahedral Fe–O geometry [84, 85] is the only Fe(III) species in As(V)-free solution. The weak 1s → 3d transition (7110–7116 eV) was not enhanced as As(V) increased (see the inset in Fig. 3.12), which implies the Fe–O coordination of Fe(III)–As(V) complexes was still octahedral and same to $[Fe(H_2O)_6]^{3+}$ In addition, the edge crest 1s → 4p becomes lower and broader as As(V) concentration increased, which shows that the Fe–O bond in Fe(III)–As(V) complexes will be inequivalent because of the common effects of complexation on distortion in octahedral geometry with shorter Fe–O distances to the anion(s) [86]. Moreover, the next higher energy feature after edge crest corresponds to single scattering of the photoelectron by the nearest neighbor O atom and shifts to higher energy with the increased As(V) concentration, implying that the nearest Fe–O distance became shorter due to the formation of Fe(III)–As(V) complexes in As(V)-contained solutions.

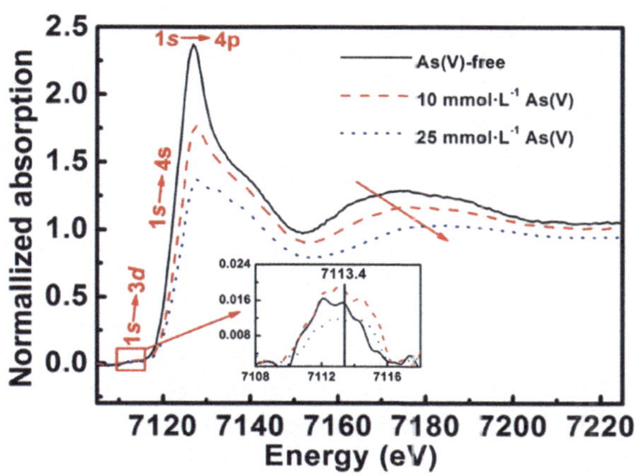


**Fig. 3.12** XANES of Fe K-edge in Fe(III)–As(V)–$HClO_4$ system (uncorrected for phase shift). Reprinted from Ref. [50] Copyright 2017, with permission from Elsevier

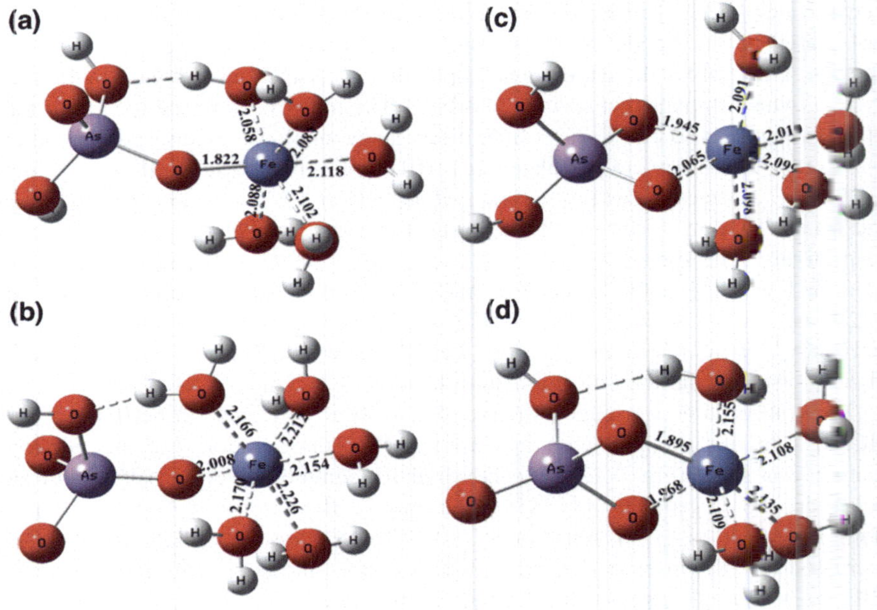


**Fig. 3.13** The optimized structures of **a** $[FeH_2AsO_4(H_2O)_5]^{2+}$, **b** $[FeH_2AsO_4(H_2O)_4]^{2+}$ **c** $[FeHAsO_4(H_2O)_5]^{+}$, and **d** $[FeHAsO_4(H_2O)_4]^{+}$. Fe–O bond distances are shown. Reprinted from Ref. [50] Copyright 2017, with permission from Elsevier

Since Fe–O coordination of Fe(III)–As(V) complexes is same as that of $Fe(H_2O)_6^{3+}$, it is feasible that $HAsO_4^{2-}$ and $H_2AsO_4^{-}$ substitute for one or two water molecules to form $[FeH_2AsO_4(H_2O)_5]^{2+}$ and $[FeHAsO_4(H_2O)_5]^{+}$, or $[FeH_2AsO_4(H_2O)_4]^{2+}$ and $[FeHAsO_4(H_2O)_4]^{+}$ (Fig. 3.13). Compared with the Fe–O bond in $Fe(H_2O)_6^{3+}$ (2.017), all Fe–O($H_2$) bonds of Fe(III)–As(V) complexes are longer than 2.017 Å. However, Fe–O(As) bonds are shorter than 2.017 Å. The average Fe–O distance of $[FeH_2AsO_4(H_2O)_5]^{2+}$, $[FeH_2AsO_4(H_2O)_4]^{2+}$, $[FeHAsO_4(H_2O)_5]^{+}$, and $[FeHAsO_4(H_2O)_4]^{+}$ are, respectively, $2.045 \pm 0.102$, $2.068 \pm 0.055$, $2.157 \pm 0.07$, and $2.065 \pm 0.096$ Å. Thus, the Fe–O bonds in Fe(III)–As(V) complexes are significantly inequivalent, which leads to irregular Fe sites and agrees with the XANES analysis. Additionally, the Fe–As distance of ~2.7 Å in $[FeH_2AsO_4(H_2O)_4]^{2+}$ and $[FeHAsO_4(H_2O)_4]^{+}$ is in accordance with that in bidentate edge-sharing ($^2E$) surface complexes [87, 88]. The Fe–As distance in $[FeHAsO_4(H_2O)_5]^{+}$ is 3.38 Å and in $[FeH_2AsO_4(H_2O)_5]^{2+}$ is 3.40 Å, which agrees with that in scorodite ($FeAsO_4{\cdot}2H_2O$) [87, 89, 90]. The extended x-ray absorption fine structure (EXAFS) of Fe K-edge were analyzed based on the four optimized structures of Fe(III)–As(V) complexes. The $k^2$-weighted $x(k)$ functions fitted by monodentate structures ($[FeHAsO_4(H_2O)_5]^{+}$ and $[FeH_2AsO_4(H_2O)_5]^{2+}$) are satisfied (Fig. 3.14a and Table 3.4). First Fe–O shell was satisfactorily fitted with a mean distance of 1.98 Å. $\Delta R = 0.10$–$0.18$ Å implies a distorted $FeO_6$ octahedra in Fe(III)–As(V)

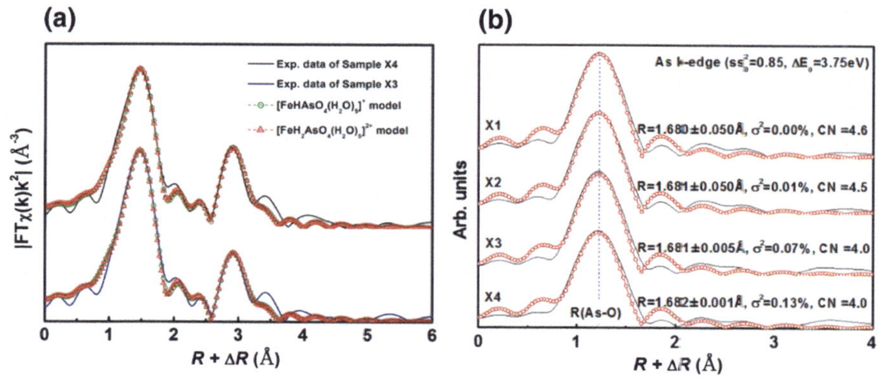


**Fig. 3.14** The Fourier transform of **a** Fe K-edge and **b** As K-edge in Fe(III)–As(V)–$HClO_4$ system (uncorrected for phase shift). Reprinted from Ref. [50] Copyright 2017, with permission from Elsevier

complexes. Similarly, $FeO_6$ octahedron is also distorted in the scorodite structure with a Fe–O bond length spread of 0.167 Å [91]. Second Fe–As shell fitting gave a mean distance of 3.25–3.26 Å, which was a little shorter than that in crystalline scorodite (3.33 ± 0.01 Å) [92]. However, bidentate structures of Fe(III)–As(V) complexes ($[FeHAsO_4(H_2O)_4]^+$ and $[FeH_2AsO_4(H_2O)_4]^{2+}$) with much shorter Fe–As distance (2.7 Å) are unlikely to be realistic due to unsatisfactory fitting. Therefore, Fe(III)–As (V) complexes are monodentate structures, i.e., $[FeHAsO_4(H_2O)_5]^+$ and $[FeH_2AsO_4(H_2O)_5]^{2+}$. Meanwhile, the EXAFS spectra of As K-edge in sample solutions were all well fitted with $AsO_4$ tetrahedra, for example, CN = 4.0 and R(As–O) = 1.68 (Fig. 3.14b). There was no As–Fe path because As(V) is great excessive and exists as $H_3AsO_4$ in sample solutions.

TDDFT calculations were carried out at the CAM-B3LYP/6-311+G** level to illuminate the absorption spectroscopic nature of $[FeHAsO_4(H_2O)_5]^+$ and $[FeH_2AsO_4(H_2O)_5]^{2+}$. The accuracy of spectra calculation was evaluated by Fe $(H_2O)_6{}^{3+}$ and the calculated feature band of $Fe(H_2O)_6{}^{3+}$ at around 240 nm is consistent with experimental spectrum. The absorptions of both $[FeHAsO_4(H_2O)_5]^+$ and $[FeH_2AsO_4(H_2O)_5]^{2+}$ are in the range of 240–340 nm (Fig. 3.15), which is consistent with our experimental results, i.e., 250–340 nm in the $HClO_4$ system and 240–300 nm in the $H_2SO_4$ system. Several strong electron excitations of $[FeH_2AsO_4(H_2O)_5]^{2+}$ and $[FeHAsO_4(H_2O)_5]^+$ with significant oscillator strength ($f \geq 0.01$) were considered to elucidate the absorption properties and listed in Table 3.5. There are three well-known types of excitations, including local excitation (LE), charge-transfer excitation (CT), and Rydberg excitation (R) [93]. The $\Delta r$ is proposed as a quantitative indicator of electron excitation type. $\Delta r \leq 2$ Å indicates LE, and $\Delta r \geq 2$ Å identifies strong CT [93]. To a great extent, S6 of $[FeH_2AsO_4(H_2O)_5]^{2+}$, S28–29 and S33 of $[FeHAsO_4(H_2O)_5]^+$ are in CT character, and other states are in LE character. Other parameters are auxiliary to identify the

**Table 3.4** Shell-fit parameters of Fe K-edge EXAFS spectra based on monodentate structures calculated by DFT

| Exp. data (fitting models) | Reduced $\chi2$ | R-factor | First Fe–O shell | | | Second Fe–As shell | | | Second Fe–O shell | |
|---|---|---|---|---|---|---|---|---|---|---|
| | | | CN | R(Å) | $\sigma^2$ (Å$^2$) | CN | R (Å) | $\sigma^2$ (Å$^2$) | CN | R (Å) |
| X3 ($[FeHAsO_4(H_2O)_5]^+$) | 56.71 | 0.0067 | 6.0 | 1.984 ± 0.186 | 0.0086 | 2.7 | 3.247 ± 0.131 | 0.0048 | 4.4 | 3.527 ± 0.003 |
| X3 ($[FeH_2AsO_4(H_2O)_5]^{2+}$) | 54.50 | 0.0065 | 6.0 | 1.979 ± 0.103 | 0.0085 | 3.3 | 3.246 ± 0.156 | 0.0058 | 4.1 | 3.500 ± 0.050 |
| X4 ($[FeHAsO_4(H_2O)_5]^+$) | 96.11 | 0.0056 | 6.0 | 1.987 ± 0.183 | 0.0083 | 2.8 | 3.259 ± 0.118 | 0.0066 | 4.7 | 3.518 ± 0.006 |
| X4 ($[FeH_2AsO_4(H_2O)_5]^{2+}$) | 58.37 | 0.0059 | 6.0 | 1.981 ± 0.102 | 0.0074 | 3.7 | 3.259 ± 0.144 | 0.0080 | 4.6 | 3.491 ± 0.059 |

*Note* Fit range = 0.9–3.7 Å. CN = coordination number (path degeneracy). $\sigma^2$ = Debye–Waller parameter. R = mean half path length. $\Delta E_0$ = the energy-shift parameter, $\Delta E_0$(Fe–O) = −3.30 eV and $\Delta E_0$(Fe–As) = 8.31 eV. $S_0{}^2$ = the passive amplitude reduction factor, $S_0{}^2$ = 0.965

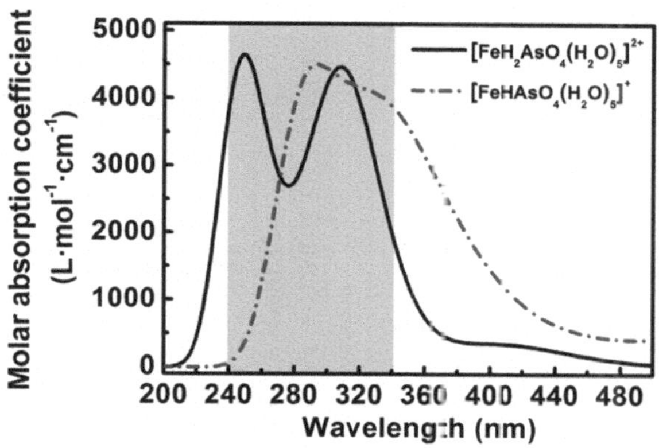


**Fig. 3.15** The calculated UV-Vis spectra of $[FeH_2AsO_4(H_2O)_5]^{2+}$, and $[FeHAsO_4(H_2O)_5]^{+}$. The grey area indicates experimental absorption range of Fe(III)–As(V) complexes. Reprinted from Ref. [50] Copyright 2017, with permission from Elsevier

excitation types, which include the distance between centroid of hole and electron (D), and the integral of overlap of hole-electron (I) (Table 3.5). As depicted in Fig. 3.16, electron and hole are severally centered on O atoms and Fe atoms, which indicate that electron transfer proceeds from O atoms to Fe atoms. However, O atoms in Fe(III)–As(V) complexes are inequivalent, which can be defined as As-bonded O, Fe–As-bridged O and water O atoms in As–O(H), Fe–O–As and H–O–H structures, respectively.

According to the contribution of molecule orbitals (MO) to electron and hole, the significant electron transitions can be symbolized as corresponding frontier molecular orbital excitations. As depicted in Figs. 3.17 and 3.18, all related frontier singly occupied molecule orbitals (SOMO) are principally characterized as d-orbital of Fe atom. However, the frontier highest occupied molecule orbitals (HOMO) are totally different, which contributes to various electron excitations. The electron excitations S28 and S29 of $[FeHAsO_4(H_2O)_5]^{+}$ have absorptions at 330–340 nm, which are symbolized as HOMO-3 → SOMO and HOMO-2 → SOMO. HOMO-2 and HOMO-3 of $[FeHAsO_4(H_2O)_5]^{+}$ are localized mainly on As-bonded O atoms. It is accordingly an electron excitation from As-bonded O atoms to d-orbital of Fe, corresponding to a ligand-to-metal charge transfer (LMCT). In the case of S6 of $[FeH_2AsO_4(H_2O)_5]^{2+}$ and S33 of $[FeHAsO_4(H_2O)_5]^{+}$, the involved HOMOs are mainly located on Fe–As-bridged O atoms, which yields an excitation around 300–310 nm. As shown in Fig. 3.16, a-S6 and b-S33, there is no significant overlap between the electron and hole, consequently LMCT type. Nearly equal excitation wavelengths for S36 and S38–39 of $[FeHAsO_4(H_2O)_5]^{+}$ is 280–290 nm. The involved HOMOs approximately spread over all O atoms, whereas contributions of Fe–As-bridged O are predominant. Accordingly, S36 and S38–39 of $[FeHAsO_4(H_2O)_5]^{+}$ are assigned to MC transitions. In a case of S20–21 of $[FeH_2AsO_4(H_2O)_5]^{2+}$, their excitation wavelengths are around 250 nm, corresponding to MC transition with the major contributions of water O atoms. Since all SOMOs are mainly characterized as d-orbital of Fe, the excitation wavelength is blue-shifted from 340 to 250 nm along with the localization of HOMOs on

**Table 3.5** Major electron excitations of (a) $[FeH_2AsO_4(H_2O)_5]^{2+}$ and (b) $[FeHAsO_4(H_2O)_5]^{+}$

| Complex | State | λ (nm) | ΔE (eV) | f | Δ*r* (Å) | I | D (Å) | MO approximation | Type |
|---|---|---|---|---|---|---|---|---|---|
| a | S6 | 309.62 | 4.0043 | 0.1058 | 2.663 | 0.184 | 1.681 | HOMO-2,3,4 → SOMO-2,3 | LMCT |
| | S20 | 248.62 | 4.9869 | 0.0154 | 1.615 | 0.098 | 0.772 | HOMO-7 → SOMO-4,3 | MC |
| | S21 | 247.39 | 5.0117 | 0.0626 | 1.609 | 0.194 | 0.616 | HOMO-8,3 → SOMO | MC |
| b | S28 | 339.32 | 3.6564 | 0.0124 | 3.890 | 0.073 | 3.847 | HOMO-3 → SOMO | LMCT |
| | S29 | 330.88 | 3.7497 | 0.0558 | 3.497 | 0.113 | 3.449 | HOMO-2 → SOMO | LMCT |
| | S33 | 303.31 | 4.0906 | 0.0124 | 2.289 | 0.084 | 2.130 | HOMO-6 → SOMO-1 | LMCT |
| | S36 | 288.90 | 4.2946 | 0.0159 | 1.297 | 0.126 | 1.067 | HOMO-8 → SOMO-2 | MC |
| | S38 | 284.75 | 4.3572 | 0.0369 | 1.420 | 0.153 | 0.472 | HOMO-6 → SOMO-2 | MC |
| | S39 | 284.21 | 4.3654 | 0.0214 | 1.047 | 0.116 | 0.977 | HOMO-9,10 → SOMO-4 | MC |

Note: $\lambda$ absorption wavelength, $\Delta E$ excitation energies, $f$ oscillator strengths, $\Delta r$ a quantitative indicator of electron excitation type, $I$ the integral of overlap of hole-electron, $D$ the distance between centroid of hole and electron. *MO Approximation* molecular orbital approximation of each electron excitation

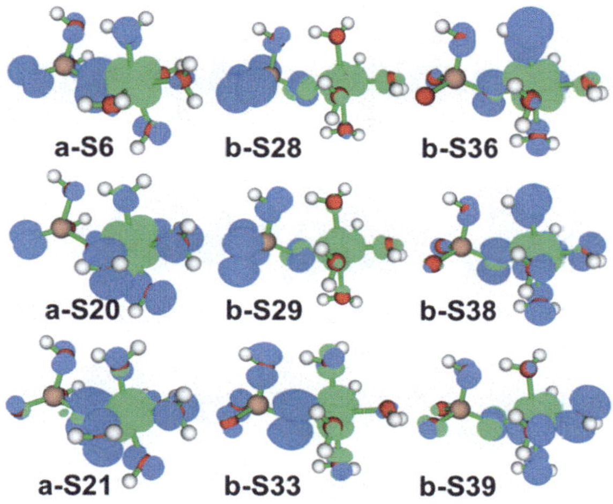


**Fig. 3.16** Isosurface of hole and electron distribution. **a** and **b** respectively, represent $[FeH_2AsO_4(H_2O)_5]^{2+}$ and $[FeHAsO_4(H_2O)_5]^+$ (e.g., a-S6 indicates S6 of $[FeH_2AsO_4(H_2O)_5]^{2+}$). Blue and green isosurface represent electron and hole distributions, respectively. Brownish, red, and white spheres correspond to arsenic, oxygen, and hydrogen atoms, respectively. Iron atoms are wrapped by the isosurface of hole distribution. Reprinted from Ref. [50] Copyright 2017, with permission from Elsevier

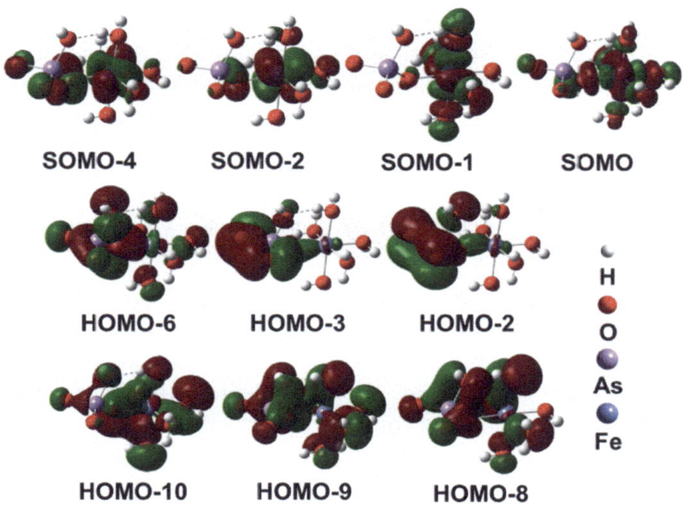


**Fig. 3.17** Related frontier molecular orbitals of $[FeHAsO_4(H_2O)_5]^+$ calculated at the TDDFT/CAM-B3LYP/6-311+G** CPCM level of theory (isosurface value = 0.02 e/Å$^3$) Reprinted from Ref. [50] Copyright 2017, with permission from Elsevier

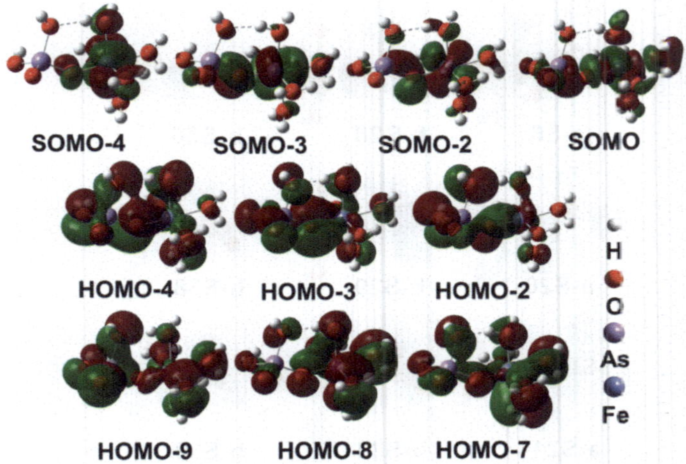


**Fig. 3.18** Related frontier molecular orbitals of $[FeH_2AsO_4(H_2O)_5]^{2+}$ calculated at the TDDFT/CAM-B3LYP/6-311+G** CPCM level of theory (isosurface value = 0.02 e/Å$^3$). Reprinted from Ref. [50] Copyright 2017, with permission from Elsevier

As-bonded O, Fe–As-bridged O and water O atoms. Therefore, the experimental spectra in the range of 250–340 nm in Fe(III)–As(V)–$HClO_4$ system and of 240–300 nm in Fe(III)–As(V)–$H_2SO_4$ system are attributed to different electron excitations of Fe(III)–As(V) complexes. Moreover, the experimental band at around 280 nm is attributed to the MC excitations chiefly from Fe–As-bridged O atoms to d-orbital of Fe, which can be considered a feature band of aqueous Fe(III)–As(V) complexes.

### 3.2.5 *Species Transformation of Fe(III)–As(V) Complexes*

The complexation of Fe(III) and As(V) was investigated spectrophotometrically in a series of Fe(III)–As(V)–$HClO_4$–$H_2O$ solutions. The absorbance at 280 nm increased proportionally to the concentrations of As(V) due to the formation of Fe (III)–As(V) complexes shown in Fig. 3.19a. In order to verify that the absorbance at 280 nm was due to ferric–arsenate complex, same method was conducted to the Fe (III)–$SO_4$ complexes and the absorbance at 305 nm corresponding to $FeSO_4^+$ [94, 95] was also shown in Fig. 3.19b. Here the correlation curve of absorbance at 305 nm versus sulfate was compared with that of absorbance at 280 nm versus arsenate in Fig. 3.20. Linear correlation with negative deviation at higher sulfate and arsenate concentrations was observed for the two correlation curves. The similar correlation trends of two curves gave indirect evidence that a peak at 280 nm corresponded to ferric–arsenate complex.

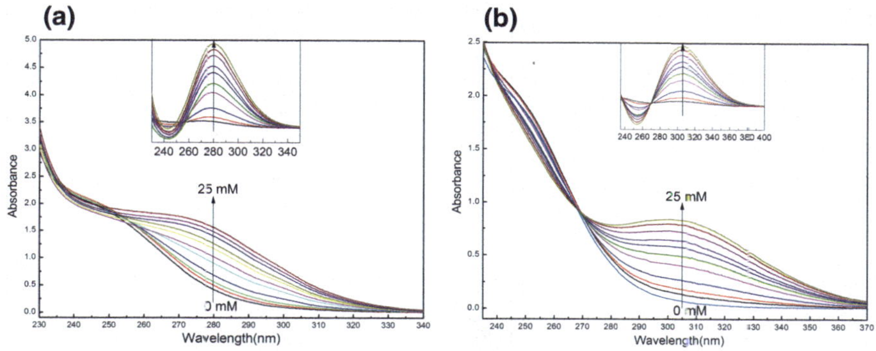


**Fig. 3.19** The UV-Vis spectra of **a** ferric–arsenate sample solutions and **b** ferric–sulfate sample solutions measured after 24 h standing at 25 ± 0.1 °C. Reprinted from Ref. [43] by permission of The Royal Society of Chemistry 2015

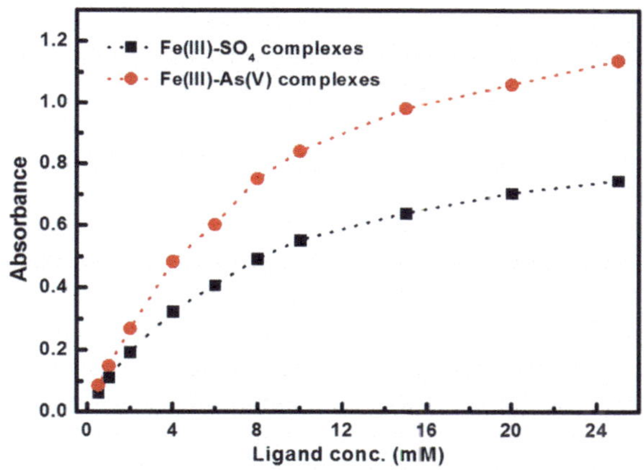


**Fig. 3.20** The correlation between absorbance of the feature band and sulfate/arsenate concentration. Reprinted from Ref. [43] by permission of The Royal Society of Chemistry 2015

To determine the equilibrium constant for Fe(III)–As(V) and Fe(III)–$SO_4$ complexes, the interaction was analyzed according to Benesi–Hildebrand equations (Eq. (3.1)) based on Fig. 3.20 [96].

$$\frac{[\text{Fe(III)}] \times l}{\Delta A} = \frac{1}{C} \times \frac{1}{\varepsilon K_f} + \frac{1}{\varepsilon} \tag{3.1}$$

where [Fe(III)] represents the concentration of the ferric ions, $l$ is the length of quartz cuvette, $\Delta A$ is the difference in absorbance between the complex and ferric ion, $C$ is the concentration of ligand; ε and $K$ are the molar absorptivity of the complex at $\lambda$ and the equilibrium constant for complex formation, respectively. The highly linear relationship ($R^2 = 0.9969$ and 0.9980) between $1/\Delta A$ and $1/C$ indicated that the absorbance of the complex followed the Benesi–Hildebrand

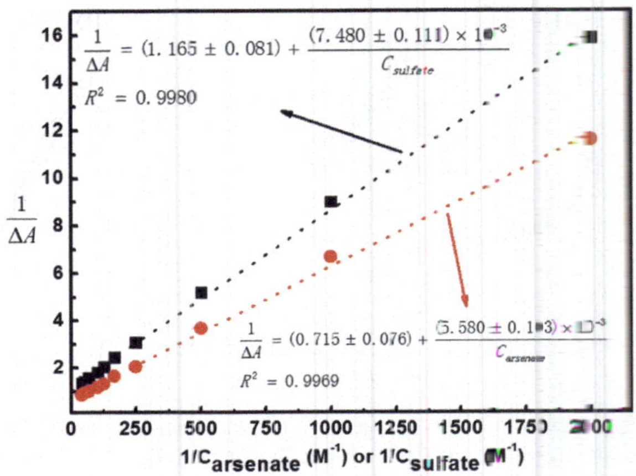


**Fig. 3.21** The Benesi–Hildebrand fitting of Fe(III)–As(V) complex (circle) and Fe(III)–$SO_4$ complex (square). Reprinted from Ref. [43] by permission of The Royal Society of Chemistry 2015

equation (Fig. 3.21). The molar absorptivity of Fe(III)–$SO_4$ and Fe(III)–As(V) complexes is $\varepsilon_{305} = 1.716 \times 10^3\ M^{-1}\ cm^{-1}$ and $\varepsilon_{280} = 2.797 \times 10^3\ M^{-1}\ cm^{-1}$ respectively. Correspondingly, the equilibrium is $K_f = 1.56 \times 10^2\ M^{-1}$ (log $K_f = 2.2$) and $K_f = 1.28 \times 10^2\ M^{-1}$ (log $K_f = 2.1$).

In order to investigate what would happen in As(V)–$SO_4^{\cdot -}$-rich system, UV-Vis spectra of two series of solutions containing both As(V) and $SO_4^{\cdot -}$ were collected for comparison. In the first series of solutions, $SO_4^{\cdot -}$ concentration was 20 mM and As(V) increased from 0 to 20 mM. Specifically, the absorbance increased significantly in the range of 240–300 nm and decreased significantly in the range of 300–380 nm (Fig. 3.22a). Therefore, Fe(III)–As(V) complexes were formed even in the presence of $SO_4^{\cdot -}$. In the second series of solutions, the concentration of As(V) was 20 mM and $SO_4^{\cdot -}$ increased from 0 to 20 mM, the absorbance of solutions showed no obvious change as $SO_4^{\cdot -}$ concentration increased (Fig. 3.22b). Accordingly, the formation of Fe(III)–$SO_4^{\cdot -}$ complexes were hindered by As(V), and thus the complexation of As(V) is stronger than $SO_4^{\cdot -}$ toward Fe(III).

The pH value is one of the key parameters for the formation of ferric arsenate solid from Fe(III)–As(V)–$H_2O$ system. In a series of solutions containing 0.5 mM Fe(III) and 20 mM As(V), perchlorate acid was added to control pH values in the range of 0.35 to 6.81. Sample solutions were limpid at pH $\leq$ 1.57 and turned turbid at higher pH of 2.38 because gel-like material was formed. These findings by naked eye were in accordance with the results of UV-Vis spectra (Fig. 3.23). The turbid solutions absorbed both ultraviolet light and visual light at pH $\geq$ 2.38, however, only ultraviolet light was absorbed at pH $\leq$ 1.57. It is well-known that the hydrolysis of Fe(III) ion and deprotonation of $H_3AsO_4$ are closely related to pH values. As(V) existed as $H_3AsO_4$ in acidic, $H_2AsO_4^-$ in weak acidic and $HAsO_4^{2-}$ in nearly neutral solutions due to $pK_1 = 2.2$ and $pK_2 = 6.96$. The first hydrolysis constant of $Fe^{3+}$ at 0.1 M $NaClO_4$ was 2.54 [97] and thus $FeOH^{2+}$ formed under

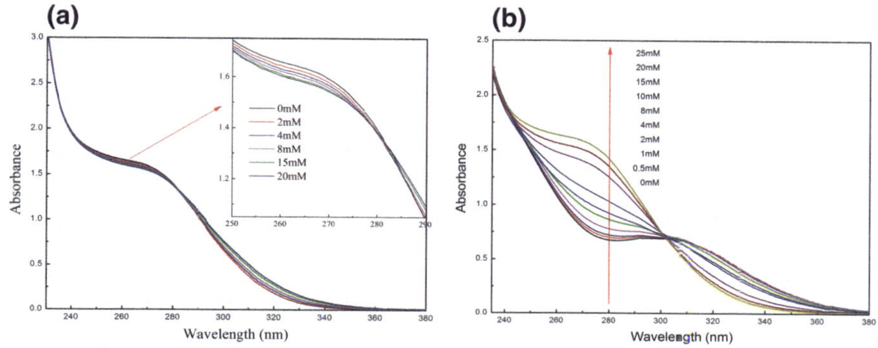


**Fig. 3.22** UV-Vis spectra of Fe(III)–As(V)–$SO_4$–$HClO_4$ solutions. **a** Fe: 0.5 mM; As(V): 20 mM; $SO_4$: 0–20 mM. **b** Fe: 0.5 mM, As(V): 0–25 mM; $SO_4$: 20 mM. Reprinted from Ref. [43] by permission of The Royal Society of Chemistry 2015

**Fig. 3.23** UV-Vis spectra at different pH values (Fe(III): 0.5 mM $L^{-1}$; As(V): 20 mM $L^{-1}$). Reprinted from Ref. [43] by permission of The Royal Society of Chemistry 2015

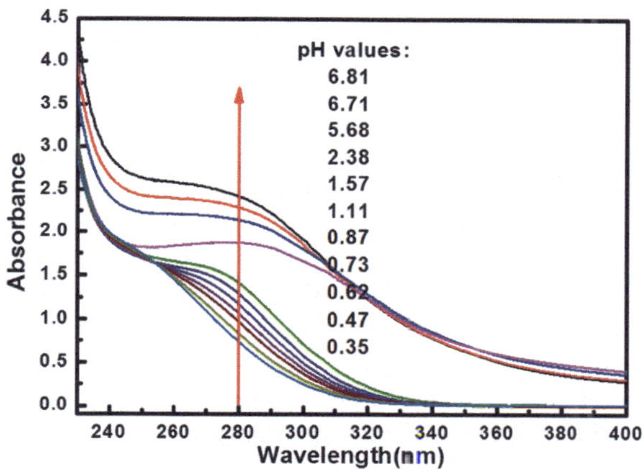


higher pH conditions, which can react with negatively charged $HAsO_4^{2-}$ to form $FeAsO_4$. Moreover, the deprotonation of $FeH_2AsO_4^{2+}$ and $FeHAsO_4^{+}$ with pH increase will also produce $FeAsO_4$. Therefore, the formed gel-like material was speculated as $FeAsO_4$ [98] under higher pH solutions.

Three solutions containing 0.5 mM Fe(III) and 10, 20 and 30 mM As(V) were heated at 90 °C for 30 min and then became turbid. The ultraviolet light was absorbed whether the solution was heated or not, and only visual light was absorbed after being heated (Fig. 3.24a). Thus, it can be concluded that higher temperature (T $\geq$ 90 °C) benefits for the formation of gel-like material. A continuous heating process was conducted for the solution containing 20 mM As(V) and the absorbance at 280 nm increased slightly at the early stage and increased dramatically later (Fig. 3.24b). An obvious boundary of the time was about 3000 s and the

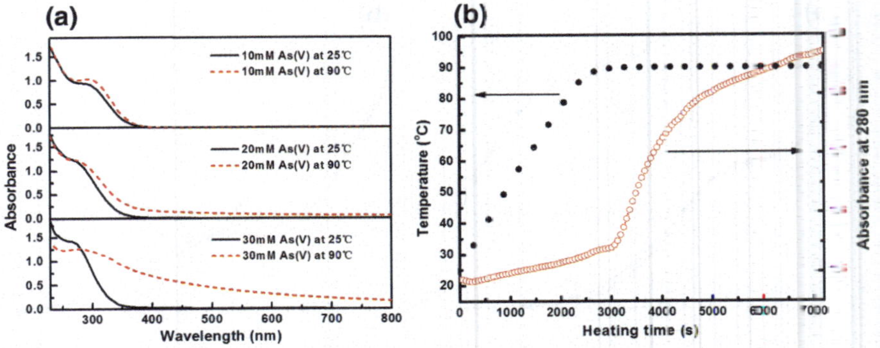


**Fig. 3.24** **a** UV-Vis spectra of sample solutions before and after heated (Fe(III): 0.5 mM $L^{-1}$; As (V): 10, 20 and 30 mM $L^{-1}$; Heating time: 20 min), **b** absorbance at 280 nm in a continuous heating process (Fe(III):0.5 mM $L^{-1}$; As(V): 20 mM $L^{-1}$). Reprinted from Ref. [43] by permission of The Royal Society of Chemistry 2015

temperature was about 90 °C. Therefore, enough heat is helpful to form the gel-like material.

Gel-like material formed in 0.1 M perchlorate acid solutions containing 45 mM As(V) and 0.5 mM Fe(III) was separated and dried at 60 °C. The molar ratio of As: Fe: O was 1: 1.11: 3.77 in this solid, which is close to the atomic ratio in $FeAsO_4$. XRD indicated that the gel-like material was poorly crystalline ferric arsenate characterized by two broad peaks centered at 2θ values of 28° and 58° (Fig. 3.25) [99]. Poorly crystalline ferric arsenate was reported as scorodite precursor with a formula of $FeAsO_4{\cdot}(2\ x)\ H_2O$ ($0 < x < 1$) [100]. The gel-like solid was further characterized using FTIR (Fig. 3.26). A strong band at 1630–1636 $cm^{-1}$ and a broad band near 3445 $cm^{-1}$ were attributed to the O–H bending and stretching vibration of water. The stretching vibration of As–O–Fe was located at 833 $cm^{-1}$

**Fig. 3.25** XRD patterns of the gel-like material in solid phase. Reprinted from Ref. [43] by permission of The Royal Society of Chemistry 2015

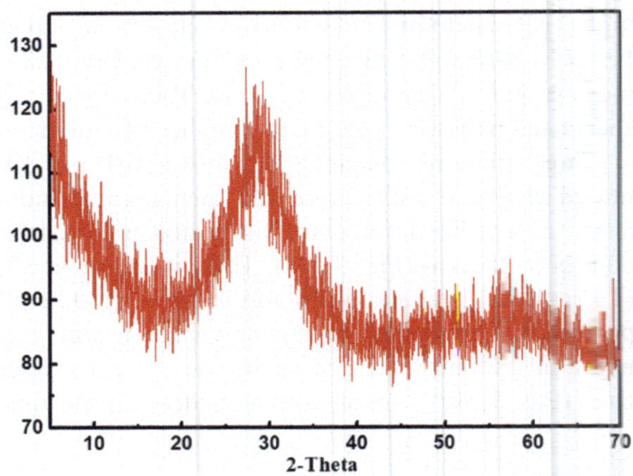

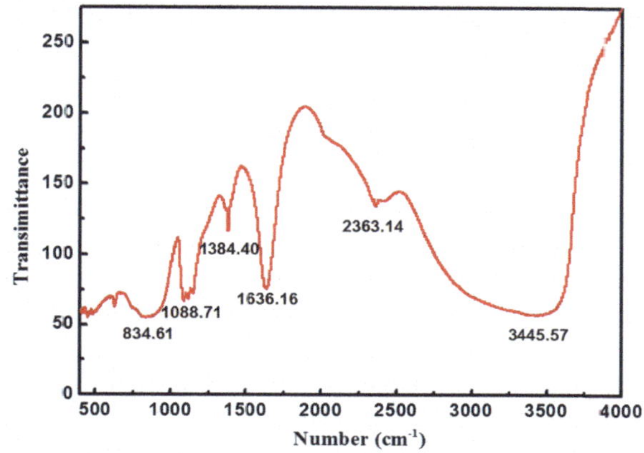


**Fig. 3.26** FTIR of the gel-like material in solid phase. Reprinted from Ref. [43] by permission of The Royal Society of Chemistry 2015

and generally with a shoulder at $\sim$750 cm$^{-1}$, which is probably caused by the hydrogen bonding between $H_2O$ and $AsO_4$. An important IR absorption at 835 cm$^{-1}$ in Fig. 3.25 with a shoulder at $\sim$750 cm$^{-1}$ belonged to the $v_3$ mode of $AsO_4$ ($\sim$833 cm$^{-1}$), identified as arsenate bonding structures of poorly crystalline ferric arsenate [99]. Other IR absorption bands at about 2363, 1384, and 1088 cm$^{-1}$ were, respectively, antisymmetric stretching mode of carbon dioxide (−OH···O=C=O) adsorbed on hydroxyl groups [101], characteristic IR absorption of $NO_3^-$ (1384 cm$^{-1}$) [102] and $ClO_4^-$ (1088 cm$^{-1}$) [103] anions. These impurities were introduced by iron nitrate and perchloric acid. The results of XRF, XRD, and FTIR demonstrated that gel-like material in our experiments was poorly crystallized ferric arsenate. Scanning electron microscopy (SEM) images showed that the sample was bulk and conchoidal (Fig. 3.27). It was different from other reported precipitates in Fe(III)–As(V)–$H_2O$ system which were aggregated particles [100]. This may be because the poorly crystalline ferric arsenate was formed from precursors of $FeHAsO_4^+$ and $FeH_2AsO_4^{2+}$, while other amorphous ferric arsenate was formed through transformation of the arsenate ions adsorption on the surface of the ferrihydrite.

**Fig. 3.27** SEM of the gel-like material in solid phase. Reprinted from Ref. [43] by permission of The Royal Society of Chemistry 2015

## 3.3 Photochemical Oxidation of As(III) and the Molecular Reaction Mechanism

### 3.3.1 The $SO_4^{\cdot -}$-Based As(III) Oxidation

Peroxydisulfate (PDS) was more effective than $H_2O_2$ for oxidizing monomethylarsonic acid (MMA) and dimethylarsinic acid (DMA) under UV irradiation [104]. PDS is advantageous over $H_2O_2$ for As(III) oxidation because the oxidation yields in PDS system were all higher than in $H_2O_2$ system, especially under Xe lamp irradiation (Fig. 3.28) [105]. The oxidation yield is 56% in PDS/Xe combination whereas only 12% and 17% in PDS/dark and PDS/daylight combination, respectively. Therefore, PDS/Xe is an effective combination for As(III) oxidation, where $SO_4^{\cdot -}$ may be generated due to the decomposition of PDS under Xe lamp irradiation [106]. Xe lamp solar simulator can emit the light of continuous band from UVA to visible range. As far as we know, UVA (320–400 nm) rays of solar radiation are natural resources on ground, which penetrate clouds more readily than UVB and UVC rays. Consequently, As(III) oxidation in PDS/Xe combination is meaningfully investigated to assess the feasibility of As(III) oxidation under solar radiation.

When As(III) oxidation under PDS/Xe combination was compared with that under Xe lamp alone and PDS alone conditions, the reaction between As(III) and PDS is very slow in absence of Xe lamp (PDS alone), and tiny As(III) (~10%) was oxidized after 30-min reaction (Fig. 3.29). However, about 55% As(III) was oxidized after 30 min with Xe lamp alone (without PDS), which is coincidentally equal to the oxidation yield in PDS/Xe combination at $R_{ox/As(III)} = 0.5$ in Fig. 3.28. When PDS was dosed in excess, i.e., $R_{ox/As(III)} = 2.5$, the reaction process in PDS/Xe combination is totally different from that of Xe lamp alone. In PDS/Xe combination at $R_{ox/As(III)} = 2.5$, a rapid oxidation was observed in the initial 2 min and the oxidation yield remained at ~82% afterwards. Similarly, a pronounced decrease of As(III) concentration was observed within the initial 2 min in Fe(II)-persulfate system, where $SO_4^{\cdot -}$ and $HO^{\cdot}$ were both determined to play key roles [106].

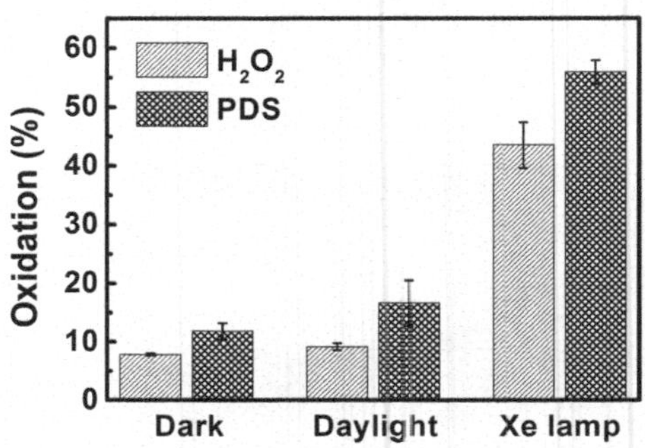


**Fig. 3.28** The oxidation yield of As(III) by PDS or $H_2O_2$ after 30-min reaction under dark, daylight and Xe lamp. $[As(III)]_0 = 20\ \mu mol\ L^{-1}$, $[PDS]_0 = [H_2O_2]_0 = 10\ \mu mol\ L^{-1}$. Reprinted from Ref. [105] Copyright 2017, with permission from Elsevier

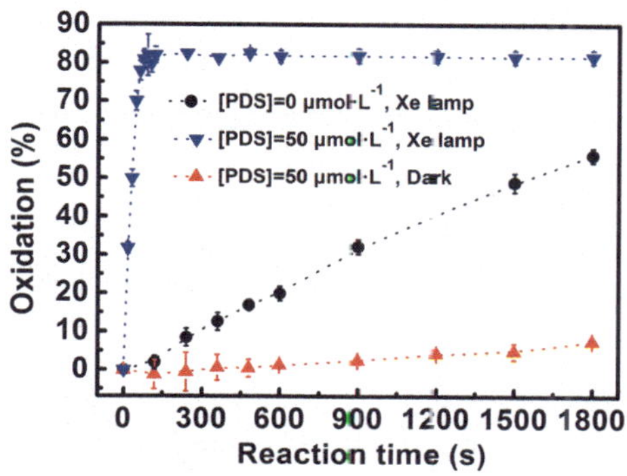


**Fig. 3.29** Comparison results in different reaction system. $[As(III)]_0 = 20\ \mu mol\ L^{-1}$. Reprinted from Ref. [105] Copyright 2017, with permission from Elsevier

$SO_4^{\cdot-}$ and $HO^{\cdot}$ were involved either individually or simultaneously during the activation of PDS due to the reaction of Eq. (3.2) or (3.3) [107].

$$SO_4^{\cdot-} + OH^- \rightarrow SO_4^{2-} + HO^{\cdot} \tag{3.2}$$

$$SO_4^{\cdot-} + H_2O \rightarrow SO_4^{2-} + HO^{\cdot} + H^+ \tag{3.3}$$

ESR was employed here to identify the radicals involved in As(III) oxidation process(Fig. 3.30). ESR signal of DMPO-HO$^{\cdot}$ adduct was obviously detected in PDS/Xe system in the presence and absence of As(III). Additionally, no DMPO-$SO_4^{\cdot-}$ adduct was detected even DMPO was added in great excess. It can be explained by the conversion from $SO_4^{\cdot-}$ to $HO^{\cdot}$ then followed by DMPO spin-trapping. In an acidic system, only $SO_4^{\cdot-}$ was expected to be the main reactive species [105]. However, only strong signal of DMPO-HO$^{\cdot}$ adduct was detected even in the presence of 2.5 mmol $H_2SO_4$ (pH = −0). Therefore, a fast disappearance of DMPO-$SO_4^{\cdot-}$ adduct and simultaneous generation of DMPO-HO$^{\cdot}$ adduct can be explained by the reaction of Eq. (3.4) [108, 109].

$$DMPO - SO_4^{\cdot-} + H_2O \leftrightarrow HSO_4^- + DMPO - HO^{\cdot} \tag{3.4}$$

Moreover, DMPO-HO$^{\cdot}$ adduct signal was not obtained in the absence of PDS, which confirmed that DMPO-HO$^{\cdot}$ adduct formation is concomitant with $SO_4^{\cdot-}$. Accordingly, $SO_4^{\cdot-}$ is the main reactive species, which simultaneously contributes to the oxidation of As(III) and the generation of $HO^{\cdot}$ in PDS/Xe system.

In quenching studies, methanol and isopropanol have been used as radical scavenger, which can readily react with both $SO_4^{\cdot-}$ and $HO^{\cdot}$ to confirm the radical and nonradical reaction [110] (Fig. 3.31). Tert-butanol was used as a selective scavenger because it reacts 1000-fold faster with $HO^{\cdot}$ ($6 \times 10^8\ M^{-1}\ s^{-1}$) [111] compared to $SO_4^{\cdot-}$ ($4–9.1 \times 10^5\ M^{-1}\ s^{-1}$) [112]. The oxidation of As(III) was almost completely stopped by 100 mmol $L^{-1}$ methanol and isopropanol, individually. Accordingly, the rapid As(III) oxidation is a radical reaction. In the presence of 1, 10 and 100 mmol $L^{-1}$ tert-butanol, the oxidation also occurred mainly in the initial few minutes, but the oxidation yield became lower in the presence of more

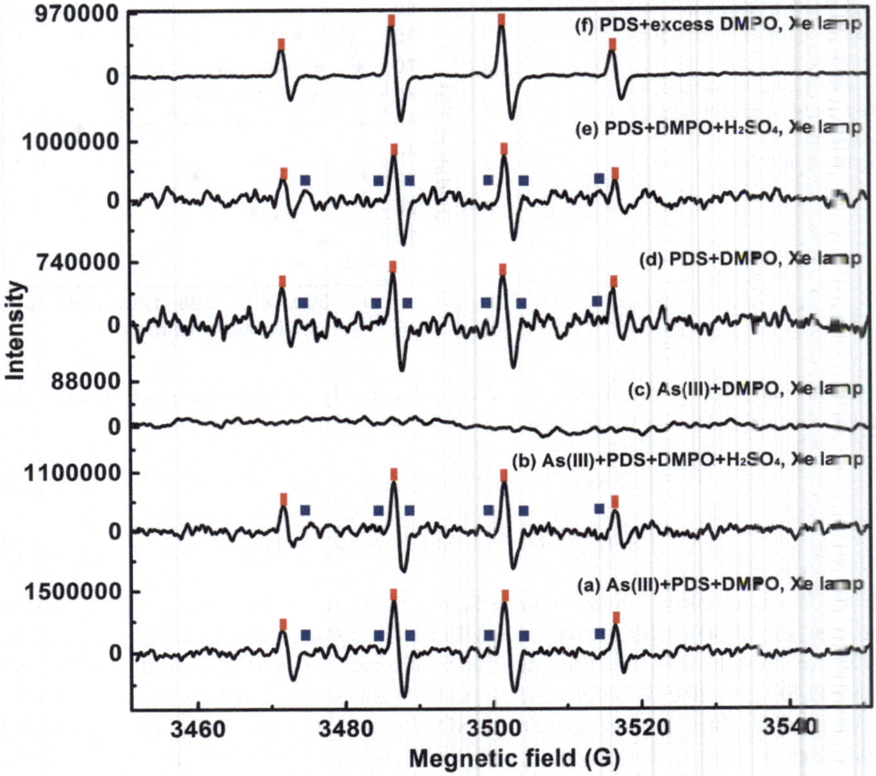


**Fig. 3.30** ESR spectra obtained from different systems within 2-min reaction. As(III) = 0.5 μmol. **a–e** PDS = 10 μmol, DMPO = 80 μmol, $H_2SO_4$ = 2.5 mmol; **f** PDS = 50 μmol, DMPO = 560 μmol. Signal of DMPO-HO˙ marked in red rectangle and DMPO-$SO_4^{\cdot-}$ marked in blue foursquare. Reprinted from Ref. [105] Copyright 2017, with permission from Elsevier

**Fig. 3.31** Quenching effect of methanol, isopropanol, and tert-butanol on the oxidation of As(III) in PDS/Xe combination system. $[As(III)]_0 = 20\ \mu mol\ L^{-1}$, $[PDS]_0 = 50\ \mu mol\ L^{-1}$. Reprinted from Ref. [105] Copyright 2017, with permission from Elsevier

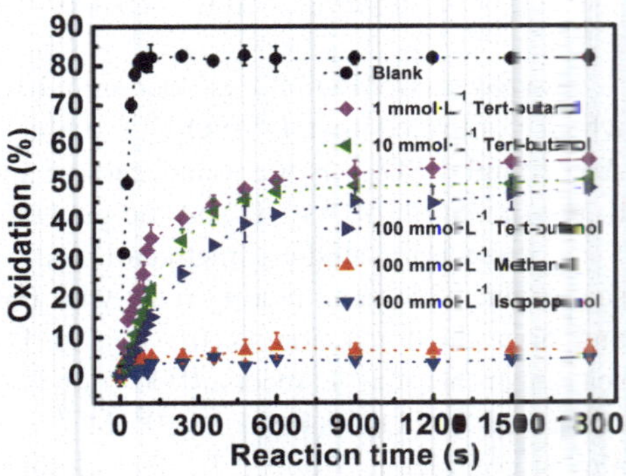

tert-butanol. Compared with the complete quenching effect of methanol and iso-propanol, the quenching effect of tert-butanol is much lower. For example, in comparison with the blank experiment (no scavenger), the oxidation yield decreased by −93% in the presence of 100 mmol $L^{-1}$ methanol or isopropanol, whereas only −42% decrease was observed in the presence of the same amount of tert-butanol. Therefore, the contribution of $SO_4^{\cdot -}$ to As(III) oxidation is over 50% and the rest contribution (not more than 42%) was attributed to $HO^{\cdot}$. Consequently, both $SO_4^{\cdot -}$ and $HO^{\cdot}$ was mainly responsible for As(III) oxidation.

### 3.3.2 *Kinetics of Oxidation Reaction*

$SO_4^{\cdot -}$ radical is a footstone for the rapid oxidation reaction occurred in the initial few minutes in PDS/Xe system and its production is controlled by PDS content but independent on light source because the power density of Xe lamp was maintained at 1.57 w $cm^{-2}$. The As(III) oxidation reaction can be simply described as Eq. (3.5), where PDS rather than $SO_4^{\cdot -}$ is associated with As(III).

$$\text{a As(III)} + \text{b } Na_2S_2O_8\text{(PDS)} \leftrightarrow \text{Products} \tag{3.5}$$

Although tetravalent As(IV) was reported as high reactive intermediate, it can be fast oxidized to As(V) and reduced to As(III) [113]. Therefore, the decrease of As (III) can represent the amount of generated As(V). Initial As(III) ($[As(III)]_0$) and residual As(III) at reaction time t ($[As(III)]_t$) were detected to study the reaction kinetics. In addition, oxidation yields of As(III) were calculated based on Eq. (3.6).

$$\text{Oxidation (\%)} = 100 \times \left([As(III)]_0 - [As(III)]_t\right)/[As(III)]_0 \tag{3.6}$$

It can be seen that As(III) oxidation yields increased faster as $[PDS]_0$ increased in the initial 2 min (Fig. 3.32a), which illustrated higher $[PDS]_0$ facilitated the rapid As(III) oxidation reaction. Moreover, when $[PDS]_0 < 50$ μmol $L^{-1}$, oxidation yield increased to −80% as prolonged reaction time. However, oxidation yield maintained unchanged when $[PDS]_0 > 50$ μmol $L^{-1}$. The cleavage of $S_2O_8^{2-}$ produces two $SO_4^{\cdot -}$ radicals, which are sufficient to oxidize As(III) to As(IV) and subsequently to As(V) [114], i.e., it is very likely that the stoichiometric ratio of As(III) and PDS is R = 1. Therefore, $[PDS]_0 > 50$ μmol $L^{-1}$ were overdosed for 20 μmol $L^{-1}$ $[As(III)]_0$ and the oxidation reaction finished quickly in the initial 2 min. However, the oxidation yields decreased as the increase of $[As(III)]_0$ in the initial 2 min. Especially, when $[As(III)]_0 > 50$ μmol $L^{-1}$, the oxidation yields decreased dramatically as the increase of $[As(III)]_0$. This is because $[PDS]_0 = 50$ μmol $L^{-1}$ is insufficient for $[As(III)]_0 > 50$ μmol $L^{-1}$. Fortunately, the lower oxidation yields can be elevated in the subsequent reaction time. For example, reaction proceeding from 2 to 30 min, the oxidation yield increased from 56 to 81% when $[As(III)]_0 = 70$ μmol $L^{-1}$ (Fig. 3.32b). This may be attributed to

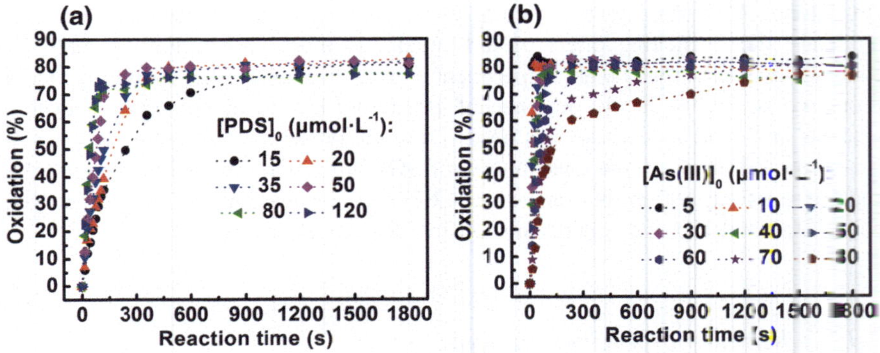


**Fig. 3.32** Oxidation yield under various initial concentrations of PDS and As(III). **a** $[As(III)]_0$ = 20 μmol L$^{-1}$ and $[PDS]_0$ = 15–120 μmol L$^{-1}$, **b** $[As(III)]_0$ = 5–80 μmol L$^{-1}$ and $[PDS]_0$ = 50 μmol L$^{-1}$. Reprinted from Ref. [105] Copyright 2017, with permission from Elsevier

superoxide radical ($HO_2^{\cdot}/O_2^{\cdot-}$), $H_2O_2$ and $O_2$, which are generated via reactions in Eqs. (3.7) and (3.8) [115].

$$As(IV) + O_2 \leftrightarrow As(V) + HO_2^{\cdot}/O_2^{\cdot -} \tag{3.7}$$

$$2O_2^{\cdot -} + 2H^+ \leftrightarrow H_2O_2 + O_2 \tag{3.8}$$

In the rapid As(III) oxidation process, PDS concentration is unambiguously known and no subproducts can be detected. The initial reaction rate method [116, 117] is chosen to study reaction kinetics. As(III) concentrations were fitted against reaction time and the derivative at zero time was determined to get the initial reaction rate ($r_0$). The reaction rate equation is constructed as Eq. (3.9).

$$r_0 = -\frac{d[As(III)]}{dt} = k \times [As(III)]_0^a \times [PDS]_0^b \tag{3.9}$$

where square brackets represent concentration, $a$ and $b$ are the apparent reaction order toward As(III) and PDS, respectively.

Because $[As(III)]_0 > 50$ μmol L$^{-1}$ was over-stoichiometrical for 50 μmol L$^{-1}$ $[PDS]_0$, the kinetics fitting was only conducted for $[As(III)]_0 < 50$ μmol L$^{-1}$. The apparent reaction order toward $[As(III)]_0$ derived from the slope of fitting of $\ln r_0$ and $[As(III)]_0$ is 0.51, i.e., $a$ = 0.51 (Fig. 3.33a). In addition, the apparent reaction order toward $[PDS]_0$ based on the fitting of $\ln r_0$ and $\ln[PDS]_0$ is 0.64, i.e., $b$ = 0.64. The apparent reaction rate equation can thus be expressed as Eq. (3.10). The apparent reaction order with respect to $[PDS]_0$ and $[As(III)]_0$ is close and the apparent reaction stoichiometric ratio of 1.25, which is close to the theoretical value R = 1. Besides, the rate constant with respect to $[As(III)]_0$ ($k_{a1}$) and with respect to $[PDS]_0$ ($k_{a2}$) is $8.4 \times 10^{-2}$ and $2.5 \times 10^{-2}$ $M^{-0.15}$ s$^{-1}$, respectively.

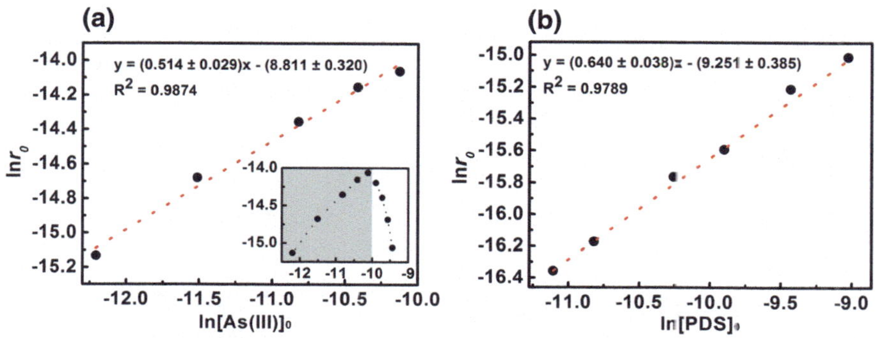


**Fig. 3.33** Linear fitting of **a** $\ln r_0$ versus $\ln[\text{As(III)}]_0$ and **b** $\ln r_0$ versus $\ln[\text{PDS}]_0$. **a** $[\text{As(III)}]_0$ = 5–40 μmol $L^{-1}$ and $[\text{PDS}]_0$ = 50 μmol $L^{-1}$, **b** $[\text{As(III)}]_0$ = 20 μmol $L^{-1}$ and $[\text{PDS}]_0$ = 15–120 μmol $L^{-1}$. Reprinted from Ref. [105] Copyright 2017, with permission from Elsevier

$$r_0 = k[\text{As}(III)]_0^{0.51}[\text{PDS}]]_0^{0.64} \tag{3.10}$$

The rapid oxidation process is different under various temperatures (293–330 K) (Fig. 3.34a). The initial reaction rates $r_0$ under different temperatures were figured out according to the apparent reaction rate equation (Eq. (3.10)). The reaction rate constant at 293 K is 6.25 × $10^{-2}$ $M^{-0.15}$ $s^{-1}$, which lies in our postulated range of 2.5–8.4 × $10^{-2}$ $M^{-0.15}$ $s^{-1}$. Following the Arrhenius equation shown in Eq. (3.11), ln*k* and 1/T was fitted linearly (Fig. 3.34b) and the apparent activation energy $E_a$ is 46.09 kJ $mol^{-1}$. It is much lower than the average activation energy of As(III) oxidation by $H_2O_2$ (96.6 kJ $mol^{-1}$) and by dissolved $O_2$ (73.2 kJ $mol^{-1}$) [118, 119].

$$k = Ae^{\frac{-E_a}{RT}} \tag{3.11}$$

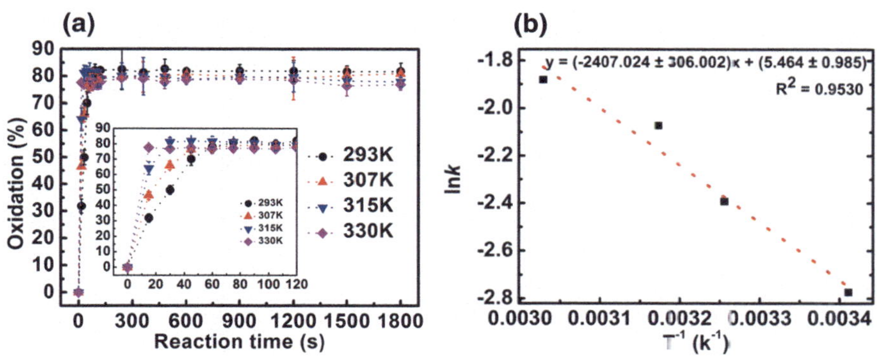


**Fig. 3.34** **a** Oxidation of As(III) under different temperatures and **b** ln*k* versus $T^{-1}$ plot. $[\text{As(III)}]_0$ = 20 μmol $L^{-1}$ and $[\text{PDS}]_0$ = 50 μmol $L^{-1}$. Reprinted from Ref. [105] Copyright 2017, with permission from Elsevier

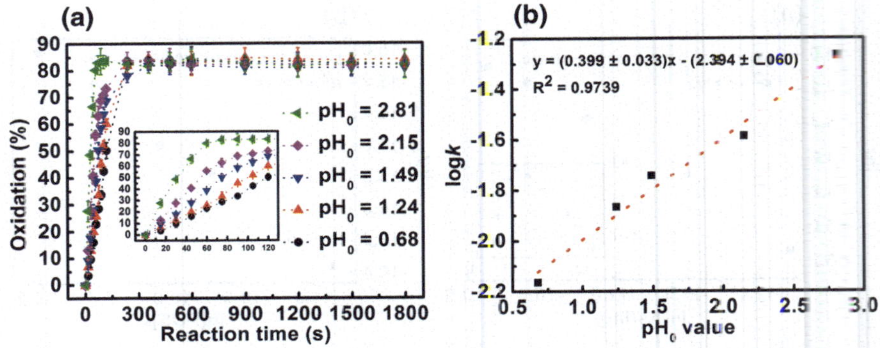


**Fig. 3.35** **a** Oxidation of As(III) under different $pH_0$ and **b** log*k* versus $pH_0$ plot. $[As(III)]_0 = 20\ \mu mol\ L^{-1}$ and $[PDS]_0 = 50\ \mu mol\ L^{-1}$. Reprinted from Ref. [105] Copyright 2017, with permission from Elsevier

**Table 3.8** The change of pH in this study and reported in reference

| | This study (10-min reaction) | | | | | | Zhou L (2013) et al. [29] | | | | |
|---|---|---|---|---|---|---|---|---|---|---|---|
| $pH_0$ | 0.68 | 1.24 | 1.49 | 2.15 | 2.81 | 6.02 | 3.0 | 4.0 | 5.0 | 6.0 | 7.0 |
| $pH_{after}$ | 0.68 | 1.25 | 1.50 | 2.12 | 2.59 | 4.28 | 3.0 | 4.0 | 4.6 | 5.3 | 5.1 |
| Change of pH | 0 | 0.01 | 0.01 | −0.03 | −0.22 | −1.74 | −0 | −0 | −0.4 | −0.7 | −1.9 |

*Note* Change of pH = $pH_{after} - pH_0$

The effect of initial pH value ($pH_0 < 3$) showed that the rapid As(III) oxidation was partly retarded under lower $pH_0$ (Fig. 3.35a). The rate constant increased from $0.7 \times 10^{-2}$ to $5.4 \times 10^{-2}\ M^{-0.15}\ s^{-1}$ with $pH_0$ increasing from 0.68 to 2.81. The linear fitting of log*k* and $pH_0$ (Fig. 3.35b) indicates that the effect of $pH_0$ on the reaction rate constant can be described as Eq. (3.12). The slight inhibition effect of $pH_0$ can be attributed to the decrease of $HO^{\cdot}$ under acidic conditions. Therefore, the mixture of $SO_4^{\cdot -}$ and $HO^{\cdot}$ is beneficial for the rapid As(III) oxidation under neutral and weak alkaline conditions. However, oxidation yields after 30-min reaction was not affected by $pH_0$. We noticed a pH drop in the 30-min reaction process, e.g., pH value went down from 6.02 to 3.96 after 30-min reaction. It is previously reported that the dropped pH is due to $H^+$ production in the reaction between $SO_4^{\cdot -}$ and solvent $H_2O$ molecule (Eq. (3.2)) [120]. It was also found that higher $pH_0$ resulted in greater pH decrease. As listed in Table 3.8, when $pH_0$ = 6.02, 2.81, 2.15 and 0.68–1.49, pH decrease was, respectively, 1.74, 0.22, 0.03 and −0 after 10-min reaction. Similarly, the greater pH decrease was reported in higher $pH_0$ reaction system of Fe(II)/PDS [29]. Additionally, the pH decrease was not limited in the initial 2 min but proceeded in the subsequent time with a slower rate. The production of $H^+$ in 2–30 min cannot be explained by the reaction of $SO_4^{\cdot -}$ and solvent $H_2O$ due to no available $SO_4^{\cdot -}$. Therefore, we postulated that deprotonation reaction of resultant $H_3AsO_4$ may also contribute to the pH decrease. Based on this, (1) no pH decrease was found at $pH_0 < 2$ because $pK_{a1}$ of $H_3AsO_4$ is 2.2 [72],

**Table 3.9** Gibbs free energy of possible elementary reactions ($\Delta G_{solv} = \Delta G_{r,\,solv} - \Delta G_r$, kcal mol$^{-1}$)

| Reactions | | $\Delta G_r$ | $\Delta G_{r,\,solv}$ | $\Delta G_{solv}$ | Reactions | | $\Delta G_r$ | $\Delta G_{r,\,solv}$ | $\Delta G_{solv}$ |
|---|---|---|---|---|---|---|---|---|---|
| A | $H_3AsO_3 + HO^{\cdot} = H_4AsO_4^{\cdot}$ | −17.59 | −14.41 | 3.18 | a | $H_3AsO_3 + SO_4^{\cdot -} = H_3AsO_3SO_4^{\cdot -}$ | −22.91 | −12.03 | 10.88 |
| **B** | **$H_3AsO_3 + HO^{\cdot} = H_2AsO_3^{\cdot} + H_2O$** | **−15.80** | **−19.99** | **−4.19** | **b** | **$H_3AsO_3 + SO_4^{\cdot -} = H_2AsO_3^{\cdot} + HSO_4^-$** | **3.93** | **−2.84** | **−6.77** |
| C | **$H_3AsO_3 + HO^{\cdot} = H_3AsO_3^{\cdot +} + HO^-$** | **186.77** | **48.15** | **−138.62** | c | **$H_3AsO_3 + SO_4^{\cdot -} = H_3AsO_3^{\cdot +} + SO_4^{2-}$** | **252.54** | **27.41** | **−225.13** |
| A1 | $H_4AsO_4^{\cdot} + HO^{\cdot} = H_5AsO_5$ | −76.54 | −73.84 | 2.70 | a1 | $H_3AsO_3SO_4^{\cdot -} = H_3AsO_4 + SO_3^{\cdot -}$ | 20.53 | 1.48 | −19.05 |
| **A2** | **$H_4AsO_4^{\cdot} + HO^{\cdot} = H_3AsO_4 + H_2O$** | **−75.25** | **−81.88** | **−6.63** | **a2** | **$H_4AsO_4^{\cdot} + SO_4^{\cdot -} = H_3AsO_4 + HSO_4^-$** | **−55.511** | **−64.73** | **−9.22** |
| **A3** | **$H_4AsO_4^{\cdot} + HO^{\cdot} = H_4AsO_4^{+} + HO^-$** | **123.73** | **−11.74** | **−135.47** | **a3** | **$H_4AsO_4^{\cdot} + SO_4^{\cdot -} = H_4AsO_4^{+} + SO_4^{2-}$** | **189.506** | **−32.49** | **−222.00** |
| B1 | $H_2AsO_3^{\cdot} + HO^{\cdot} = H_3AsO_4$ | −77.03 | −76.30 | 0.73 | b1 | $H_2AsO_3^{\cdot} + SO_4^{\cdot -} = H_2AsO_3SO_4^-$ | −88.77 | −71.76 | 17.01 |
| **B2** | **$H_2AsO_3^{\cdot} + HO^{\cdot} = HAsO_3 + H_2O$** | **−48.68** | **−51.38** | **−2.70** | **b2** | **$H_2AsO_3^{\cdot} + SO_4^{\cdot -} = HAsO_3 + HSO_4^-$** | **−28.95** | **−34.23** | **−5.28** |
| **B3** | **$H_2AsO_3^{\cdot} + HO^{\cdot} = H_2AsO_3^{+} + HO^-$** | **176.78** | **35.41** | **−141.37** | **b3** | **$H_2AsO_3^{\cdot} + SO_4^{\cdot -} = H_2AsO_3^{+} + SO_4^{2-}$** | **242.55** | **14.66** | **−227.89** |
| C1 | $H_3AsO_3^{\cdot +} + HO^{\cdot} = H_4AsO_4^{+}$ | −80.63 | −74.31 | 6.32 | c1 | $H_3AsO_3^{\cdot +} + SO_4^{\cdot -} = H_3AsO_3SO_4$ | −178.64 | −69.06 | 109.58 |
| **C2** | **$H_3AsO_3^{\cdot +} + HO^{\cdot} = H_2AsO_3^{+} + H_2O$** | **−25.80** | **−32.74** | **−6.94** | **c2** | **$H_3AsO_3^{\cdot +} + SO_4^{\cdot -} = H_2AsO_3^{+} + HSO_4^-$** | **−6.06** | **−15.59** | **−9.53** |
| **C3** | **$H_3AsO_3^{\cdot +} + HO^{\cdot} = H_3AsO_3^{2+} + HO^-$** | **333.04** | **83.88** | **−249.16** | **c3** | **$H_3AsO_3^{\cdot +} + SO_4^{\cdot -} = H_3AsO_3^{2+} + SO_4^{2-}$** | **398.81** | **63.14** | **−335.67** |
| **D** | **$H_5AsO_5 = H_3AsO_4 + H_2O$** | **1.30** | **−8.04** | **−9.34** | **d** | **$H_2AsO_3SO_4^- = H_2AsO_4^- + SO_3$** | **53.88** | **35.32** | **−18.56** |
| E | $H_4AsO_4^{+} + HO^- = H_5AsO_5$ | −200.28 | −62.09 | 138.19 | e | $H_2AsO_4^- + H^+ = H_3AsO_4$ | −319.83 | −166.47 | 153.36 |
| F | $HAsO_3 + H_2O = H_3AsO_4$ | −28.35 | −24.92 | 3.43 | **f** | **$H_3AsO_3SO_4 = H_3AsO_4 + SO_3$** | **14.58** | **9.46** | **−5.12** |
| G | $H_2AsO_3^{+} + HO^- = H_3AsO_4$ | −253.81 | −111.71 | 142.10 | X | $H_2AsO_3^{\cdot} + H^+ = H_3AsO_3^{\cdot +}$ | −190.65 | −143.32 | 47.33 |
| H | $H_3AsO_3^{2+} + 2HO^- = H_5AsO_5$ | −613.94 | −220.29 | 393.65 | | | | | |

**Table 3.10** The possible pathways and the controlling step (kcal mol$^{-1}$)

| | Pathway | Total Gibbs free energy | Step number | The controlling step | Gibbs free energy | $SO_4^{\cdot -}$-based step |
|---|---|---|---|---|---|---|
| *Only $HO^{\cdot}$ system* | | | | | | |
| 1 | $H_3AsO_3 \rightarrow H_4AsO_4^{\cdot} \rightarrow H_5AsO_5 \rightarrow H_3AsO_4$ | −96.29 | 3 | $H_5AsO_5 \rightarrow H_3AsO_4$ | −8.04 | None |
| 2 | $H_3AsO_3 \rightarrow H_4AsO_4^{\cdot} \rightarrow H_3AsO_4$ | | 2 | $H_3AsO_3 \rightarrow H_4AsO_4^{\cdot}$ | −14.41 | |
| 3 | $H_3AsO_3 \rightarrow H_4AsO_4^{\cdot} \rightarrow H_4AsO_4^{\cdot +} \rightarrow H_5AsO_5 \rightarrow H_3AsO_4$ | | 4 | $H_5AsO_5 \rightarrow H_3AsO_4$ | −8.04 | |
| 4 | $H_3AsO_3 \rightarrow H_2AsO_3^{\cdot} \rightarrow H_3AsO_4$ | | 2 | $H_3AsO_3 \rightarrow H_2AsO_3^{\cdot}$ | −19.99 | |
| 5 | $H_3AsO_3 \rightarrow H_2AsO_3^{\cdot} \rightarrow HAsO_3 \rightarrow H_3AsO_4$ | | 4 | | | |
| 6 | $H_3AsO_3 \rightarrow H_2AsO_3^{\cdot} \rightarrow H_3AsO_3^{\cdot +} \rightarrow H_4AsO_4^{\cdot +} \rightarrow H_5AsO_5 \rightarrow H_3AsO_4$ | −307.75 | 5 | $H_5AsO_5 \rightarrow H_3AsO_4$ | −8.04 | |
| 7 | $H_3AsO_3 \rightarrow H_2AsO_3^{\cdot} \rightarrow H_3AsO_3^{\cdot +} \rightarrow H_2AsO_3^{+} \rightarrow H_3AsO_4$ | | 4 | $H_3AsO_3 \rightarrow H_2AsO_3^{\cdot}$ | −19.99 | |
| *Mixed $SO_4^{\cdot -}$ and $HO^{\cdot}$ system* | | | | | | |
| 1 | $H_3AsO_3 \rightarrow H_4AsO_4^{\cdot} \rightarrow H_3AsO_4$ | −79.14 | 2 | $H_3AsO_3 \rightarrow H_4AsO_4^{\cdot}$ | −14.41 | $H_4AsO_4^{\cdot} \rightarrow H_3AsO_4$ |
| 2 | $H_3AsO_3 \rightarrow H_4AsO_4^{\cdot} \rightarrow H_4AsO_4^{\cdot +} \rightarrow H_5AsO_5 \rightarrow H_3AsO_4$ | −117.03 | 4 | $H_5AsO_5 \rightarrow H_3AsO_4$ | −8.04 | |
| 3 | $H_3AsO_3 \rightarrow H_2AsO_3^{\cdot} \rightarrow H_3AsO_4$ | −78.78 | 2 | $H_3AsO_3 \rightarrow H_2AsO_3^{\cdot}$ | −2.48 | $H_3AsO_3 \rightarrow H_2AsO_3^{\cdot}$ |
| 4 | $H_3AsO_3 \rightarrow H_2AsO_3^{\cdot} \rightarrow HAsO_3 \rightarrow H_3AsO_4$ | −61.63 | 3 | | | |
| 5 | $H_3AsO_3 \rightarrow H_2AsO_3^{\cdot} \rightarrow H_3AsO_3^{\cdot +} \rightarrow H_4AsO_4^{\cdot +} \rightarrow H_5AsO_5 \rightarrow H_3AsO_4$ | −290.24 | 5 | | | |
| 6 | $H_3AsO_3 \rightarrow H_2AsO_3^{\cdot} \rightarrow H_3AsO_3^{\cdot +} \rightarrow H_2AsO_3^{+} \rightarrow H_3AsO_4$ | −273.10 | 4 | | | $H_3AsO_3 \rightarrow H_2AsO_3^{\cdot}$ $H_3AsO_3^{\cdot +} \rightarrow H_4AsO_4^{\cdot +}$ |

(2) more $H^+$ can be generated under higher pH conditions due to the significant deprotonation of $H_3AsO_4$ under higher pH conditions, (3) the quick pH decrease in the initial 2 min is due to the rapid formation of resultant $H_3AsO_4$.

$$lnk = 0.40 \times pH_0 - 2.39 \tag{3.12}$$

### 3.3.3 Molecular Reaction Mechanism

$H_3AsO_3$ is the main species of As(III) under acidic and weak alkaline conditions due to its $pK_{a1} = 9.2$ [59]. $SO_4^{\cdot-}$ is mainly responsible for As(III) oxidation in PDS/Xe combination system, however, $HO^{\cdot}$ induced by $SO_4^{\cdot-}$ can also be involved in As(III) oxidation process, especially under neutral and weak alkaline conditions. The oxidation of $H_3AsO_3$ by $SO_4^{\cdot-}$ and $HO^{\cdot}$ was both envisaged to occur via reactions displayed in Table 3.9. The calculations were conducted by Gaussian 09 software at IEFPCM (water)/M06-2X/6-311++G**//M06-2X/6-31G* level. Seven pathways were supposed for the oxidation from $H_3AsO_3$ to $H_3AsO_4$ under the single function of $HO^{\cdot}$ (Table 3.10) and the second and forth pathway has least elementary

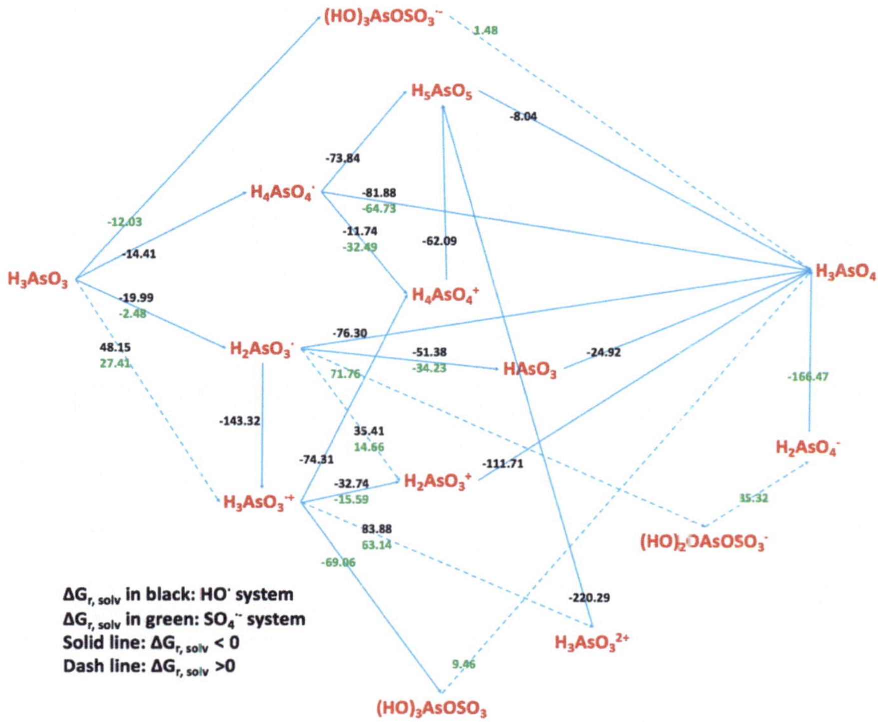


**Fig. 3.36** Possible pathways of $H_3AsO_3$ oxidation to $H_3AsO_4$ by $SO_4^{\cdot-}$ and $HO^{\cdot}$ and the Gibbs free energy (kcal $mol^{-1}$). Reprinted from Ref. [105] Copyright 2017, with permission from Elsevier

reactions and the lowest Gibbs free energy in the controlling step. Although $SO_4^{\cdot-}$ can play a role in several elementary reactions, there is no available pathway for $H_3AsO_3$ oxidized by only $SO_4^{\cdot-}$ to $H_3AsO_4$. In mixed $SO_4^{\cdot-}$ and $HO^{\cdot}$ system six different pathways were proposed for the oxidation from $H_3AsO_3$ to $H_3AsO_4$ and the first pathway has least elementary reactions and the lowest Gibbs free energy in the controlling step (Fig. 3.36). Therefore, the most favorable pathways for $H_3AsO_3$ oxidation to $H_3AsO_4$ is $H_3AsO_3 \rightarrow H_4AsO_4^{\cdot} \rightarrow H_3AsO_4$ and $H_3AsO_3 \rightarrow H_2AsO_3^{\cdot} \rightarrow H_3AsO_4$ in $HO^{\cdot}$-based system, and $H_3AsO_3 \rightarrow H_4AsO_4^{\cdot} \rightarrow H_3AsO_4$ in mixed $SO_4^{\cdot-}$ and $HO^{\cdot}$ system where $SO_4^{\cdot-}$ plays a role in the first step.

## 3.4 Formation Mechanism and Characteristics of Tooeleite from High-Arsenic Acid Wastewater

The sequestering of arsenic in tooeleite is considered to be the best environmentally friendly possible waste form for stable outdoor storage.

### *3.4.1 Direct Removal of As(III) by the Formation of Tooeleite*

The formation of tooeleite was investigated in pH range from 1.8 to 9. XRD patterns (Fig. 3.37b [121]) confirmed that the precipitate obtained in pH range of 1.8 to 4.5 was tooeleite, consistent with PDF44-1468. This pH range for tooeleite formation is wider than the previous report (2.0–3.5) [122]. However, tooeleite would transform to amorphous ferric arsenite when pH > 4.5. As(III) and Fe(III) removal efficiency increased with the increase of pH, for example, As(III) removal sharply increased from 30% at pH 1.8 to around 90% at pH 2.8 (Fig. 3.37a). This is because that the

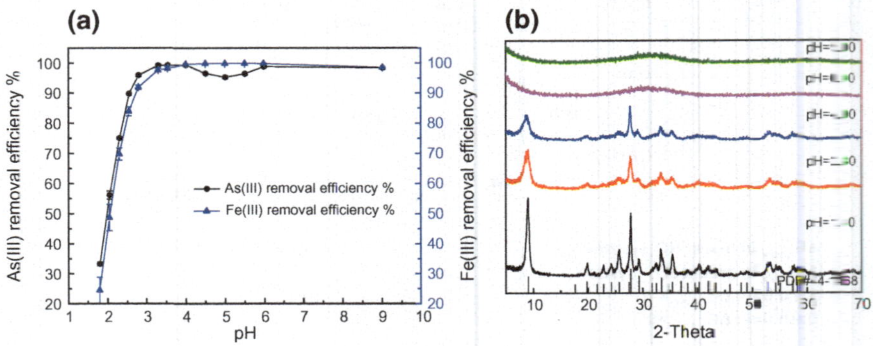


**Fig. 3.37** As(III) and Fe(III) removal efficiency (**a**) and XRD patterns of As-Fe precipitates (**b**) at pH 1.8–9. Reprinted from Ref. [121] Copyright 2017, with permission from Elsevier

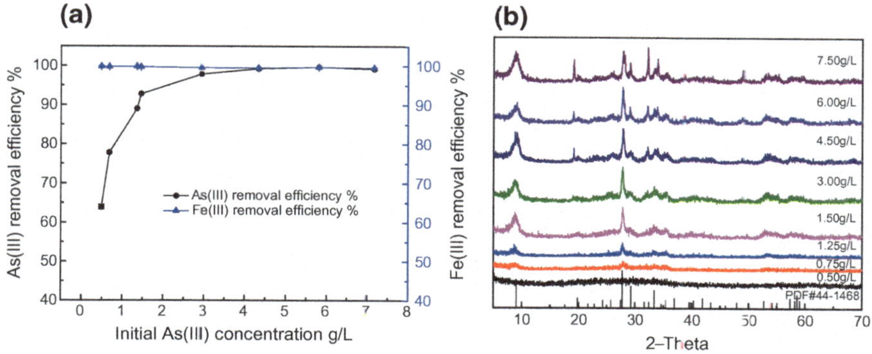


**Fig. 3.38** As(III) and Fe(III) removal (**a**) and XRD patterns of As–Fe precipitates (**b**) as affected by initial As(III) concentration varied from 0.5 to 7.5 g $L^{-1}$. Reprinted from Ref. [121] Copyright 2017, with permission from Elsevier

mixed solution was in the metastable state at pH 1.8, and primary nucleation of tooeleite may carry a portion of As(III) and Fe(III) after a period of inducing time [123]. Considering hydrolysis reaction, Fe(III) is supposed to start precipitating at pH 1.76 and complete at pH 3.17 and thus the majority of Fe(III) will be removed with the increase of pH from 1.8 to 2.8. The formation of iron hydroxides makes contributions to the removal of As(III) as well as the formation of tooeleite. However, As(III) removal decreased at pH 4–5 and then increase again with continuous pH increase. The decrease can be explained by As releasing from the partial transform of tooeleite to amorphous FeOOH(As) [122] and the formation of $FeAsO_3$ at pH 5–9 lead to the reprecipitation of As(III) [124]. The removal of As(III) can reach above 99% at pH 4 by forming well-crystallized tooeleite. Therefore, pH 4 is suggested for the treatment of acid wastewater containing a high concentration of As(III).

The As(III) concentration from 0.75 to 7.5 g $L^{-1}$ is suitable for forming tooeleite but higher As(III) facilitates the crystallization of tooeleite (Fig. 3.38b). The higher As(III) also facilitates As(III) removal efficiency but has no significant effect on iron removal. Since the solution pH is 4, precipitation of iron leads to persistent high iron removal but different As(III) removal due to different initial As(III) concentration (Fig. 3.38a). When As(III) concentration lowed to 0.5 g $L^{-1}$, As(III) removal efficiency was 60% but no tooeleite was found. Adsorption of As(III) on iron hydroxides benefits As(III) removal [125], with the increase of As(III) initial concentration, As(III) was incorporated into the growing iron hydroxides phase and finally co-precipitated as tooeleite [126].

The formation of tooeleite can be realized at the Fe/As from 0.8 to 2 and the crystallinity degree increased with Fe/As (Fig. 3.39b). For all Fe/As, Fe(III) removal maintained over 99%, however, As(III) removal linearly increased from 46 to 97% against Fe/As from 0.8 to 2 (Fig. 3.39a). As(III) removal exhibited a good Fe-dependency and we hypothesized that iron compounds like ferrihydrite were formed due to the introduction of NaOH and increased As(III) promoted the transformation of ferrihydrite to tooeleite [127].

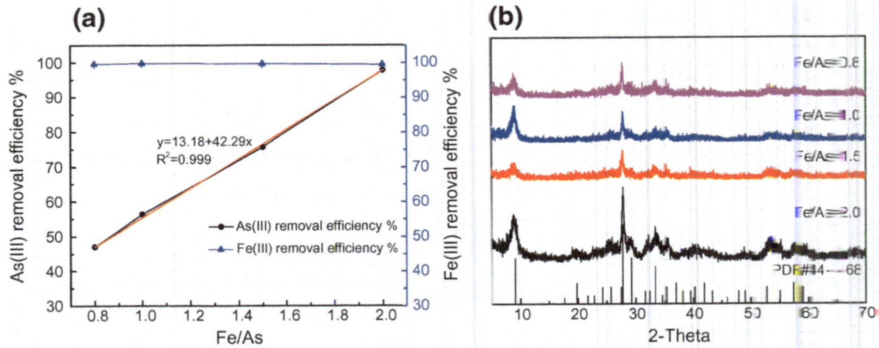


**Fig. 3.39** As(III) and Fe(III) removal (**a**) and XRD patterns of As–Fe precipitates (**b**) as affected by Fe/As varied from 0.8 to 2. Reprinted from Ref. [121] Copyright 2017, with permission from Elsevier

### 3.4.2 *Characterization of Tooeleite*

SEM images of tooeleite obtained at pH 4 exhibited a flower structure of platy particles (1 μm) aggregating together (Fig. 3.40a). Quantitative chemical analysis shows the As content of tooeleite is 23.38%, which is much higher than normal arsenic-bearing stabilized slag (5–15%) [128]. The XPS As 3d spectrum confirmed that arsenic exists as As(III) in tooeleite precipitate (Fig. 3.40b). This provides strong evidence that As(III) can be directly removed and solidified in tooeleite from high-As(III) acid wastewater. IR spectrum (Fig. 3.40c) confirmed stretching modes and bending vibrations of adsorbed water at 3396 and 1632 $cm^{-1}$, three unique $SO_4^{2-}$ absorption bands in tooeleite (sorbed $\nu_3(SO_4)$ at 1112 $cm^{-1}$, outer sphere bound $\nu_1(SO_4)$ at 990 $cm^{-1}$ and structural v4($SO_4$) at 618 $cm^{-1}$) [129], band of $AsO_3^{3-}$ at 778 $cm^{-1}$ and Fe–O–As band at 513 $cm^{-1}$ [130].

### 3.4.3 *Safety Evaluation*

The short-term stability experiments (Fig. 3.41a) showed that As leaching at pH 1.8, 2.8 and 4 was far less than at pH 6 and 9. This is because tooeleite was formed at pH below 4, while As–Fe compound with poor crystal quality at pH 6 and 9 (Fig. 3.41b). Therefore, tooeleite is much more stable than poor-crystallized precipitates obtained at higher pH. However, much lower pH is unfavorable for the stability of tooeleite due to the adsorption of some As on the surface of tooeleite. Tooeleite prepared at pH 1.8 showed the lowest arsenic leaching, however, it still exceeds the national criteria (5 mg $L^{-1}$). This puts us heavy pressure to find some ways to improve the stability of tooeleite.

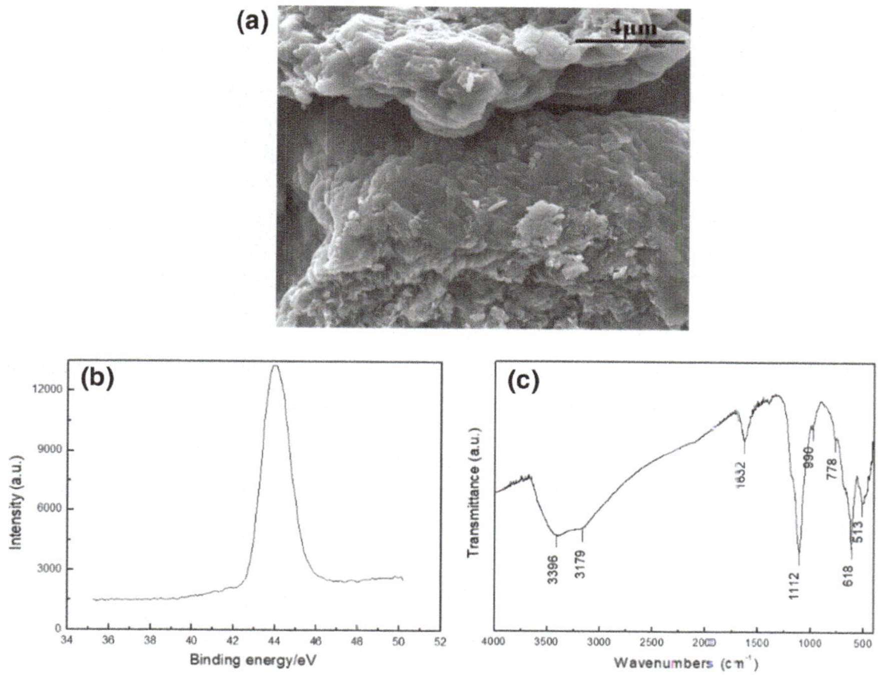


**Fig. 3.40** SEM image (**a**), XPS analysis (**b**) and FTIR analysis (**c**) of precipitate obtained at pH 4.0. Reprinted from Ref. [121] Copyright 2017, with permission from Elsevier

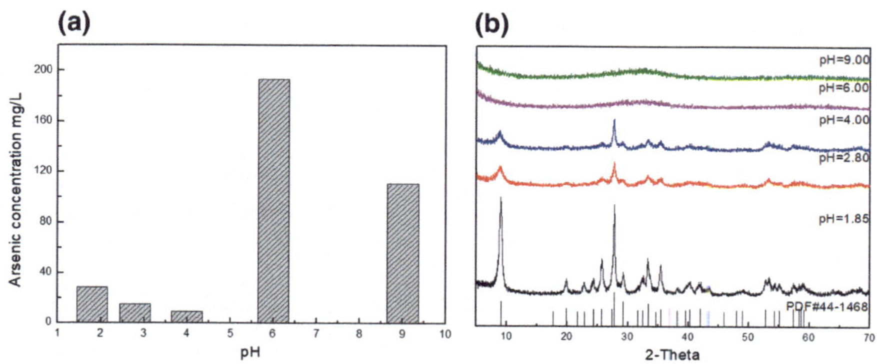


**Fig. 3.41** Leaching concentration of arsenic (**a**) and XRD patterns after leaching experiments (**b**) at pH 1.85, 2.80. 4.00, 6.00 and 9.00

### 3.4.4 Tooeleite Formation Mechanism

Samples were collected during the formation process of tooeleite and their color changed from reddish brown to lemon yellow as time goes by (Fig. 3.42). XRD (Fig. 3.43b) of the sample at 1 h exhibited two broad bands at 2θ of 31° (d ≈ 2.86 Å) and 60° (d ≈ 1.55 Å). Comparing with ferrihydrite [131], a little shift toward higher d-spacing was ascribed to the adsorption of As(III) [129]. Samples obtained after 17 h showed the enhanced diffraction peaks at 9.06, 27.78, and 29.28°, respectively, marked by the indices (020), (200) and (061) of tooeleite. In addition, FTIR spectra (Fig. 3.44) of samples obtained at 1, 6, 17 and 24 h showed weakened characteristic peaks of ferrihydrite at 1193 and 1046 $cm^{-1}$ and enhanced Fe–O–As bond at 510 $cm^{-1}$ [130]. During the transformation process, Fe(III) was almost removed and As(III) removal increased from 90% at beginning to ~100% at 5 day (Fig. 3.43a).

Samples were obtained after 24-h reaction in solutions with Fe/As = 0.8 and after another 24-h reaction when Fe/As was adjusted by iron sulfate to 1.5. The two group samples were all confirmed as tooeleite (Fig. 3.45b). However, the removal

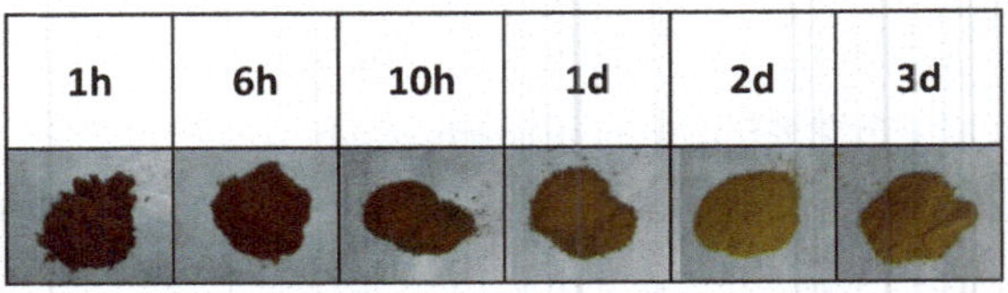


**Fig. 3.42** Color change of obtained samples as affected by reaction time. Reprinted from Ref. [121] Copyright 2017, with permission from Elsevier

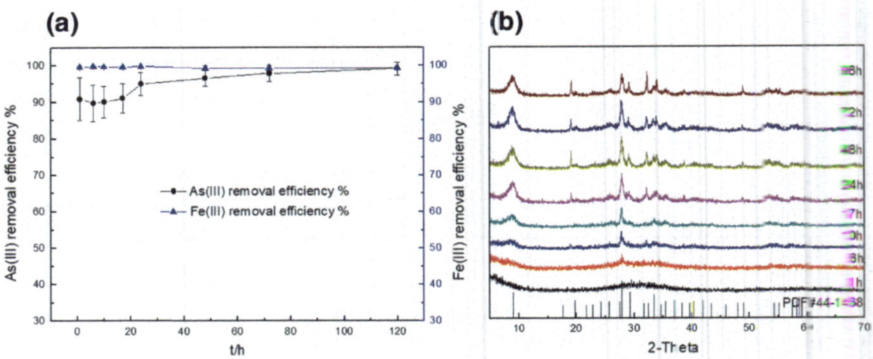


**Fig. 3.43** As(III) and Fe(III) removal efficiency (**a**) and XRD patterns of As–Fe precipitates (**b**) as affected by reaction time varied from 1 to 96 h. Reprinted from Ref. [121] Copyright 2017, with permission from Elsevier

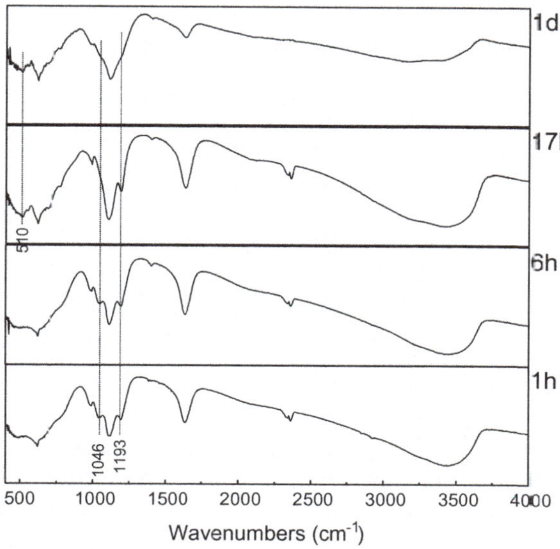


**Fig. 3.44** FTIR of As–Fe precipitates (1, 6, 17, 24 h). Reprinted from Ref. [121] Copyright 2017, with permission from Elsevier

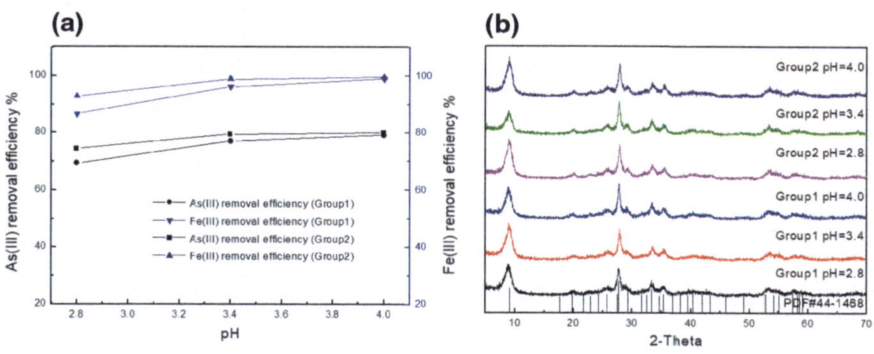


**Fig. 3.45** As(III) and Fe(III) removal efficiency (**a**) and XRD patterns (**b**) on the two Groups. Reprinted from Ref. [121] Copyright 2017, with permission from Elsevier

of As(III) and Fe in the latter group are a little higher than in the former group (Fig. 3.45a). The formation mechanism of tooeleite can be described as that Fe(III) precipitates as ferrihydrite at pH up to 4.0 and adsorbs As and sulfate, and then transforms to tooeleite (Fig. 3.46).

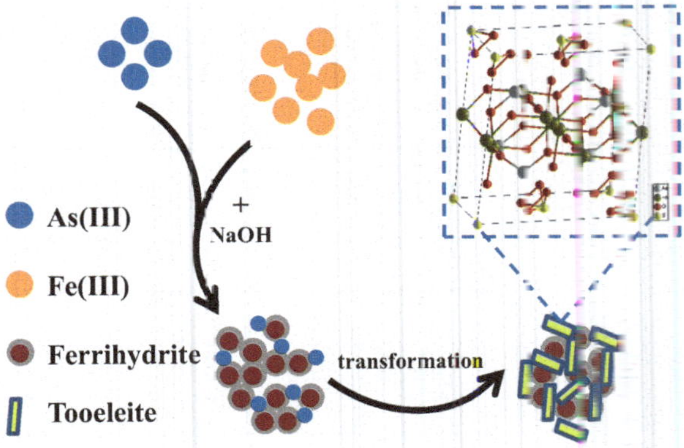


**Fig. 3.46** Schematic diagram of tooeleite formation process. Reprinted from Ref. [121] Copyright 2017, with permission from Elsevier

## 3.5 Cascade Sulfide Precipitation and Separation of Copper and Arsenic from Acidic Wastewater

Many industrial processes, especially in the mining and metallurgical processing industry, discharge acidic effluents containing significant amounts of metals such as copper, nickel, zinc, lead, and arsenic [132, 133]. Copper is often associated with arsenic in the mixed sulfide minerals such as enargite ($Cu_3AsS_4$) and tennantite ($Cu_{12}As_4S_{13}$) [134], therefore, among these heavy metals, copper and arsenic are found to behave similarly and exist simultaneously in the wastewater [135].

Arsenic contamination has greatly threatened water safety due to its high toxicity and carcinogenicity [136, 137]. In the case of liquid effluents, there are many methods available for the removal of heavy metals, such as chemical precipitation, adsorption, coagulation, ion exchange, microbial reduction, and so on [138–140]. On the other hand, dissolved air flotation was used to separate and recover heavy metals. Stalidis et al. [141] separated copper, zinc and arsenic ions from dilute aqueous solutions by the dissolved air technique via the production of fine gas bubbles. It seems to be complex and expensive for the treatment of acidic wastewater. Sulfide precipitation is indeed an effective process for the treatment of toxic heavy metals ions [142, 143]. One of the primary advantages of sulfide precipitation is that the process can achieve a high degree of metal removal over a broad pH range. Metal sulfide sludge also exhibits better thickening and dewatering characteristics than the corresponding metal hydroxide sludge [144]. Bhattacharyya et al. [145] separated arsenic and other heavy metals using sodium sulfide. The removals of cadmium, zinc, and copper from the actual wastewaters are greater than 99%, and those of arsenic and selenium are 98% and > 92%, respectively. However, it is $H_2S$ that reacts with heavy metals by gas–liquid reaction as sulfur exists almost in the form of $H_2S$ rather than $S^{2-}$ and $HS^-$ in acid conditions. It takes 5–10 times excess of theoretical amount of sodium sulfide, moreover, a lot of hydrogen sulfide gas escapes, which leads to serious secondary pollution [146].

A novel process of treating acidic wastewater containing copper and arsenic is proposed, involving primary separation of copper and arsenic by sulfide precipitation and further separation by the replacement of arsenic in the precipitates by copper in the solution [147].

### *3.5.1 Thermodynamics for Separation of Copper and Arsenic*

With the introduction of $S^{2-}$, $Cu^{2+}$, and As(III) precipitated in the form of CuS and $As_2S_3$, respectively. The solubility equilibrium reaction of CuS and $As_2S_3$ can be expressed as follows:

$$\mathrm{CuS} = \mathrm{Cu}^{2+} + \mathrm{S}^{2-} \tag{3.13}$$

$$K_{sp1} = \left[\mathrm{Cu}^{2+}\right]\left[\mathrm{S}^{2-}\right] = 6.3 \times 10^{-36} \tag{3.14}$$

$$\mathrm{As_2S_3} + 6\mathrm{H_2O} = 2\mathrm{H_3AsO_3} + 3\mathrm{H_2S} \tag{3.15}$$

$$K_{sp2} = [\mathrm{H_3AsO_3}]^2[\mathrm{H_2S}]^3 = 10^{-11.9} \tag{3.16}$$

The sulfur-containing aqueous solution consists of $[H^+]$, $[OH^-]$, $[H_2S]$, $[HS^-]$, $[S^{2-}]$, and the total sulfur ion of the system can be expressed as follows:

$$[\mathrm{S}]_T = [\mathrm{H_2S}] + [\mathrm{HS^-}] + [\mathrm{S^{2-}}] \tag{3.17}$$

The stepwise stability constants for each ion at 298.15 K are used as follows:

$$\mathrm{H_2S} = \mathrm{H^+} + \mathrm{HS^-} \tag{3.18}$$

$$K_{\alpha1} = \frac{[\mathrm{HS^-}][\mathrm{H^+}]}{[\mathrm{H_2S}]} = 1.3 \times 10^{-7} \tag{3.19}$$

$$\mathrm{HS^-} = \mathrm{H^+} + \mathrm{S^{2-}} \tag{3.20}$$

$$K_{\alpha2} = \frac{\left[\mathrm{S^{2-}}\right]\left[\mathrm{H^+}\right]}{\left[\mathrm{HS^-}\right]} = 7.1 \times 10^{-15} \tag{3.21}$$

It can be deduced from Eqs. (3.17) and (3.19) to Eq. (3.21):

$$[\mathrm{S^{2-}}] = \frac{\mathrm{S}_T}{[\mathrm{H^+}]^2/K_{\alpha1}K_{\alpha2} + [\mathrm{H^+}]/K_{\alpha2} + 1} \tag{3.22}$$

$$[H_2S] = \frac{S_T[H^+]^2}{K_{\alpha1}K_{\alpha2} + K_{\alpha1}[H^+] + [H^+]^2} \tag{3.23}$$

The total copper of the solution system can be expressed as follows:

$$[Cu]_T = \frac{K_{sp1}}{[S]_T}\left(\frac{[H^+]^2}{K_{\alpha1}K_{\alpha2}} + \frac{[H^+]}{K_{\alpha2}} + 1\right)\times \left(1 + \frac{K_1}{[H^+]} + \frac{K_2}{[H^+]^2} + \frac{K_3}{[H^+]^3} + \frac{K_4}{[H^+]^4}\right) \tag{3.24}$$

Similarly, arsenic in the water has the following chemical equilibrium:

$$H_3AsO_3 \Leftrightarrow H_2AsO_3^- + H^+$$

$$K_{\alpha1}' = \frac{[H_2AsO_3^-][H^+]}{[H_3AsO_3]} = 10^{-9.17} \tag{3.25}$$

$$H_2AsO_3^- \Leftrightarrow HAsO_3^{2-} + H^+$$

$$K_{\alpha2}' = \frac{[HAsO_3^{2-}][H^+]}{[H_2AsO_3^-]} = 10^{-14.1} \tag{3.26}$$

$$[As]_T = [H_3AsO_3] + [H_2AsO_3^-] + [HAsO_3^{2-}] \tag{3.27}$$

It can be deduced from Eq. (3.25) to Eq. (3.27):

$$\frac{[As]_T}{[H_3AsO_3]} = 1 + \frac{K_{\alpha1}'}{[H^+]} + \frac{K_{\alpha1}'K_{\alpha2}'}{[H^+]^2} \tag{3.28}$$

Combined Eqs. (3.17) and (3.23) with Eq. (3.28), the total arsenic can be expressed as follows:

$$[As]_T = \left(1 + \frac{K_{\alpha1}'}{[H^+]} + \frac{K_{\alpha1}'K_{\alpha2}'}{[H^+]^2}\right)\times \sqrt{\frac{K_{sp2}}{[S]_T^3}\left(\frac{K_{\alpha1}K_{\alpha2}}{[H^+]^2} + \frac{K_{\alpha1}}{[H^+]} + 1\right)^3} \tag{3.29}$$

For $[H^+] = 10^{-pH}$, it can be deduced from Eqs. (3.24) and (3.29):

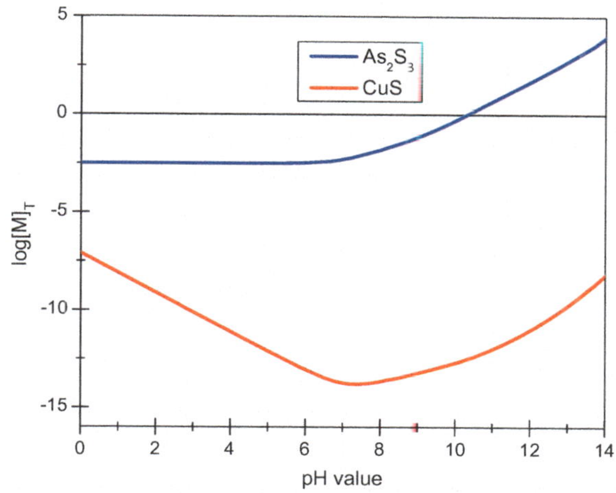


**Fig. 3.47** $Log[M]_T$ – pH curves of CuS and $As_2S_3$. Reprinted from Ref. [147] Copyright 2017, with permission from Elsevier

$$
\begin{aligned}
[\mathrm{Cu}]_T = {} & \frac{6.3\times10^{-36}}{[\mathrm{S}]_T}\times\left(1.08\times10^{21-2\mathrm{pH}}+1.41\times10^{14-\mathrm{pH}}+1\right) \\
& \times\left(1+10^{\mathrm{pH}-7}+10^{2\mathrm{pH}-14.32}+10^{3\mathrm{pH}-25}+10^{4\mathrm{pH}-37.5}\right)
\end{aligned} \tag{3.30}
$$

$$
\begin{aligned}
[\mathrm{As}]_T = {} & \left(1+10^{\mathrm{pH}-9.17}+10^{2\mathrm{pH}-23.27}\right) \\
& \times\sqrt{\frac{10^{-11.9}}{[\mathrm{S}]_T^3\left(9.23\times10^{2\mathrm{pH}-22}+1.3\times10^{\mathrm{pH}-7}+1\right)^3}}
\end{aligned} \tag{3.31}
$$

Hypothesizing that CuS and $As_2S_3$ stoichiometrically achieved dissolution equilibrium, then $[Cu]_T = [S]_T$, $3[As]_T = 2[S]_T$, $\log[Cu]_T$ and $\log[As]_T$ as a function of pH value are shown in Fig. 3.47. With the increase of pH value, the solubility of CuS decreases first and then increases. The minimum solubility of CuS is $1.733 \times 10^{-14}$ mol $L^{-1}$ when the pH value is 7.75. The solubility of $As_2S_3$ is greater than that of CuS and increases with the increasing pH value, which favors the possibility of cascade sulfidation of copper and arsenic in the solution.

### 3.5.2 *Cascade Sulfide Precipitation and Separation Technique*

The experimental program was divided into two stages. In the first stage, copper and arsenic were primarily separated by sulfide precipitation. The schematic diagram of the experimental apparatus is shown in Fig. 3.48. In the second stage, precipitates obtained in the first stage were recycled to the initial solution to improve the grade of copper in the precipitates through the replacement of arsenic by copper.

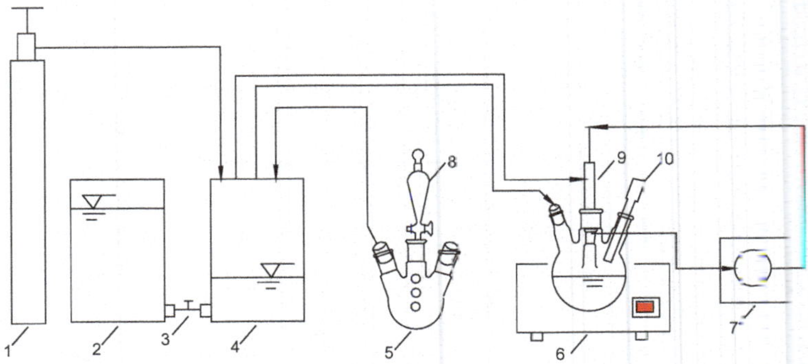


Symbol: 1 nitrogen cylinder, 2 fluid reservoir, 3 two-way valve, 4 gasholder, 5 gas generating bottle, 6 temperature controller, 7 peristaltic pump, 8 separating funnel, 9 Reactor, 10 pH meter

**Fig. 3.48** Schematic diagram of the experimental apparatus. Reprinted from Ref. [14] Copyright 2017, with permission from Elsevier

Gaseous sulfide source ($H_2S$) generated was collected by displacement of water. Two-way valve was turned off at first. The gas tank was filled with the saturated $H_2S$ solution and sealed with paraffin. Gaseous sulfide generated in the three-necked flask was then introduced into the tank. Solution in the gasholder was pushed into the fluid reservoir when the two-way valve was opened. A certain volume fraction of $H_2S$ gas was obtained by introducing nitrogen into the gasholder when the pressure gauge of cylinder was adjusted to the specified pressure with nitrogen cylinder open. The operating temperature and pH were controlled at a desired level. $H_2S$ gas was then delivered to the reactor and the simulated copper sulfate was pumped to the reactor by peristaltic pump.

### 3.5.3 Removal Efficiencies of Copper and Arsenic

(1) Reaction time

Arsenic and copper removal efficiency, barely influenced by reaction time (Fig. 3.49), are constant at 10% and 30%, respectively, as the reaction time increases from 0.5 to 4 min with addition of 10 mmol sulfide. However, it becomes 100% for both arsenic and copper when 70 mmol sulfide was added into the solution, which implies that sulfidation of arsenic and copper has finished within 0.5 min.

(2) Temperature

The copper removal efficiency increases from 60 to 80% when temperature increases from 24 to 60 °C, whilst arsenic removal efficiency decreases from 25 to 15%, indicating significant effect of temperature on arsenic and copper removal

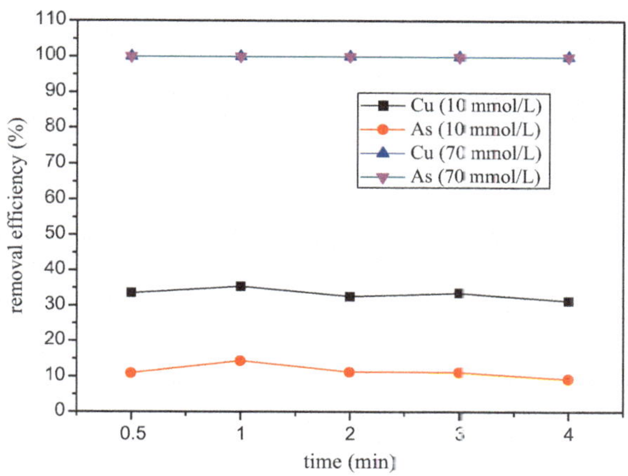


**Fig. 3.49** Arsenic and copper removal efficiency at different reaction times (50 °C). Reprinted from Ref. [147] Copyright 2017, with permission from Elsevier

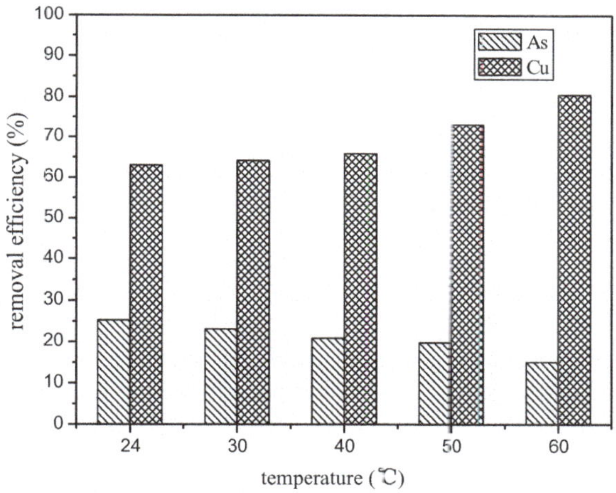


**Fig. 3.50** Arsenic and copper removal efficiency at different temperatures. Reprinted from Ref. [147] Copyright 2017, with permission from Elsevier

efficiency (Fig. 3.50). The improvement of separation efficiency of copper and arsenic can be achieved via raising the temperature. Besides, the temperature of acidic wastewater discharged from the industry process is as high as 55 °C, so the separation of arsenic and copper under the high temperature could be realized practically.

(3) Sulfide dosage

Arsenic and copper removal efficiencies increase as the amount of sulfide increases (Fig. 3.51), which are consistent with results previously reported by van Hille and coworkers [148] for copper sulfide precipitation in a fluidized bed reactor. In all cases, the copper removal efficiency increases more significantly than the arsenic removal efficiency does. Both arsenic and copper removal efficiency are as high as nearly 100% when the addition of sulfide is more than 70 mmol. Compared the

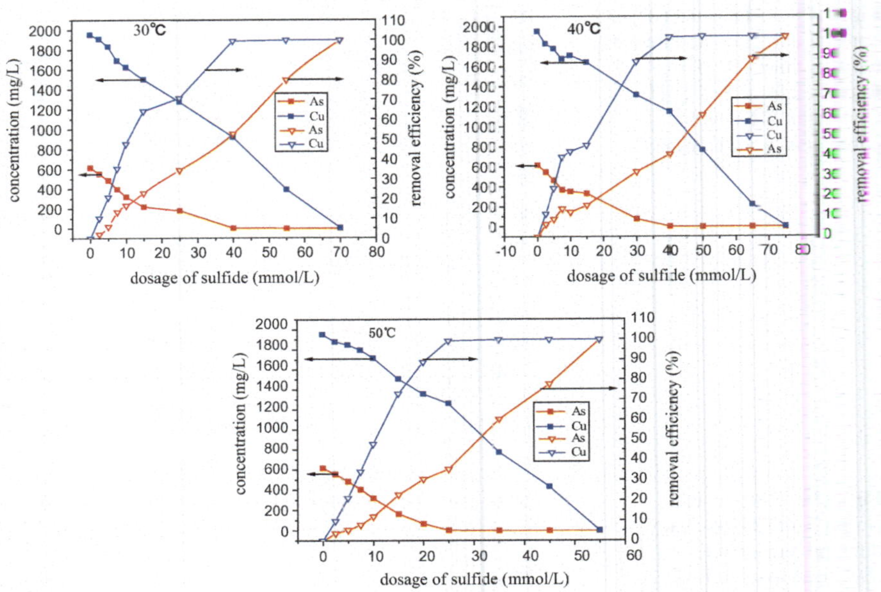


**Fig. 3.51** Arsenic and copper removal efficiency and concentration as a function of sulfide dosage at different temperatures (arsenic concentration of 600 mg $L^{-1}$, copper concentration of 2000 mg $L^{-1}$). Reprinted from Ref. [147] Copyright 2017, with permission from Elsevier

results obtained at 30, 40, and 50 °C, the difference between copper and arsenic removal efficiency increases with the temperature increasing. The copper removal efficiency is higher than 80% and arsenic removal efficiency is only 20% when a certain amount of sulfide is added into the solution at 50 °C. The results indicate that separation efficiency of copper and arsenic become higher with the temperature increases which consists of the results mentioned above.

In the first stage of primary separation, more than 80% copper and nearly 20% arsenic are precipitated with 20 mmol $L^{-1}$ $H_2S$ at 50 °C within 0.5 min.

### 3.5.4 Replacement of Arsenic by Copper

(1) Solid to liquid ratio

Arsenic and copper concentration in acidic wastewater when precipitates generated in the first stage are recycled to the initial solution with different solid to liquid ratios (Fig. 3.52). The results show that arsenic concentration in the initial solution increases as solid to liquid ratio increases, while copper concentration becomes lower at the similar conditions. It attributes to the replacement reaction between $As_2S_3$ and copper ions; it could be described as follows:

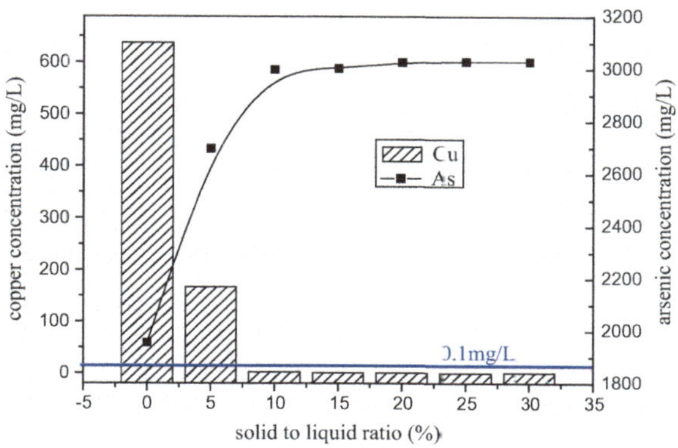


**Fig. 3.52** Arsenic and copper concentration in acidic wastewater at different solid/liquid ratios. Reprinted from Ref. [147] Copyright 2017, with permission from Elsevier

$$3Cu^{2+} + As_2S_3 + 4H_2O = 3CuS + HAsO_2 + 6H^+ \quad (3.32)$$

Most of the copper in the initial acidic wastewater is distributed into precipitates and the copper concentration in the initial acidic wastewater is lower than 0.1 mg $L^{-1}$ when the solid to liquid ratio is more than 10%. Arsenic was precipitated into the residue before dissolves into the solution. Therefore, the addition of precipitates generated in the first separation stage into the initial acidic wastewater with certain solid to liquid ratio can promote the separation of arsenic and copper effectively.

(2) Reaction time

The data of arsenic and copper concentration in acidic wastewater as a function of reaction time (Fig. 3.53) show that arsenic concentration increases within the first 10 min and then becomes constant at about 3200 mg $L^{-1}$ as time increasing, while copper concentration decreases rapidly from 110 to 0.5 mg $L^{-1}$ within the first 5 min and further decreases to lower than 0.1 mg $L^{-1}$ when time is longer than 10 min. The separation efficiency of arsenic and copper is more than 99%.

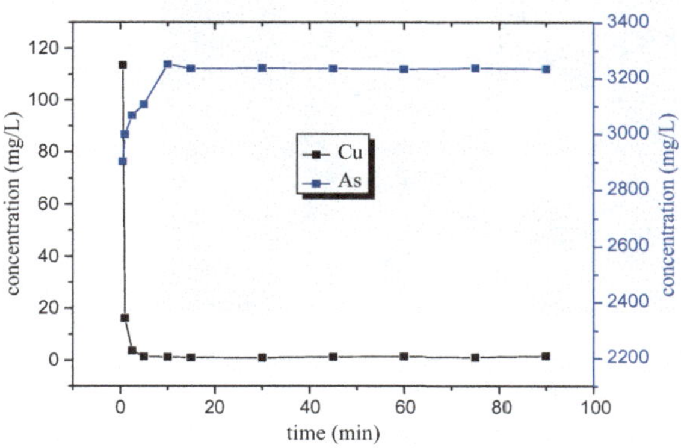


**Fig. 3.53** Arsenic and copper concentration in acidic wastewater with different reaction time. Reprinted from Ref. [147] Copyright 2017, with permission from Elsevier

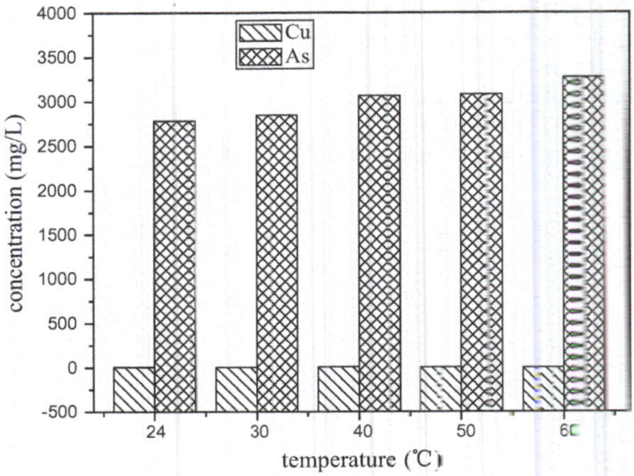


**Fig. 3.54** Arsenic and copper concentration in acidic wastewater at different temperatures. Reprinted from Ref. [147] Copyright 2017, with permission from Elsevier

It indicated that separation of arsenic and copper is finished within 10 min which is favorable to the application in the industry.

(3) Temperature

Second separation of arsenic and copper is temperature-independent represented by the results that copper concentration in the solution scarcely changes when temperature increases (Fig. 3.54). However, arsenic concentration increases slightly as temperature increasing, which attributes to the redissolution of arsenic sulfide in the precipitates at high temperature.

In the second stage of separation, the separation efficiencies of copper and arsenic are more than 99% with precipitates generated in the first stage recycled to the initial solution when solid to liquid ratio is not less than 10% at 20 °C within 10 min.

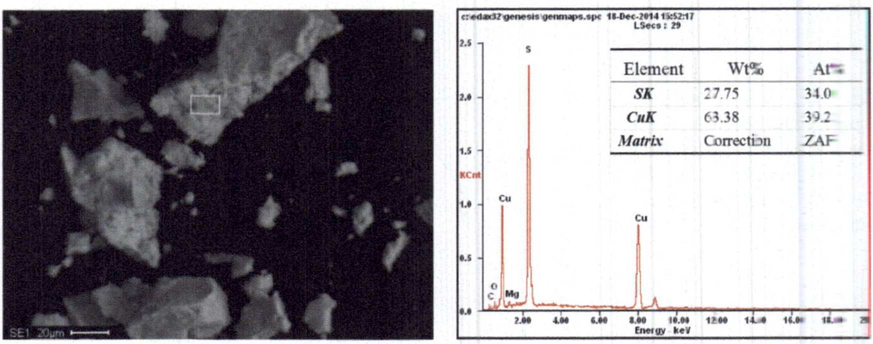


**Fig. 3.55** SEM-EDS image of the precipitate. Reprinted from Ref. [147] Copyright 2017, with permission from Elsevier

Generally, the morphology of the precipitate (Fig. 3.55) exhibits agglomeration of dense, amorphous particles in the micron range. The mole ratio of Cu to S is 1.15, which is approximate to 1, indicating that the precipitate mainly consists of CuS. The copper content of precipitate is as high as 63.38% in mass fraction.

## 3.6 Arsenic Removal by $Fe_3O_4$-Based Nanomaterials via Adsorption

### *3.6.1 Controllable Synthesis of Hierarchical Porous $Fe_3O_4$ Particles for Arsenic Removal*

In terms of the application of $Fe_3O_4$ as an adsorbent, decreasing the $Fe_3O_4$ particle size from micrometers to nanometers would increase the available adsorptive areas by 100–1000 times [149–151]. However, as the $Fe_3O_4$ particle size decreases to nanometers, its response to an external magnetic field undesirably decreases, which will not be large enough to overcome Brownian motion and no efficient magnetic separation will occur [149, 152, 153]. To tackle this problem, one practical strategy is to prepare magnetic hierarchical structures, which are constructed with building blocks of nano units. The hierarchical nanostructures not only exhibit high specific surface area because of the abundant interparticle spaces or intraparticle pores, but also possess satisfactory magnetic response because of their larger size and weaker Brownian motion, which therefore show great superiority to individual nanometer- and micrometer-sized materials [154–159]. Hence, we controllably prepare $Fe_3O_4$ hierarchical particles via modulating the grain property for arsenic adsorption [160].

**(1) Morphology, structure, and properties of $Fe_3O_4$**

The SEM images and size distribution of the $Fe_3O_4$-i (i = 1–6, representing the dosage of PDDA from 1 to 6 g) particles are presented in Fig. 3.56. The size of monodispersed hierarchical particles monotonously decreases from (A) 420 nm to (F) 100 nm, as increasing the poly (diallyl dimethyl ammonium chloride) (PDDA) dosage from 1 to 6 g. Correspondingly, the morphology of hierarchical particles gradually becomes coarse and porous, with the increase of PDDA dosage. HRTEM images were also conducted to give further insight into the grain assembly (Fig. 3.57). Taking $Fe_3O_4$-4 as an example, the particle shows pineal-like morphology with fringe spacing of 0.48 nm, corresponding to the (111) lattice planes of $Fe_3O_4$. The result indicates the possible oriented assembly of grain along (111) plane, which is the crystallographic plane with the highest energy and preferential for oriented attachment [161]. The structures and grain size of $Fe_3O_4$ were further measured by XRD, as shown in Fig. 3.58. All the diffraction peaks at $18.32 \pm 0.03$, $30.10 \pm 0.05$, $35.48 \pm 0.03$, $43.10 \pm 0.02$, $53.40 \pm 0.04$, $57.02 \pm 0.05$, and $62.58 \pm 0.08°$ can be indexed to the indices (111), (220), (311), (400), (422), (511), and (440) of $Fe_3O_4$. Briefly speaking, PDDA-modulated

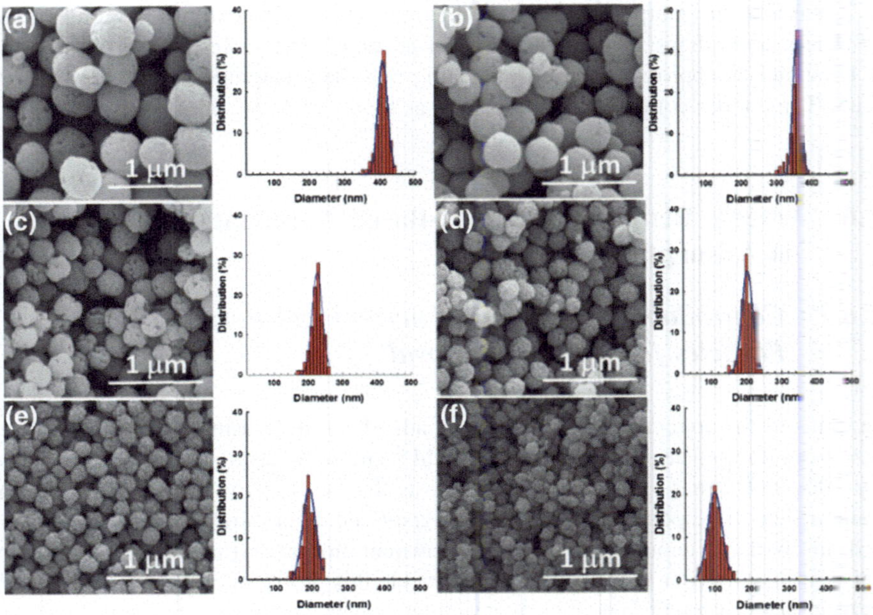


**Fig. 3.56** SEM images and the corresponding hierarchical particle size distribution of **a** $Fe_3O_4$-1, **b** $Fe_3O_4$-2, **c** $Fe_3O_4$-3, **d** $Fe_3O_4$-4, **e** $Fe_3O_4$-5, and **f** $Fe_3O_4$-6 at initial PDDA dosage varying from 1 to 6 g. Reprinted with the permission from Ref. [160] Copyright 2013, American Chemical Society

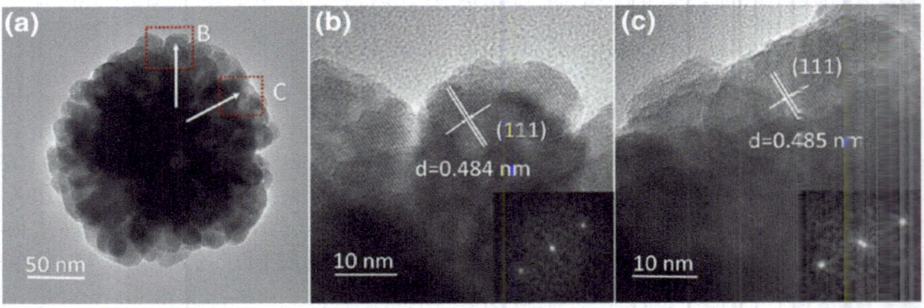


**Fig. 3.57** HRTEM images (**a–c**) of $Fe_3O_4$-4; **b** and **c** represent the magnification of the dotted area in (**a**). Reprinted with the permission from Ref. [160] Copyright 2013, American Chemical Society

solvothermal method successfully modulate products morphology, particle size, grain size, and facilitate the oriented grain assembly.

On the other hand, the surface area and pore size distribution of as-synthesized $Fe_3O_4$-i (i = 2, 4, 5, 6) were determined by nitrogen adsorption–desorption measurements (Fig. 3.59). $Fe_3O_4$-6 (Fig. 3.59d) synthesized with the highest PDDA

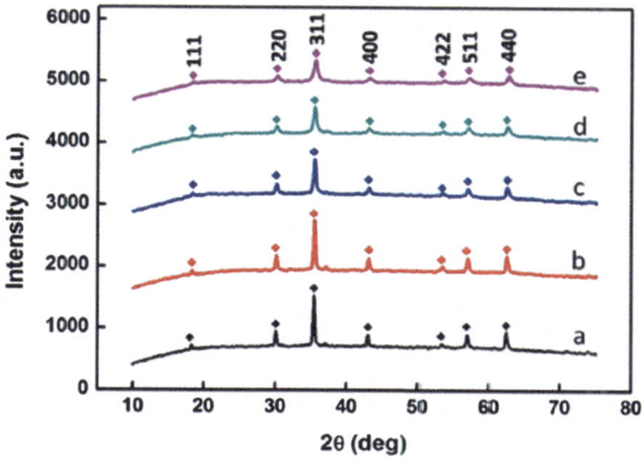


**Fig. 3.58** XRD patterns of $Fe_3O_4$ particles obtained at different PDDA dosage: (a) 2 g, (b) 3 g, (c) 4 g, (d) 5 g, (e) 6 g. Reprinted with the permission from Ref. [160] Copyright 2013, American Chemical Society

dosage possesses surface area and pore volume of 32.75 $m^2\ g^{-1}$ and 0.12 $cm^3\ g^{-1}$, respectively, both of which are higher than that of the $Fe_3O_4$-5 (31.16 $cm^2\ g^{-1}$ and 0.117 $cm^3\ g^{-1}$, Fig. 3.59c), $Fe_3O_4$-4 (19.13 $m^2\ g^{-1}$ and 0.07 $cm^3\ g^{-1}$, Fig. 3.59b) and $Fe_3O_4$-2 (7.05 $m^2\ g^{-1}$ and 0.015 $cm^3\ g^{-1}$, Fig. 3.59a). All the samples pose pore sizes in the range of 7–12 nm. The results above can be ascribed to the fact that smaller grain assembly possesses more channels, leading to the increased surface area and pore amount. Hence, increasing the PDDA dosage yields $Fe_3O_4$ hierarchical particles composing of smaller grain, which exhibit higher surface area and porosity.

The magnetic property of $Fe_3O_4$ hierarchical particles was evaluated, as shown in Fig. 3.60. The $M_s$ for the $Fe_3O_4$-i (i = 2–6) is in the range of 50–80 emu $g^{-1}$, which is comparable with many other magnetic particles [154].

Briefly speaking, hierarchical porous $Fe_3O_4$ particles with high magnetism were synthesized by facile PDDA-modulated solvothermal method, which is achieved in one-pot solution reaction and avoids the time/energy consuming precursor calcination process. Furthermore, PDDA-induced grain size tunable strategy has been proved to be an efficient way to enhance the surface area and porosity of particles.

**(2) Mechanism for the formation of $Fe_3O_4$ hierarchical particles mediated by PDDA**

The morphology and structure of the products with initial PDDA dosage of 4 g at various reaction time were examined by TEM, FTIR and XRD to preliminarily understand the morphology and structure evolution of $Fe_3O_4$ hierarchical particles (Fig. 3.61).

TEM results give insight into the morphology evolution of mesoporous $Fe_3O_4$ (Fig. 3.61a–f), three typical stages were observed for the formation of $Fe_3O_4$, namely, the formation of spindle precursor with length of 5–10 nm (0–1.5 h), the formation and assembly of grain to sphere particles (1.5–4 h), and the oriented assembly/Ostwald ripening process of preformed sphere into porous particles (4–8 h).

The XRD patterns (Fig. 3.61g) of $Fe_3O_4$-4 at 1.5 h depict a strong peak at 7.55° along with a broad weak one at 25° probably originated from (001) and

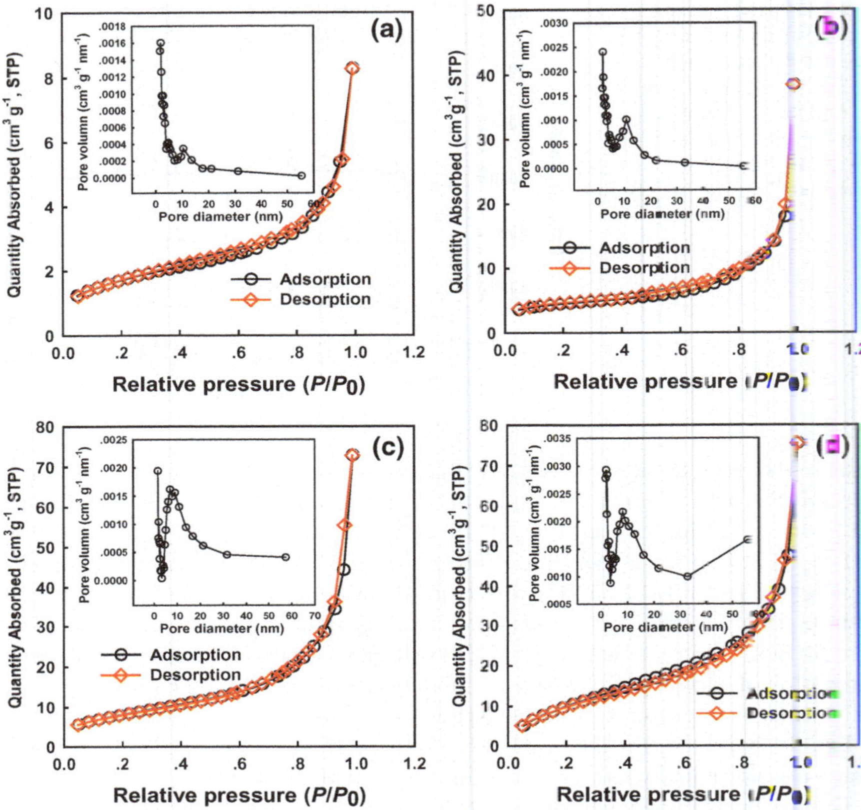


**Fig. 3.59** Nitrogen adsorption–desorption isotherms and pore size distribution curves (the corresponding insert) of **a** as-obtained $Fe_3O_4$-2, **b** $Fe_3O_4$-4, **c** $Fe_3O_4$-5, and **d** $Fe_3O_4$-6 respectively. Reprinted with the permission from Ref. [160] Copyright 2013, American Chemical Society

(013) planes of an iron oxide acetate hydroxide hydrate with a formula of $Fe_2O$ $(CH_3COO)$ $(OH)_3 \cdot H_2O$ according to JCPDS. Then, XRD patterns of the samples obtained at the time from 2.5 to 4 h show gradually enhanced peaks at 30.00°, 35.48°, 43.14°, 53.44°, 57.04°, and 62.58°, marked by the indices (220), (311), (400), (422), (511), and (440) of $Fe_3O_4$ phases. When the reaction time was 6–8 h, the produced aggregates were pure $Fe_3O_4$.

The FT-IR spectra (Fig. 3.61h) of $Fe_3O_4$-1.5 h and $Fe_3O_4$-2.5 h show reduced absorption peaks at 1578 and 1445 $cm^{-1}$ due to the asymmetric and symmetric stretching of COO– group, band at 1090 $cm^{-1}$ owing to C–O stretching of the COO– group, band at 887 $cm^{-1}$ due to the OH bending [162]. The spectra of the samples within 4–8 h show broad strong band at 591 $cm^{-1}$ due to the Fe–O lattice mode of $Fe_3O_4$ [157]. Except for the anticipated typical peak for iron composite

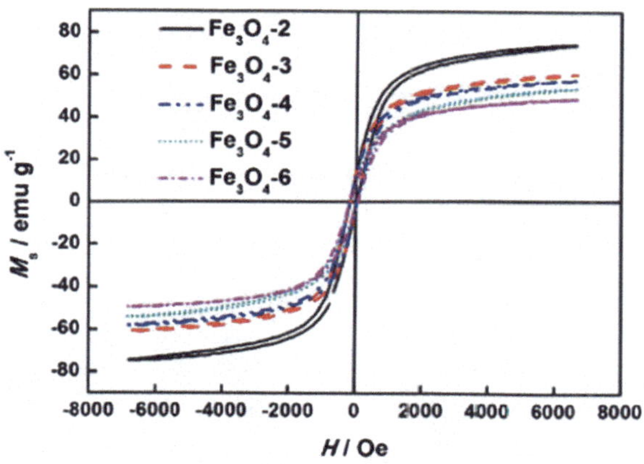


**Fig. 3.60** Room temperature hysteresis loops for the $Fe_3O_4$-*i* (*i* = 2–6) with initial PDDA dosage increasing from 2 to 6 g. Reprinted with the permission from Ref. [160] Copyright 2013, American Chemical Society

($Fe_2O(CH_3COO)(OH)_3 \cdot H_2O$ or $Fe_3O_4$), the peak at 1125 $cm^{-1}$ was ascribed to the C–N symmetric stretching vibration of PDDA. The PDDA also exhibits weak $CH_2$ bending vibrations (around 1474, 1326, and 960 $cm^{-1}$), C-H asymmetric, and C-H symmetric stretching frequencies (2918 and 2867 $cm^{-1}$) [163, 164]. Thus, both the XRD and FT-IR indicate the gradual formation of $Fe_3O_4$ phase at the expense of preformed $Fe_2O(CH_3COO)(OH)_3 \cdot H_2O$ phase.

Thus, a possible mechanism was proposed to elucidate the PDDA-induced grain size tunable strategy for the controllable synthesis of porous $Fe_3O_4$ hierarchical particles. A mixture composed of $FeCl_3$, EG, NaAc, and PDDA was first obtained and the viscosity of mixture was greatly enhanced by PDDA (Fig. 3.62). Spindle particles were then obtained with PDDA as capping agents, which improved the particle dispersibility. As time goes on, hierarchical $Fe_3O_4$ particles were eventually produced, and meanwhile PDDA function on capping effect and increasing viscosity declines particle and grain size, facilitates oriented assembly, thus synchronously enhancing surface area and porosity. Briefly speaking, a PDDA-modulated solvothermal method can controllably prepare porous $Fe_3O_4$ hierarchical particles.

### (3) Arsenic adsorption performance of $Fe_3O_4$

The adsorption capacity of $Fe_3O_4$-i (i = 2–6) for As(V) and As(III) was evaluated using the equilibrium adsorption isotherm by varying the initial As(V) and As(III) concentrations. The adsorption capacity of As(V) and As(III) monotonously increased from 1.93 and 1.57 mg $g^{-1}$ for $Fe_3O_4$-2 to 7.23 and 6.77 mg $g^{-1}$ for $Fe_3O_4$-6 (Fig. 3.63A, C), indicating that the morphology mediated by PDDA greatly facilitates the absorption performance of particles. $Fe_3O_4$-5 and $Fe_3O_4$-6 exhibit higher adsorption capacity than commercial $Fe_3O_4$ particles (1.35 mg $g^{-1}$ for As(V) and 0.76 mg $g^{-1}$ for As(III)). The adsorption of As(V) onto $Fe_3O_4$ fits the Langmuir isotherm model (Fig. 3.63B), which interprets the adsorption process as a monolayer adsorption on a homogeneous surface. In contrast, the adsorption of As (III) onto $Fe_3O_4$ fits well with the Freundlich isotherm model (Fig. 3.63D),

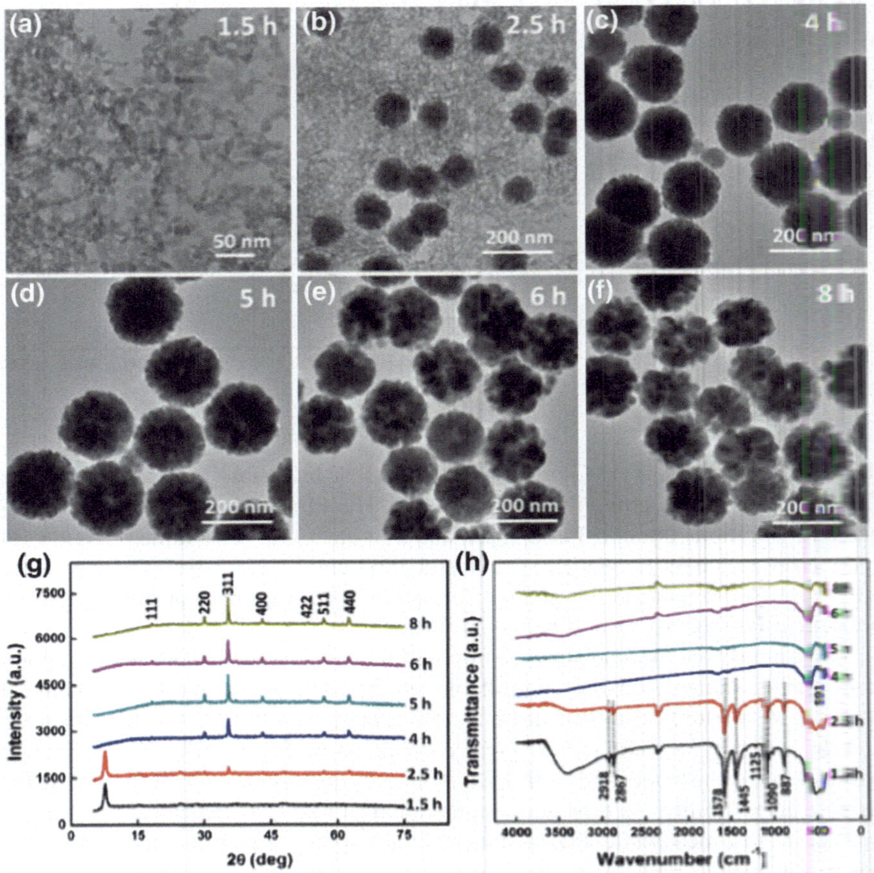


Fig. 3.61 **a–f** TEM, **g** XRD, and **f** FT-IR spectra of $Fe_3O_4$-4-xh samples (x = 1.5, 2.5, 4, 5, 6, 8). Reprinted with the permission from Ref. [160] Copyright 2013, American Chemical Society

indicating that the adsorption process is a multilayer adsorption on a homogeneous surface.

The two different adsorption isotherm models may be attributed to the different surface charge effects of As(V) and As(III) species under the environment of pH 5 [165]. For As(V), there exists an electrostatic attraction between positively charged $Fe_3O_4$ samples and negatively charged As(V) species. However, As(III) exists predominantly as noncharged $H_3AsO_3$ [166]. The interaction between $Fe_3O_4$ samples and noncharged As(III) species is little so that the adsorption of As(III) should continue to increase with the increase of As(III) concentration [165].

The adsorption of As(V) or As(III) is rapid at first and then slows considerably (Fig. 3.64). The rapid adsorption at first is ascribed to the process of arsenic adsorption on the exterior surface of the $Fe_3O_4$ particles. The slower adsorption rate

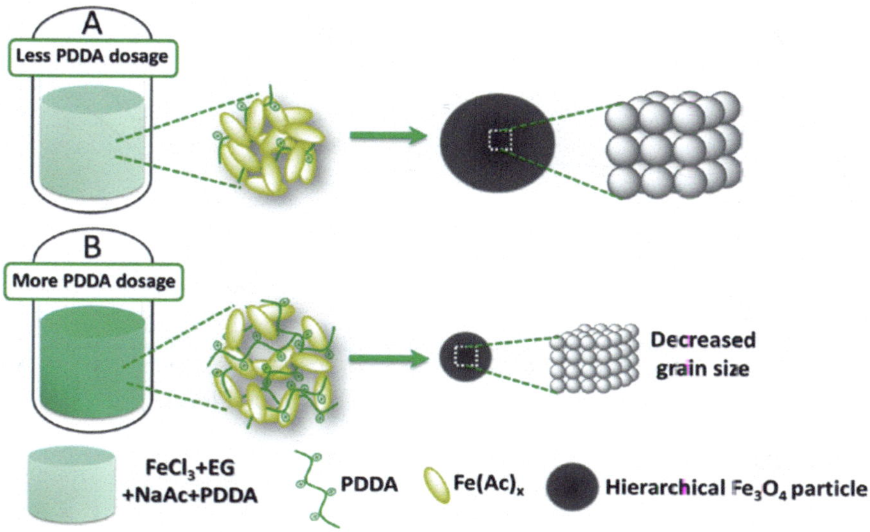


**Fig. 3.62** Scheme of the formation of hierarchical $Fe_3O_4$ particles mediated by PDDA. Reprinted with the permission from Ref. [160] Copyright 2013, American Chemical Society

followed might be partially due to the higher diffusion resistance as the arsenic begins to enter and move into the interior of the $Fe_3O_4$ particles via the nanopores [165]. All the above adsorption kinetic experimental data can be best fitted into a pseudo-second-order rate kinetic model.

The full-range XPS spectra (Fig. 3.65a) show the appearance of arsenic species and the increase of O intensity after arsenic adsorption, validating the arsenic adsorption. XPS of Fe2p of all samples (Fig. 3.65b) exhibit the binding energies of Fe 2p1/2 at 724.4 eV and Fe 2p3/2 at 710.5 eV [167, 168]. XPS of As3d (Fig. 3.65c) in $Fe_3O_4$ adsorbed As(V) shows a peak located at 45.1 eV, attributing to As(V)–O bonding, and that of As3d in $Fe_3O_4$ adsorbed As(III) indicates fitted peak located at 43.9 eV, corresponding to As(III)–O, respectively [169–173]. The results confirmed no major differences in the valence state of the Fe and As species in arsenic adsorption. O1s XPS spectrum (Fig. 3.65d) can be deconvoluted into peaks located at 530.0, 531.5, and 533.0 eV, which are attributed to oxygen in the lattice (e.g., Fe–O or As–O), oxygen atoms in the surface hydroxyl groups (H–O), and oxygen in the outermost layer of $H_2O$ or $CO_2$ adsorbed [169, 171, 174–176]. The high peak intensity of H–O species of $Fe_3O_4$ confirms the existence of many hydroxyl groups on the surface of $Fe_3O_4$ spheres, which plays a vitally important role in the arsenic removal [169]. Moreover, after arsenic adsorption, the shift of the O 1s binding energy to low energy, the proportion decrease of the H–O (531.5 eV) and the proportion increase of O in the lattice (530.0 eV) suggest that the adsorption mechanism was mainly ascribed to the substitution of Fe–OH groups by arsenic species.

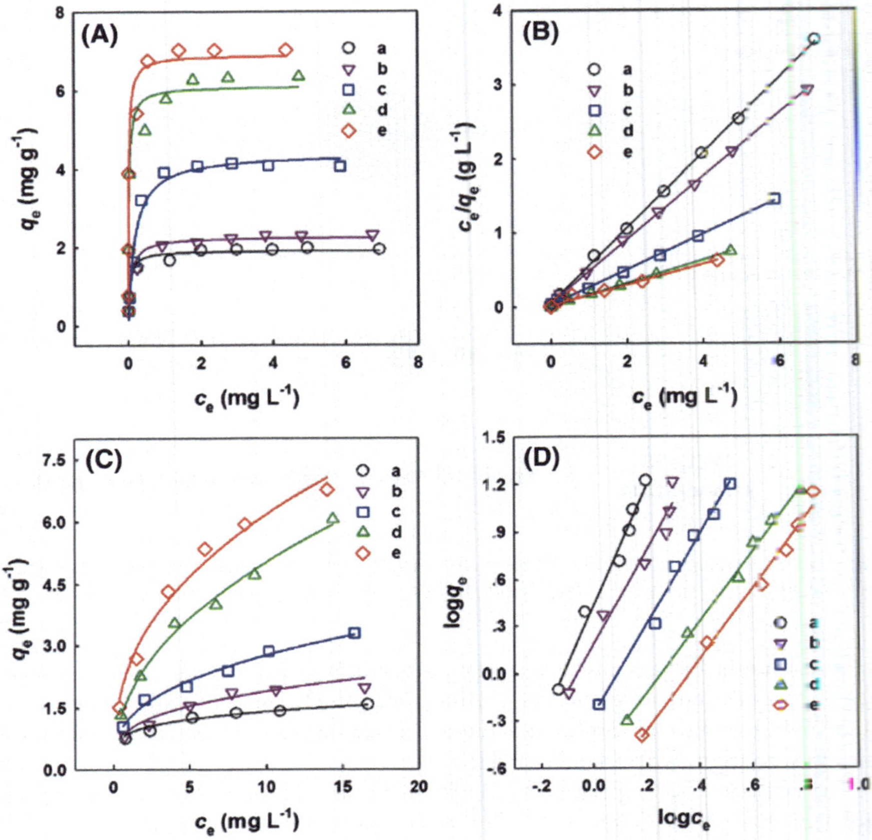


**Fig. 3.63** Adsorption isotherms of **A** As(V) and **C** As(III) onto (a) $Fe_3O_4$-2, (b) $Fe_3O_4$-3, (c) $Fe_3O_4$-4, (d) $Fe_3O_4$-5, and (e) $Fe_3O_4$-6 samples with the initial PDDA dosage of 2 g, 3 g, 4 g, 5 g, and 6 g, respectively. **B** Linearized Langmuir isotherm for As(V) adsorption and **D** linearized Freundlich isotherm for As(III) adsorption by (a) $Fe_3O_4$-2, (b) $Fe_3O_4$-3, (c) $Fe_3O_4$-4, (d) $Fe_3O_4$-5, and (e) $Fe_3O_4$-6 samples (T = 25 °C; adsorbent doses = 0.5 g L$^{-1}$; pH = 5 ± 0.2). Reprinted with the permission from Ref. [160] Copyright 2013, American Chemical Society

### 3.6.2 Arsenic Adsorption by Cu Doped $Fe_3O_4$ Magnetic Adsorbent

The transformation of As(III) to As(V) is necessary for arsenic removal [177–180]. Doping is an effective approach to modify particle structure, such as increased amount of surface defects [181, 182], increased hydroxyl amount [183, 184], tunable surface charge [185] and high electronic conduction properties [186], which could enhance adsorption performance and broaden applications [187]. Adding oxidant to convert As(III) to As(V) is normal and efficient route. Most recently,

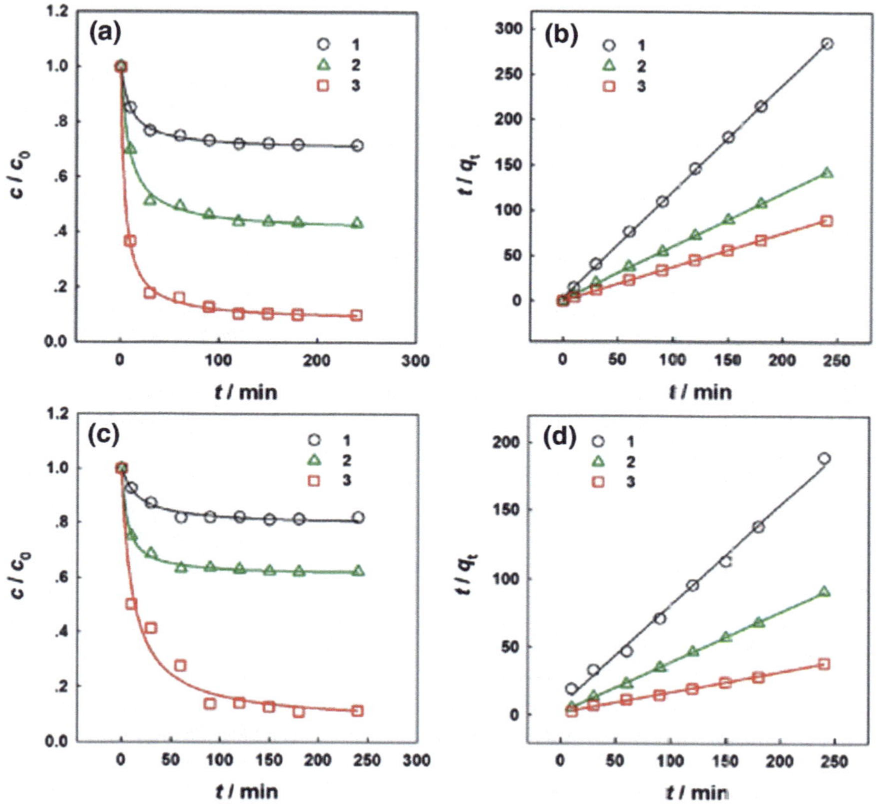


**Fig. 3.64** (**a, c**) Adsorption rate of As(V) and As(III) by $Fe_3O_4$-3 (1), $Fe_3O_4$-4 (2), and $Fe_3O_4$-5 (3) samples. (**b, d**) Pseudo-second-order kinetic plots for the adsorption of As(V) and As(III). (T = 25 °C; adsorbent doses = 0.5 g $L^{-1}$; pH = 5 ± 0.2). Reprinted with the permission from Ref. [160] Copyright 2013, American Chemical Society

Cu doping has attracted great attention, copper ions were able to act as electron mediation centers to accelerate the oxidation reaction based on Cu(II)/Cu(I) couple [188, 189]. It is rational to dope copper ions in magnetic adsorbents, which are expected to simultaneously possess good adsorption performance, catalytic oxidation and readily magnetic separation property.

**(1) Morphology and structure of $Fe_3O_4$:Cu**

As the initial amount of $Cu^{2+}$ increased from 0 to 2.5 mM, the sphere size decreased from −560 to −120 nm (Fig. 3.66). The particle structure transformed from solid spheres to porous spheres with decreased size accordingly. EDS-mapping images of $Fe_3O_4$:Cu-2.5 in Fig. 3.75g–i revealed that Cu, Fe, and O elements distributed uniformly on the $Fe_3O_4$:Cu-2.5 particles. Microscopic images

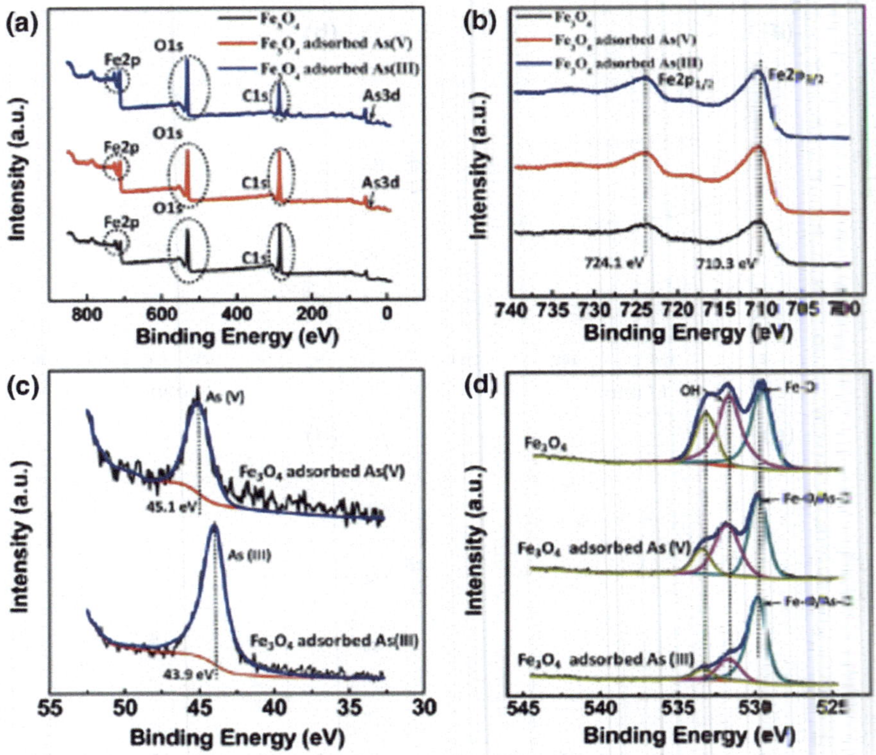


**Fig. 3.65** **a** Full-range, **b** Fe 2p, **c** As 3d, and **d** O 1s XPS spectra of several samples of interests including the $Fe_3O_4$, $Fe_3O_4$ adsorbed As(V), and $Fe_3O_4$ adsorbed As(III) particles. Reprinted with the permission from Ref. [160] Copyright 2013, American Chemical Society

indicated that indicated that porous $Fe_3O_4$:Cu particles with Cu, Fe, O uniformly distributed shows particle size and structure dependent on initial $Cu^{2+}$ dosage.

The XRD spectra of all $Fe_3O_4$:Cu particles in Fig. 3.67 show well-defined diffraction lines suggesting the well-crystallized structure. As expected, the typical crystal planes (111), (220), (311), (400), (511), and (440) of $Fe_3O_4$ (JCPDS no. 86-1344) were observed for all samples. With the increase of initial $Cu^{2+}$ dosage, the intensity of diffraction peaks at 43.31, 50.44, and 74.10 correspond to the crystal planes (111), (200), and (220) of Cu increase, indicating the augment in the amount of metal Cu (JCPDS no. 85-1326). The metal Cu might be produced due to the reduction of $Cu^{2+}$ by ethylene glycol [190]. No peaks attributed to CuO or $Cu_2O$ were detected [191, 192].

The combined microscopic and XRD results indicated the concurrent presence of doped $Cu^{2+}$ and the Cu nanocrystal on $Fe_3O_4$:Cu particles. The mixture of $Cu^0$ and $Cu^{2+}$ doped $Fe_3O_4$ nanograin constituted $Fe_3O_4$:Cu particles with Fe, Cu, and O uniformly distributed. The Cu loading decreased the particle size, which should be ascribed to the nucleation effect of the doped $Cu^{2+}$ [193, 194].

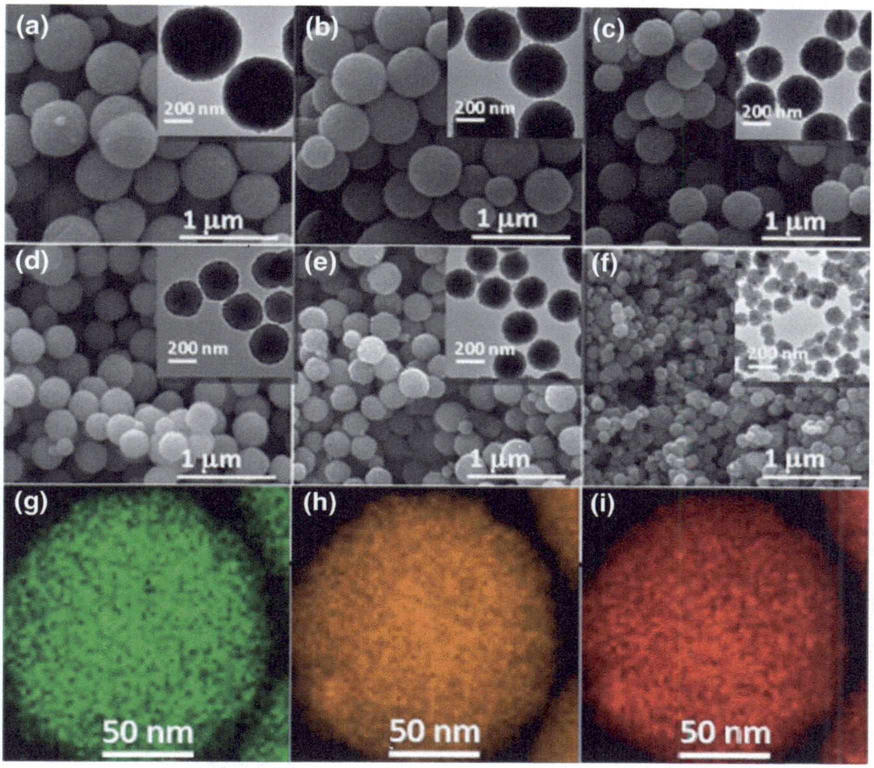


**Fig. 3.66** SEM of $Fe_3O_4$:Cu-0 (**a**), $Fe_3O_4$:Cu-0.05 (**b**), $Fe_3O_4$:Cu-0.25 (**c**), $Fe_3O_4$:Cu-0.5 (**d**), $Fe_3O_4$:Cu-1.0 (**e**), $Fe_3O_4$:Cu-2.5 (**f**) particles, EDS-mapping images Cu (**g**), Fe (**h**) and O (**i**) of $Fe_3O_4$:Cu-2.5 particles. The inset shows the corresponding TEM image. Reprinted from Ref. [136] by permission of The Royal Society of Chemistry 2015

The pH point of zero charged ($pH_{PZC}$) of $Fe_3O_4$:Cu increased from 6.97 to 8.86 as increasing initial dosage of copper ions from 0 to 2.5 mM (Fig. 3.68). The function of Cu incorporation in enhancing $pH_{PZC}$ should be that the doping of $Cu^{2+}$ caused more defects in $Fe_3O_4$ crystals, which was favorable for the adsorption of hydroxyl on particles and then increased the affinity to proton, eventually leading to the increase in $pH_{PZC}$ [195]. The increased $pH_{PZC}$ indicates that the adsorbent has a high tendency to be positive and thus has the potential of high affinity to negative arsenic species [196, 197].

**(2) Arsenate adsorption performance**

As the initial dosage of copper ions increase from 0 to 2.5 mM, the adsorption capacity for As(V) and As(III) increases from 7.32 to 42.90 mg $g^{-1}$ and from 8.12 to 32.97 mg $g^{-1}$ (Fig. 3.69a and c), respectively. The adsorption capacity of $Fe_3O_4$: Cu-2.5 particle is much higher than many reported related adsorbents. Adsorbents

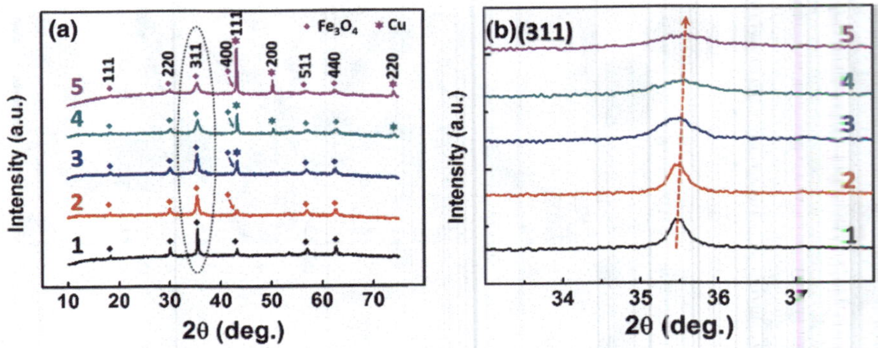


**Fig. 3.67** XRD spectra (**a**) and the magnified peak (311) (**b**) of $Fe_3O_4$:Cu-0 (1), $Fe_3O_4$:Cu-0.05 (2), $Fe_3O_4$:Cu-0.25 (3), $Fe_3O_4$:Cu-1.0 (4), and $Fe_3O_4$:Cu-2.5 (5) particles. Rhombic and stellate icons represent the crystal place of $Fe_3O_4$ and Cu, respectively. Reprinted from Ref. [136] by permission of The Royal Society of Chemistry 2015

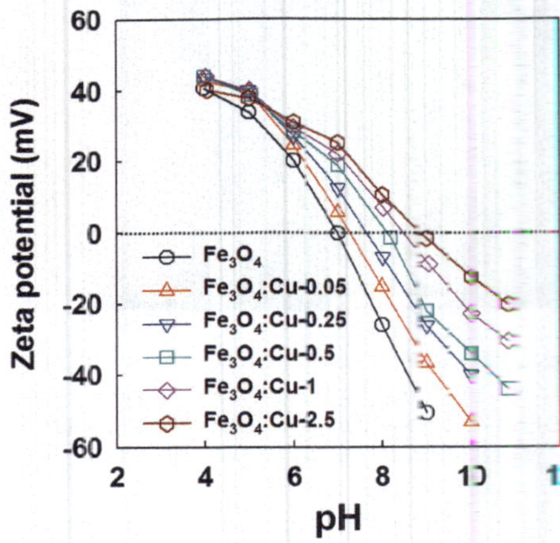


**Fig. 3.68** Zeta potential of $Fe_3O_4$:Cu particles as a function of pH. The dashed line represents zero zeta potential. Reprinted from Ref. [136] by permission of The Royal Society of Chemistry 2015

with higher zeta potential possessed positive-charged surface and thus high affinity toward negative As(V) species. However, the modified surface charged state could not explain increased affinity for As(III). The speciation of adsorbed As(III) was then characterized in detail to explain the phenomenon.

The Langmuir models can well describe the adsorption behavior of As(V) on $Fe_3O_4$:Cu (Fig. 3.69b), while Freundlich models well describe the adsorption behavior of As(III) on $Fe_3O_4$:Cu (Fig. 3.69d). The two different adsorption isotherm models may be attributed to the different surface charges of arsenic species, namely negative-charged As(V) and noncharged As(III) [197, 198]. With the increase of Cu amount, the calculated Langmuir maximum adsorption capacities for As(V) increased from 7.90 to 43.37 mg $g^{-1}$.

The regeneration and reuse of the adsorbent are important in considering the practical applicability. The $Fe_3O_4$:Cu adsorbent was magnetically separated within

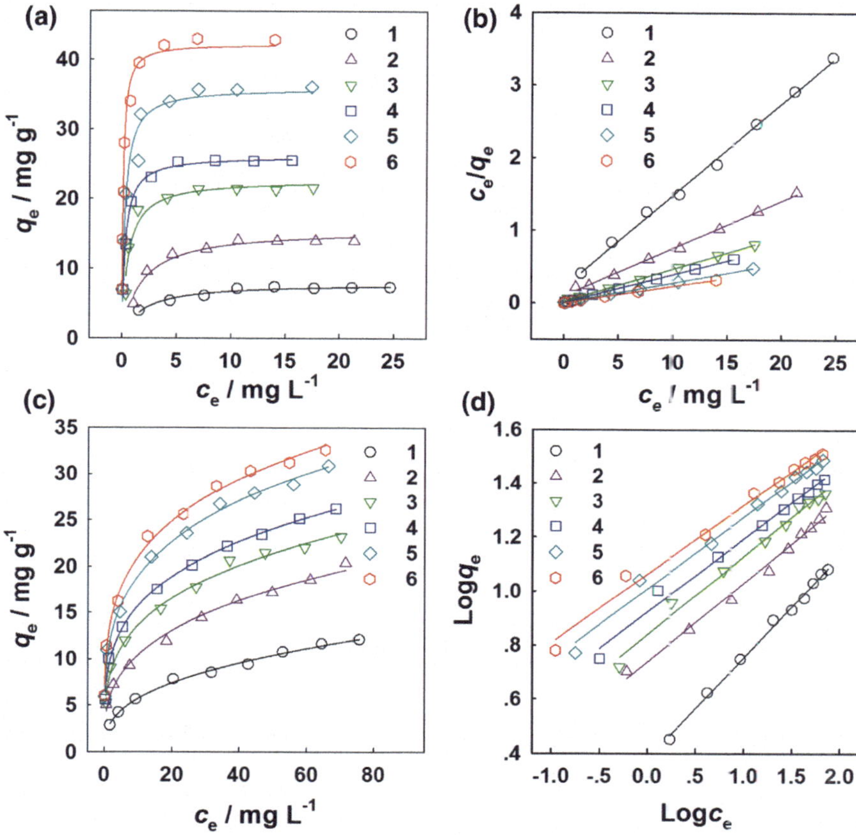


**Fig. 3.69** Adsorption isotherms of As(V) (**a** and **b**) and As(III) (**c** and **d**) on $Fe_3O_4$:Cu-0 (1), $Fe_3O_4$:Cu-0.05 (2), $Fe_3O_4$:Cu-0.25 (3), $Fe_3O_4$:Cu-0.5 (4), $Fe_3O_4$:Cu-1 (5), and $Fe_3O_4$:Cu-2.5 (6) particles at 298 K. Reprinted from Ref. [136] by permission of The Royal Society of Chemistry 2015

10 s and readily re-dispersed by slightly stirring. The regeneration of $Fe_3O_4$:Cu was conducted using 0.1 M NaOH solution as eluent. It was found that the removal efficiency remained 80% after six cycles (82.4% for As(V) and 81.8% for As(III)), which indicated the feasibility of regenerating the $Fe_3O_4$:Cu adsorbent (Fig. 3.70).

### (3) Arsenate adsorption mechanism

Full-range XPS spectra of $Fe_3O_4$:Cu-2.5 exhibited the information of Fe, Cu and O element (Fig. 3.71). As 3d peaks appeared after As(V) and As(III) adsorption. High-resolution XPS spectra of the As 3d peak are shown in Fig. 3.71a. $Fe_3O_4$:Cu-2.5 with As(V) adsorbed shows the peak at 45.3 eV ascribed to As(V)–O, indicating no valence change for As(V) [169, 199]. The adsorption of As(III) on $Fe_3O_4$:Cu was conducted under atmosphere, $N_2$ protection, and $O_2$ bubbling

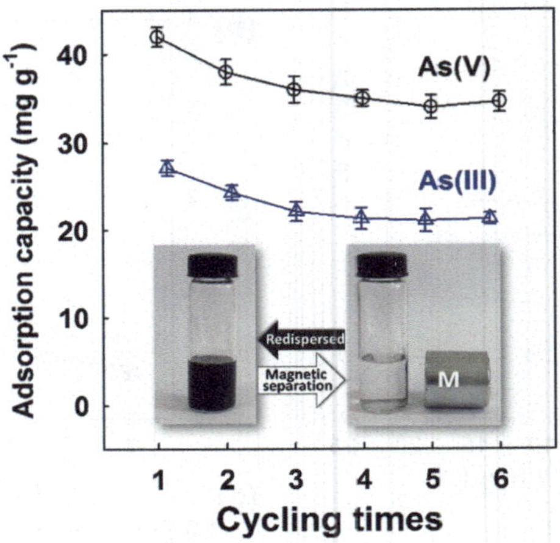


**Fig. 3.70** Arsenic removal efficiency of $Fe_3O_4$:Cu-2.5 particles for As(V) (1) and As(III) (2) in different cycle numbers. The inset: separation/redispersion property of $Fe_3O_4$:Cu-2.5 under external magnet (M). Reprinted from Ref. [136] by permission of The Royal Society of Chemistry 2015

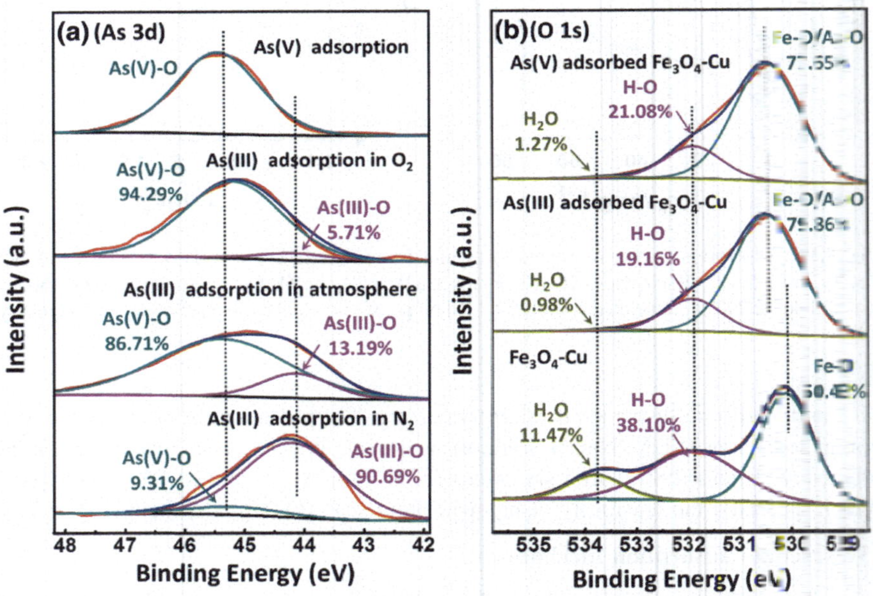


**Fig. 3.71** As XPS 3d spectra (**a**) of $Fe_3O_4$:Cu-2.5 with arsenic adsorbed and O XPS 1s spectra (**b**) of $Fe_3O_4$:Cu-2.5 before and after arsenic adsorption. Reprinted from Ref. [136] by permission of The Royal Society of Chemistry 2015

condition, respectively. The As(V)–O were estimated as 9.31%, 86.71%, and 94.29% for $N_2$ protection, atmosphere, and $O_2$ bubbling condition, respectively. The results suggested that As(III) was efficiently oxidized to As(V) by $O_2$ on $Fe_3O_4$:Cu particles. A few oxidation efficiencies for $N_2$ protection were attributed to trace rudimental oxygen in solution.

The O 1s spectra of $Fe_3O_4$:Cu-2.5 (Fig. 3.71b) can be deconvoluted into three overlapped peaks corresponding to X–O (Fe–O, As–O), hydroxyl groups (–OH) and adsorbed water ($H_2O$), respectively. After arsenic adsorption, the peak of the X–O has shifted from 530.11 to 530.50 eV and the area ratio attributed to X–O increased from 50.43% to 79.86% and 77.65% after As(V) and As(III) adsorption, respectively. The variation for X–O may be due to: (i) the formation of Fe-O on the surface after the reaction between adsorbents and adsorbates; (ii) the As–O in adsorbed arsenic species on the surface. The intensity of the –OH peak slightly decreased from 38.10 to 19.16% and 21.08% after arsenic adsorption, implying that the hydroxyl groups have exchanged with arsenic species.

XANES spectra of As(V) adsorbed and As(III) adsorbed $Fe_3O_4$:Cu particles were recorded in Fig. 3.72. For comparison, the XANES spectrum of $Na_3AsO{\cdot}12H_2O$ and $NaAsO_2$ were shown, which exhibited peaks at 11874.1 eV for As(V)–O and peaks at 11870.3 eV for As(III)–O, respectively [200]. The XANES spectrum of all Cu–$Fe_3O_4$ samples after As(V) adsorption exhibited a typical peak for As(V)–O at 11874.1 eV (Fig. 3.72a). After As(III) adsorption (Fig. 3.72b), two peaks were observed, a peak at 11870.3 eV ascribed to As(III)–O and a peak at 11874.1 eV ascribed to As(V)–O. The intensity of As(V)–O peak increased with the increase of Cu amount, higher than the negligible peak for As (V)–O on $Fe_3O_4$:Cu-0. The results indicate that the oxidation efficiency of As(III) was highly dependent on the amount of Cu on adsorbent. The mechanism is $Cu^{2+}$ acting as electron transfer center between reaction agents. Based on Cu(II)/(I), the proposed pathway is that $Cu^{2+}$ first acted as electron receptor, producing oxidation

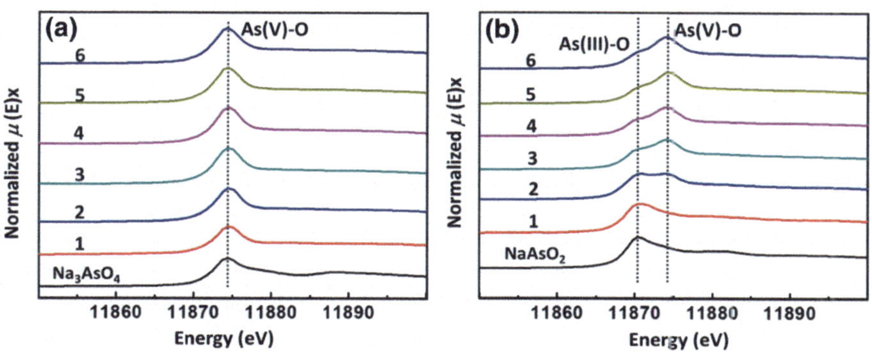


**Fig. 3.72** Normalized absorbance of As K-edge XANES spectra for As(V) absorbed (**a**) and As (III) absorbed (**b**) $Fe_3O_4$:Cu-0 (1), $Fe_3O_4$:Cu-0.05 (2), $Fe_3O_4$:Cu-0.25 (3), $Fe_3O_4$:Cu-0.5 (4), $Fe_3O_4$:Cu-1 (5), and $Fe_3O_4$:Cu-2.5 (6) under atmosphere condition. Reprinted from Ref. [136] by permission of The Royal Society of Chemistry 2015

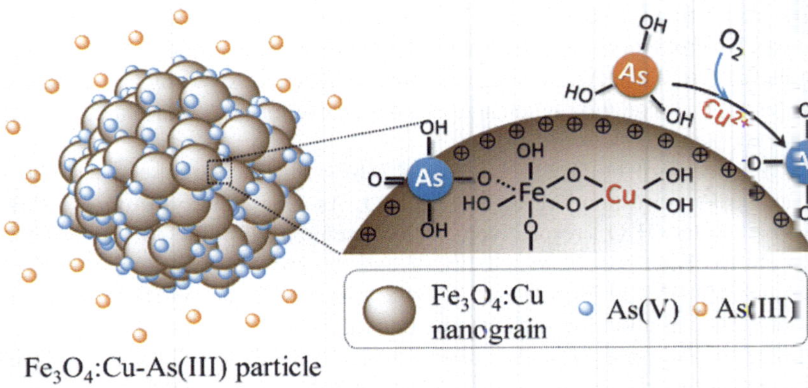


**Fig. 3.73** The As(III) redox and adsorption on $Fe_3O_4$:Cu particle. Reprinted from Ref. [132] by permission of The Royal Society of Chemistry 2015

product and $Cu^+$ as transient species. The active $Cu^+$ then serves as electron donor, which readily transfers electron to oxidants, yielding reduction product and $Cu^{2+}$ [201, 202].

The proposed mechanism of arsenic adsorption on $Fe_3O_4$:Cu particle was given in Fig. 3.73. The mixture of $Cu^0$ and $Cu^{2+}$ doped $Fe_3O_4$ nanograin consisted $Fe_3O_4$:Cu particles. Due to the Cu incorporation, the $Fe_3O_4$:Cu particles have high surface area, high porosity, abundant hydroxyl groups, and positive-charged surface, which are beneficial for As(V) adsorption. When $Fe_3O_4$:Cu adsorbent was applied for As(III) adsorption, As(III) was efficiently oxidized to As(V) by $O_2$ based on $Cu^{2+}$ as electron mediation agent between As(III) and $O_2$. The produced negative As(V) species was then in situ adsorbed on $Fe_3O_4$:Cu particles based on the electrostatic interaction and hydroxyl exchange. Thus, the $Fe_3O_4$:Cu particles were efficient adsorbents for both As(III) and As(V).

### 3.6.3 *Arsenic Adsorption by $Fe_3O_4$@$Cu(OH)_2$ Composites*

Magnetic nanoparticles (MNPs) have aroused great attention because of the unique merits of facile magnetic separation property [203, 204], such as porous $Fe_3O_4$ [157], flower-like $Fe_3O_4$ adsorbent [203], superparamagnetic high-surface-area $Fe_3O_4$ nanoparticles [205]. Nevertheless, these materials always suffer from undesirable adsorption performances because of high tendency of self-aggregation, less quantity of functional groups, and weak affinity to contaminants [206]. To address the problems, surface functional coating provides a prospective solution for MNPs performance improvement by combining multiple functionalities of introduced chemicals or special structures [207], which attracted increasing interests in specific drug targeting, magnetic cell separation, magnetic sealing, etc. Copper-based nanomaterials have received great attention because of their potential application in optoelectronic devices, catalysis, and superconductors [208, 209]. Thus, $Fe_3O_4$@Cu $(OH)_2$ composites were synthesized and used for arsenic adsorption.

**(1) Characterization of $Fe_3O_4$@$Cu(OH)_2$ composites**

From the TEM images, bare $Fe_3O_4$ particles apparently showed regular microspheres of size around 250 nm without Cu addition (Fig. 3.74a) [210]. When Cu was added initially, some tiny flocculent like structures can be observed obviously along the surface of $Fe_3O_4$ (Fig. 3.74b), implying obtained copper hydroxide was preferably assembled on the particles. It is more interesting that as the amount of $CuCl_2$ increased successively, the initial dendrites gradually grew up to constitute final leaflike nanostructures around the $Fe_3O_4$ particles with ca. 300–500 nm in length and at most 200 nm in width (Fig. 3.74c–e).

The phase and composition of the $Fe_3O_4$ and $Fe_3O_4$@$Cu(OH)_2$ composites were characterized by XRD (Fig. 3.75). All the diffraction peaks located at about 30.24°, 35.52°, 43.12°, 57.04°, and 62.66° can be assigned to the diffraction of $Fe_3O_4$ crystal with inverse spinel structure from the (220), (311), (400), (511), and (440) planes, respectively (PDF#75-1610). Compared with curves 1, 2, 3, 4, and 5, $Fe_3O_4$@$Cu(OH)_2$ exhibited the typical magnetic $Fe_3O_4$ peak patterns indicating that the prepared $Fe_3O_4$@$Cu(OH)_2$ retained the magnetic property of $Fe_3O_4$.

Figure 3.76 presents the FTIR spectra of the $Fe_3O_4$@$Cu(OH)_2$ composites with various Cu addition. The broad adsorption between 3500 and 3000 $cm^{-1}$ represents the stretching/bending vibration of –OH belonging to free surface water and metal oxides (M–OH, M = Fe, Cu). The strong –OH absorption bands at 1618 $cm^{-1}$ arise from physically adsorbed water molecules [211]. The characteristic peaks at 1060 $cm^{-1}$ can be assigned to the bending vibration of hydroxyl groups on metal oxides [212, 213]. Profoundly, as Cu/Fe ratios rose, the bands intensity of hydroxyl groups in the spectrum all showed increasing tendency, strongly demonstrating that $Cu(OH)_2$ coating could significantly enhance the –OH group concentration in surface and lattices of obtained composites.

**(2) Magnetic property of $Fe_3O_4$@$Cu(OH)_2$**

The saturation magnetization (Ms) of $Fe_3O_4$ (1), $Fe_3O_4$@$Cu(OH)_2$-1 (2), $Fe_3O_4$@$Cu(OH)_2$-2 (3), $Fe_3O_4$@$Cu(OH)_2$-3 (4), and $Fe_3O_4$@$Cu(OH)_2$-4 (5) was

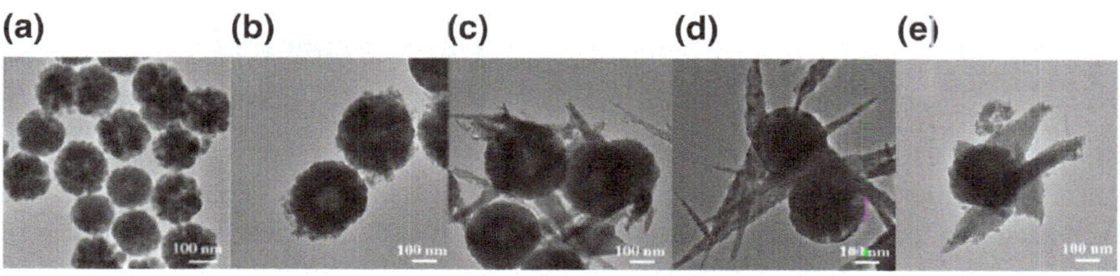


**Fig. 3.74** TEM images of $Fe_3O_4$ (**a**), $Fe_3O_4$@$Cu(OH)_2$-1 (**b**), $Fe_3O_4$@$Cu(OH)_2$-2 (**c**), $Fe_3O_4$@Cu $(OH)_2$-3 (**d**) and $Fe_3O_4$@$Cu(OH)_2$-4 (**e**). Reprinted from Ref. [210] Copyright 2016, with permission from Elsevier

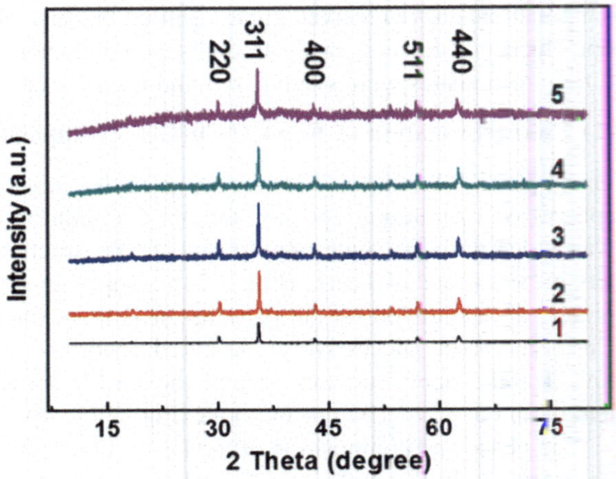


**Fig. 3.75** X-ray diffraction pattern of $Fe_3O_4$ (1), $Fe_3O_4$@$Cu(OH)_2$-1 (2), $Fe_3O_4$@$Cu(OH)_2$-2 (3), $Fe_3O_4$@$Cu(OH)_2$-3 (4) and $Fe_3O_4$@$Cu(OH)_2$-4 (5). Reprinted from Ref. [210] Copyright 2016, with permission from Elsevier

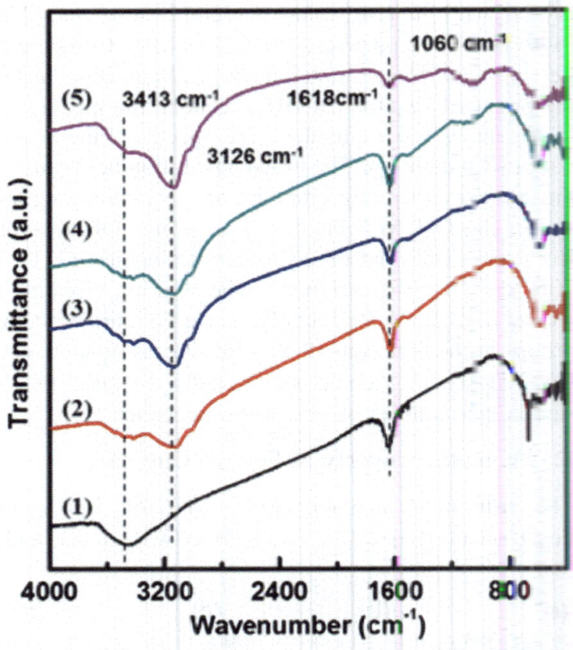


**Fig. 3.76** FTIR spectra of $Fe_3O_4$ (1), $Fe_3O_4$@$Cu(OH)_2$-1 (2), $Fe_3O_4$@$Cu(OH)_2$-2 (3), $Fe_3O_4$@$Cu(OH)_2$-3 (4) and $Fe_3O_4$@$Cu(OH)_2$-4 (5). Reprinted from Ref. [210] Copyright 2016, with permission from Elsevier

98.84, 97.74, 91.42, 67.38 and 56.53 emu $g^{-1}$, respectively (Fig. 3.77). It is evident that Cu coating has a negative effect on the $Fe_3O_4$ magnetic characteristic. The decreased Ms. can be attributed to the coating of $Cu(OH)_2$ shell on the surface of $Fe_3O_4$ microsphere cores, which reduced the magnetite fraction in each microsphere [214–217]. However, the saturation magnetization value achieved with $Fe_3O_4$ and $Fe_3O_4$@$Cu(OH)_2$ composites was high enough for magnetic separation [218]. The

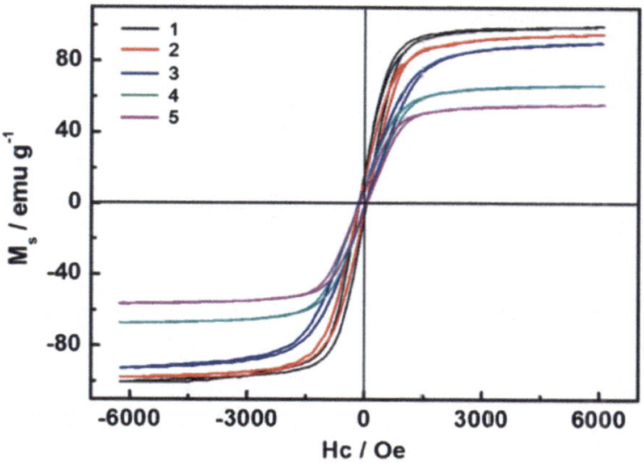


**Fig. 3.77** Magnetic hysteresis loop of $Fe_3O_4$ (1), $Fe_3O_4@Cu(OH)_2$-1 (2), $Fe_3O_4@Cu(OH)_2$-2 (3), $Fe_3O_4@Cu(OH)_2$-3 (4), and $Fe_3O_4@Cu(OH)_2$-4 (5). Reprinted from Ref. [210] Copyright 2016, with permission from Elsevier

superior magnetic property guaranteed the convenient separation in adsorption application.

**(3) Arsenate adsorption isotherms**

The arsenate adsorption capacity of $Fe_3O_4@Cu(OH)_2$-1 to $Fe_3O_4@Cu(OH)_2$-4 increased from 11.11 to 35.71 mg g$^{-1}$ with the increase of $Cu(OH)_2$ loading ratio (Fig. 3.78), indicating that Cu coating on bare magnetic particles played a significant role in enhancing adsorption performance of composites. Obviously, $Fe_3O_4@Cu(OH)_2$ composites with high magnetic have a powerful arsenic removal capacity, thus making them promising as adsorbents for arsenic removal.

**(4) Arsenate adsorption kinetics**

The arsenate adsorption on the $Fe_3O_4@Cu(OH)_2$ composites was rapid at first due to the large number of available adsorption sites at the initial stage (Fig. 3.79). With the increase of adsorption time, it had a slower removal rate and then reached

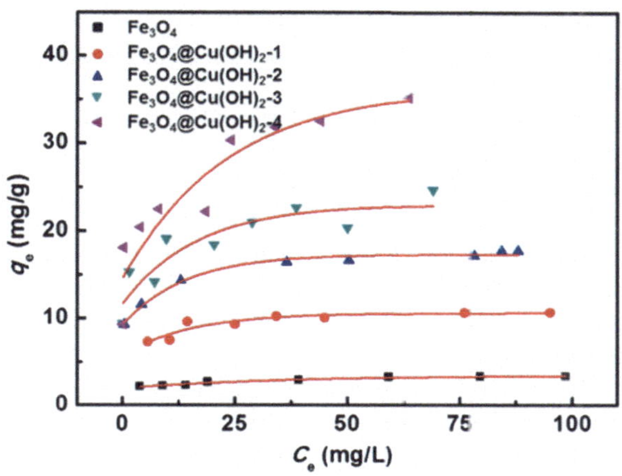


**Fig. 3.78** Adsorption performance of $Fe_3O_4@Cu(OH)_2$ composites toward As(V). Reprinted from Ref. [210] Copyright 2016, with permission from Elsevier

Fig. 3.79 Adsorption rate of As(V) by $Fe_3O_4$@$Cu(OH)_2$-2. Reprinted from Ref. [210] Copyright 2016, with permission from Elsevier

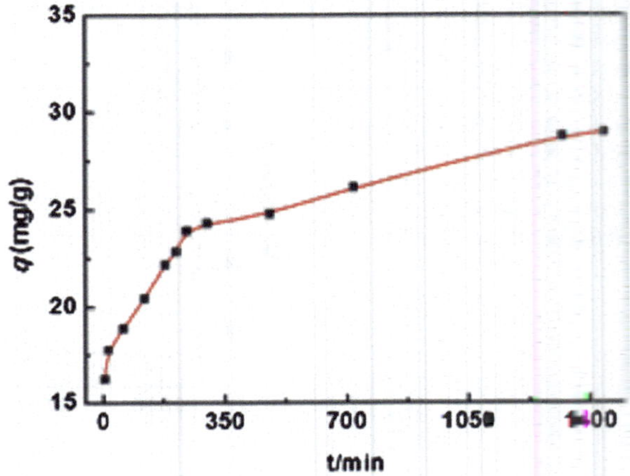


equilibrium which was ascribed to the gradual occupation of available adsorption sites on $Fe_3O_4$@$Cu(OH)_2$ composites by arsenate at the later time. The results further revealed that the $Fe_3O_4$@$Cu(OH)_2$ composites had an excellent percentage of the arsenic adsorption.

**(5) Arsenate adsorption mechanism**

XPS spectra of the adsorbent before and after As(V) adsorption at pH 5.0 were analyzed in Fig. 3.80. Figure 3.80A depicts two intense bands with binding energies of 711.7 eV and 724.0 eV that are assigned to Fe $2p_{3/2}$ and Fe $2p_{1/2}$ respectively, demonstrating the existence of the typical $Fe_3O_4$ structures [205, 206] The Cu 2p XPS spectrum is shown in Fig. 3.80B. The peaks located at 934.5 and 954.2 eV are assigned to Cu $2p_{3/2}$ and Cu $2p_{1/2}$ [219, 220]. After arsenate adsorption, the binding energy of Fe and Cu has no changed, indicating that $Fe_3O_4$@$Cu(OH)_2$ composites are stable in arsenate adsorption. XPS of As 3d peak appears after adsorption (Fig. 3.80D). The As 3d binding energies is 45 5 eV which is attributing to As(V)–O bonding [168, 207, 221–223], meaning that arsenate species have been loaded on the surface of the $Fe_3O_4$@$Cu(OH)_2$ composites after the adsorption.

As for O 1s spectra in Fig. 3.80C, after arsenate adsorption, the O 1s peak locates at 531.5 eV decreased from 56.18 to 26.17% suggesting that the surface hydroxyl groups have surely important effects on arsenate adsorption, while the increased proportions of oxygen in the lattice at 530.0 eV from 39.33 to 67.12% maybe originate from the formation of M–O on the surface and As–O groups in the adsorbed arsenate species after arsenate adsorption [224]. That is to say, the M–OH groups on the $Fe_3O_4$@$Cu(OH)_2$ composites surface are responsible for arsenic adsorption.

The possible mechanism of arsenate adsorption on $Fe_3O_4$@$Cu(OH)_2$ composites was shown in Fig. 3.81. The M–OH groups on the $Fe_3O_4$@$Cu(OH)_2$ composites surface played a key role in As(V) adsorption through the formation of coordination

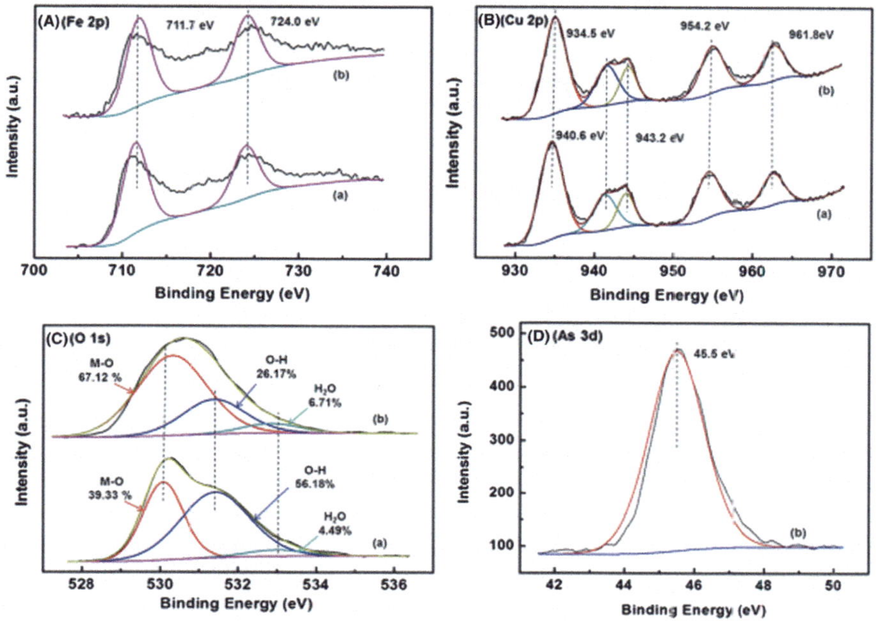

**Fig. 3.80** Fe 2p (**A**), Cu 2p (**B**), O 1s (**C**) and As 3d (**D**) XPS spectra of $Fe_3O_4@Cu(OH)_2$-2 (a), $Fe_3O_4@Cu(OH)_2$-2 with arsenic adsorbed (b). Reprinted from Ref. [210] Copyright 2016, with permission from Elsevier

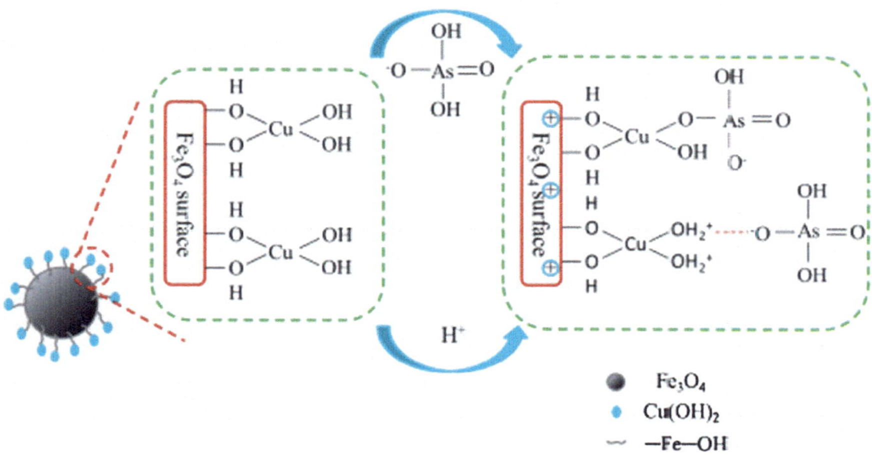

**Fig. 3.81** As(V) adsorption mechanism on $Fe_3O_4@Cu(OH)_2$ composites. Reprinted from Ref. [210] Copyright 2016, with permission from Elsevier

compounds between As–OH and M–OH and the corresponding M–O groups formed. Meanwhile, $Fe_3O_4@Cu(OH)_2$ composites can be protonated in acidic solution to form the positively charged surface, the negative arsenate species can be directly adsorbed on the positive sites of the adsorbent via electrostatic attraction.

In summary, there were two different mechanisms contributing to adsorption the formation of coordination compounds and electrostatic attraction between As(V) species and $Fe_3O_4@Cu(OH)_2$ composites.

### 3.6.4 *Arsenic Removal by Hollow Cu-Loaded Poly (m-Phenylenediamine) Particles*

As a diamine derivative of polyaniline, poly (phenylenediamine) has many important applications in sensors, catalysis, electrodes, actuators, etc. [225–227]. In particular, widespread attention has been attracted to investigate the adsorption performance of poly (phenylenediamine) over recent decades due to its superior redox reversibility and chelation ability [228–231], and it can be used for removing various metals ions from water, such as As(V). A sustainable synthesis of poly (m-phenylenediamine) microparticles with hollow structures was developed via Cu-catalyzed air oxidation. The conversion efficiency of monomer in this research can reach close to 100%, which is higher than the highest yield reported (93.1%) [232]. Moreover, the nanoscaled Cu-loaded PmPD particle is a hopeful material in arsenic removal in aqueous solution.

**(1) Morphology of PmPD particles**

The PmPD particles were named as PmPD-Cux (x: 1:1, 1:0.5, 1:0.25, 1:0.125 and 1:0.1), where x corresponds to the mPD:$Cu^{2+}$ molar ratio. Obviously, amount of Cu addition has a significant influence on the micro-morphology of PmPD particles. As shown in Fig. 3.82g, h, the product synthesized in $n_{mPD}$:$n_{Cu}$ = 1:0.125 just consisted of abundant irregular solid particles. However, great variation emerged by increasing Cu content. When mPD/Cu ratio became 1:0.5, a lot of distinguishable solid microspheres appeared actually in the product, though their surfaces are very rough (Fig. 3.82c, d). It is noted that some hollow nanostructures could be identified apparently in Fig. 3.82d. Especially, in a further experiment, the obtained PmPD-Cu1:1 particles were mainly composed of nanosized smooth spheres with the diameters of 300–500 nm (Fig. 3.82a, b). More importantly, the microspheres are hollow with a wall thickness of about 60–80 nm, as confirmed by the TEM image. Up to our knowledge, that is a new sustainable alternate to constitute hollow structure of PmPD particles.

The self-formation mechanism of the hollow structures can be described as follows. When in contact with $Cu^{2+}$, mPD instantly reacted to generate the positively charged complex, which would increase the molecules amphiphilicity to constitute micelles in water. As higher amount of Cu(II) salt was added into the

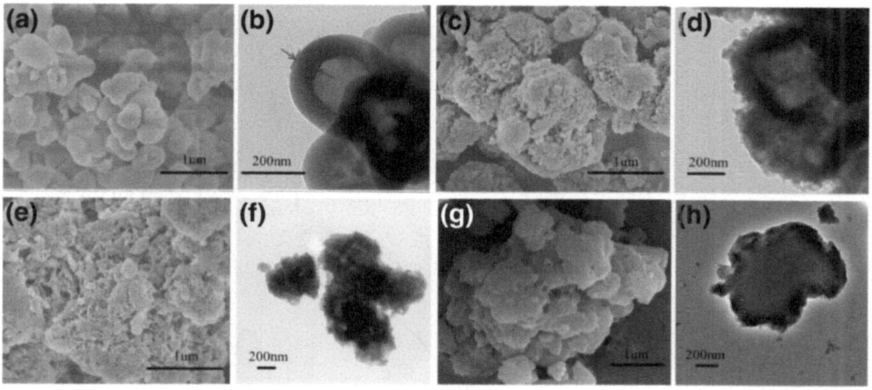


**Fig. 3.82** SEM and TEM images: **a** and **b** PmPD-Cu1:1; **c** and **d** PmPD-Cu1:0.5; **e** and **f** PmPD-Cu1:0.25 and **g** and **h** PmPD-Cu1:0.125 (reaction time was 24 h)

system, the counter ion concentration also increased, which can compress the electrical double layer and reduce charge repulsion, allowing the micelles to come closer to each other [233–235]. At the same time, the increased ionic strength can cause the molecules' polar groups undergo more "dehydration" to enhance their hydrophobic character, which will greatly reduce monomer molecules exchange between the micelles and the bulk phase [236]. Hence, the effects worked together to drive micellar molecules to be rearranged to form enlarged stable micellar cluster or bilayer aggregates. When air was introduced into the bulk solution, the oxidation chain-propagation will principally take place around the surface of the aggregates to make them growing up [234]. However, when the reaction reached to a certain extent, the dynamic equilibrium cannot be maintained between the grown aggregates and polymer molecules. Particularly, if the total Cu addition is not enough, low ionic strength will go against keeping hollow aggregates stable and finally formed irregular solid particles.

**(2) Structure of PmPD particles**

The FTIR spectra of PmPD-Cux (x: 1:1, 1:0.5, 1:0.25, 1:0.125, 1:0.1) were similar (Fig. 3.83). The broad absorption centered between 3500 and 3000 $cm^{-1}$ should be due to the stretching mode of –NH– [237–239]. The peak at $\sim$1620 $cm^{-1}$ was associated with phenazine and quinoid imine. And the peak at $\sim$1500 $cm^{-1}$ was attributed to benzenoid amine structures [240, 241]. Meanwhile, the peak at $\sim$1250 $cm^{-1}$ corresponded to the C–N stretching mode in the PmPD [242]. Moreover, it was found that the relative content of the two peaks changed obviously with the variation of the Cu/monomer molar ratio. With the increase of $Cu^{2+}$, the peak at $\sim$1620 $cm^{-1}$ turned to stronger which indicated that content of phenazine and quinoid structure were increased.

Figure 3.84 clearly indicates that the PmPD particles with Cu loaded are basically made up of carbon, oxygen, nitrogen, copper, and a small amount of chlorine. With the decrease of the initial dosage of $Cu^{2+}$, the peak of Cu became weaker.

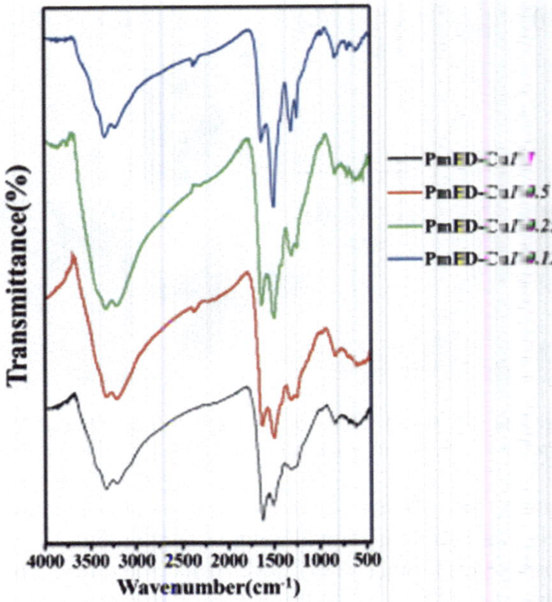


**Fig. 3.83** FTIR spectra of PmPD-Cu1:1, PmPD-Cu1:0.5, PmPD-Cu1:0.25 and PmPD-Cu1:0.125

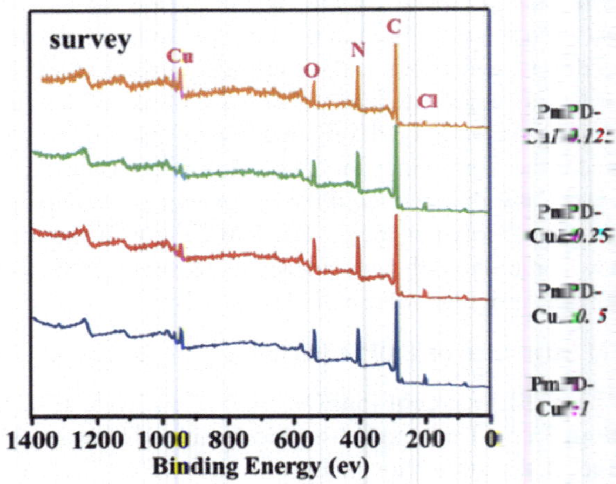


**Fig. 3.84** Wide energy range surface spectra of PmPD-Cu1:1; PmPD-Cu1:0.5; PmPD-Cu1:0.25 and PmPD-Cu1:0.125

The XPS spectra of N element of various PmPD particles (Fig. 3.85a–d) were investigated to study the relative molar contents of N-containing segments. The peak at 399.2 eV was associated with the neutral –N= in both quinoid imine and phenazine while the one at 400.2 eV was due to the –NH– in the benzenoid amine units. Besides these, there was a weak peak at 401.2 eV assigned to $-N^{+}=$ [243–245]. With the increase of $Cu^{2+}$, the area of –N= and $-N^{+}=$ increased, indicating the enhancement of oxidation state. This is in agreement with the analysis of FTIR. Based on XPS analysis, it was interesting to see that the oxidation state of PmPD increased with the increase of Cu/monomer molar ratio.

The XPS spectra of Cu element of PmPD particles (Fig. 3.85e–h) indicates that copper existed in two oxidation states on the PmPD particles, $Cu^{2+}$ and $Cu^{+}$ [246, 247]. $Cu^{+}$ contents tended to rise with the increase of Cu/mPD ratio. The variation of $Cu^{+}$ contents proved that $Cu^{2+}$ was involved in the mPD oxidation polymerization. It was the reduction of $Cu^{2+}$ to $Cu^{+}$ achieving the oxidation and the increased amount of $Cu^{+}$ was possibly a direct reason for the increased oxidation state of PmPD. Therefore, it could be further correlated to the increase of conversion rate.

**(3) Arsenate adsorption properties**

When initial As(V) concentration is 70 mg $L^{-1}$, PmPD-Cu0 exhibited a certain arsenic absorbance as low as about 8 mg $g^{-1}$ (Fig. 3.86). As the Cu/mPD ratios rose, the absorbance of PmPD particles loaded with Cu significantly increased. The maximal absorbance (about 27.4 mg $g^{-1}$) occurred at Cu/mPD ratios 0.25:1, 4 times as much as that of PmPD-Cu0. Subsequently, further enhancing Cu/mPD ratio had no great impacts upon the As(V) adsorption of PmPD particles with just a slight decrease to about 23–24 mg $g^{-1}$. It is indicative that Cu played a crucial role in the arsenic removal of PmPD particles. Consequently, PmPD-Cu1:0.25 is more preferable adsorbent for As(V) removal. The Freundlich can better describe the adsorption since the correlation efficiency of Freundlich is all much higher than that of Langmuir. This suggests that PmPD particles adsorption of arsenic is a multi-component adsorption isotherm.

The adsorption kinetics of PmPD-Cu1:0.25 (Fig. 3.87) indicate the adsorption process can be roughly divided into fast and slow steps. The fast step lasted for about 10 min; the residue percentage of arsenic in the filtrate was sharply decreased to 18.5% for. This rapid process was caused by the adsorption of arsenic to the functional groups on the surface of PmPD particles. As prolonging the time, the adsorption became slow apparently, owing to the gradual diffusion of arsenic molecules from the surface to the inner structures of the PmPD particles. This rapid attainment of adsorption equilibrium is of great importance to the practical process of the obtained PmPD particles.

Arsenic removal property of PmPD-Cu1:0.25 was measured with initial pH value from 3 to 11 (Fig. 3.88). The removal rate was calculated according to the equations described in the literature [229]. When decreasing the pH from 5.0 to 3.0, the removal rate declined obviously from only 53.2 to 16.1%. This means that the acidic condition with pH lower than 5 is not beneficial for arsenic removal. While at pH 5.0–7.0, the removal rate varies little (<2%), implying that the weak acidic condition influences slightly on arsenic adsorption. As pH decreased from 5 decrease to 3, negative-charged $H_2AsO_4^-$ began to turn to neutral molecular state $H_3AsO_4$, which is difficult to be adsorbed, thus leading to the decrease of removal rate. As pH increased to 11, the removal rate of arsenic rapidly dropped to 0%, which is mainly due to $OH^-$ competition in alkaline condition. Consequently, the solution pH at 5.0–7.0 is suitable for arsenic adsorption of PmPD particles (Fig. 3.88).

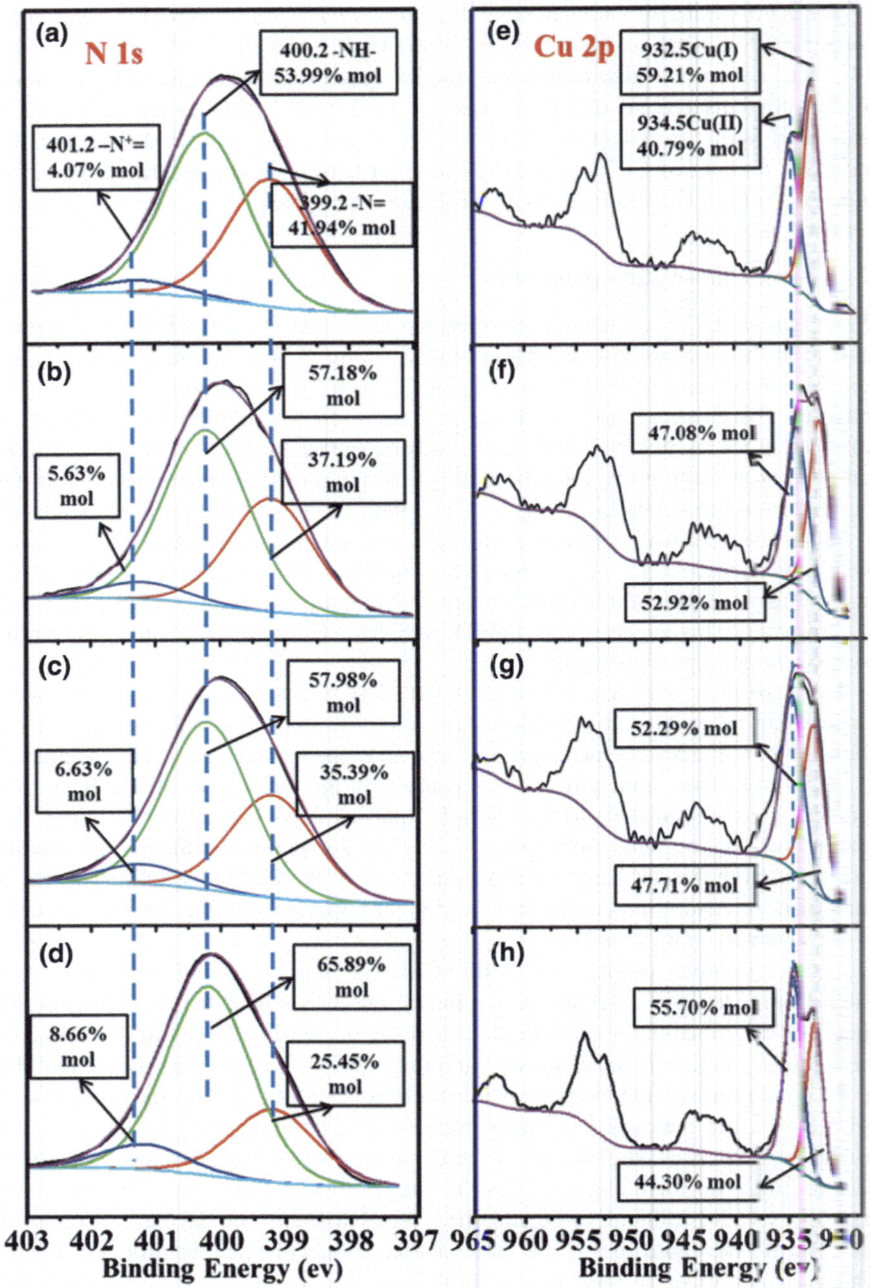


**Fig. 3.85** Deconvolution results of N 1s (**a–d**) and Cu 2p (**e–h**) XPS spectra of PmPD-Cu1:1 (**a–e**), PmPD-Cu1:0.5 (**b–f**), PmPD-Cu1:0.25 (**c–g**) and PmPD-Cu1:0.125

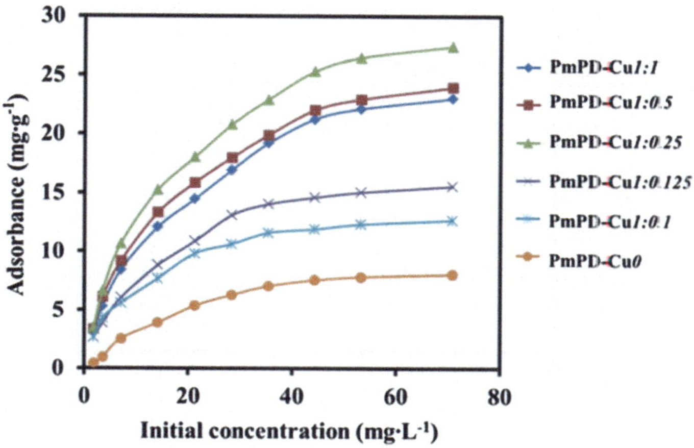


Fig. 3.86 Effects of initial concentration of PmPD particles (T = 35 °C; adsorbent dosage = 0.5 g $L^{-1}$; pH = 5)

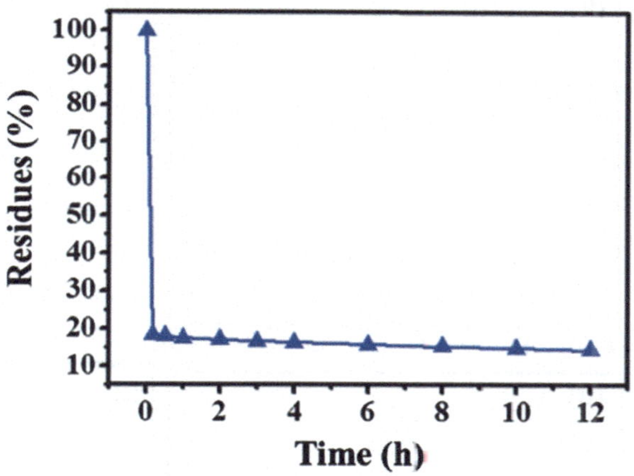


Fig. 3.87 Effects of time on the adsorption performance of PmPD-Cu1:0.25 particles (T = 35 °C; adsorbent doses = 0.5 g $L^{-1}$; pH = 5)

### (4) Adsorption mechanism

The As 3d and N 1s data (XPS) of PmPD particles and relevant calculation are given in Fig. 3.89 [248]. Based on the arsenic 3d data, the adsorption of arsenic on PmPD particles existed in two forms, 58.14% mol As(V) and 41.86% mol As(III). That means an oxidation process took place to make As(V) reduced to As(III) during the adsorption. As compared the data of N 1s before and after adsorption (in Fig. 3.88), it can be found that the content of –NH– declined by 5.7% after treating As(V), in turn the content of =N– and $=N^{+}$– increased correspondingly.

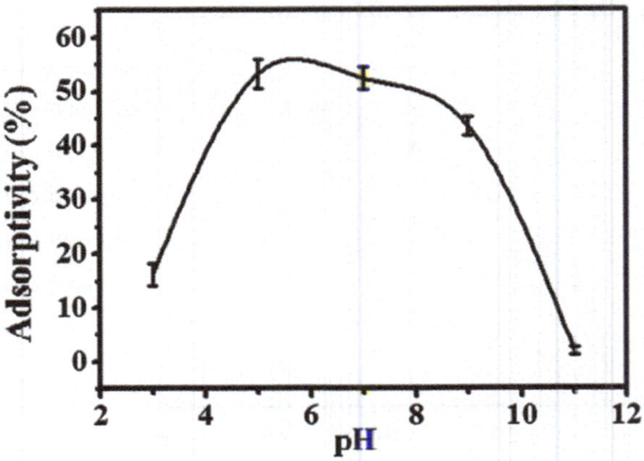


**Fig. 3.88** Effect of pH on the adsorption performance of PmPD-Cu1:0.25 particles (T = 3[illegible] °C, adsorbent doses = 20 mg $L^{-1}$)

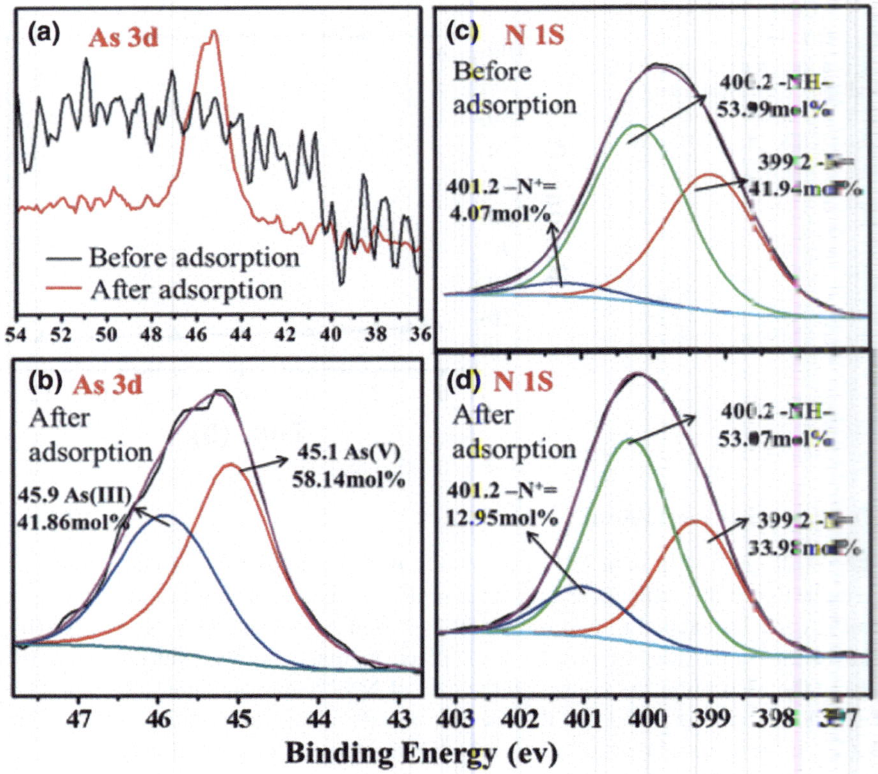


**Fig. 3.89** Deconvolution results of N 1s and As 3d XPS spectra of PmPDCu1:0.25 before and after adsorption

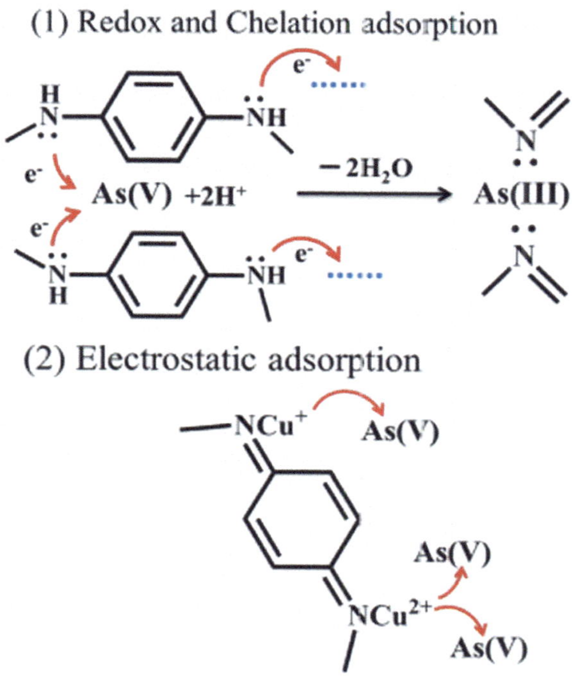


**Fig. 3.90** Possible mechanism for arsenic adsorption with PmPD particles

As is known to all, –NH– of conjugated polymers can be oxidized readily and then hydrogen left, which was finally transformed to =N– [244]. 0.92% of –NH– was oxidized by As(V) form to =N–. On the other hand, the electron density of =N– is much higher than that of –NH– due to its different molecular configuration [249, 250], which allows it to interact with As(III) cations through coordination. The increased amount of $=N^{+}-$ after adsorption also strongly verifies this chelation between As(III) and =N–. It should be noted that part of the =N– in PmPD before adsorption has been already chelated by Cu ions to generate $=N^{+}-$ which can interact with negative ionic As(V) through static manner. Based on the discussion of XPS, the corresponding interaction was illustrated in Fig. 3.90 [244, 251, 252]. As soon as the addition of PmPD-Cu nanoparticles into the As(V) solution, abundant As(V) was adsorbed by $=N^{+}-$ components via electrostatic attraction. Meanwhile, the redox reaction between part of As(V) and nearby imine group (–NH–) occurred which produced the =N– and As(III). Therein, As(III) can readily tend to interact with =N– through coordination.

## References


1. Luo, T., Cui, J., Hu, S., et al.: Arsenic removal and recovery from copper smelting wastewater using $TiO_2$. Environ. Sci. Technol. **44**(23), 9094–9098 (2010)
2. Baig, J.A., Kazi, T.G., Shah, A.Q., et al.: Speciation and evaluation of arsenic in surface water and groundwater samples: a multivariate case study. Ecotoxicol. Environ. Saf. **73**(5), 914–923 (2010)
3. Mandal, B.K., Suzuki, K.T.: Arsenic round the world: a review. Talanta **58**(1), 201–235 (2002)
4. Riveros, P.A., Dutrizac, J.E., Spencer, P.: Arsenic disposal practices in the metallurgical industry. Can. Metall. Q. **40**(4), 395–420 (2001)
5. Emett, M.T., Khoe, G.H.: Photochemical oxidation of arsenic by oxygen and iron in acidic solutions. Water Res. **35**(3), 649–656 (2001)
6. Kim, M.J., Nriagu, J.: Oxidation of arsenite in groundwater using ozone and oxygen. Sci. Total Environ. **247**(1), 71–79 (2000)
7. Jia, Y.F., Zhang, D.N., Pan, R.R., et al.: A novel two-step coprecipitation process using Fe (III) and Al (III) for the removal and immobilization of arsenate from acidic aqueous solution. Water Res. **46**(2), 500–508 (2012)
8. Cui, J.L., Jing, C.Y., Che, D.S., et al.: Groundwater arsenic removal by coagulation using ferric(III) sulfate and polyferric sulfate: a comparative and mechanistic study. J. Environ. Sci. **32**, 42–53 (2015)
9. Mertens, J., Casentini, B., Masiond, A., et al.: Polyaluminum chloride with high Al30 content as removal agent for arsenic-contaminated well water. Water Res. **46**(1), 53–62 (2012)
10. An, B., Liang, Q.Q., Zhao, D.Y.: Removal of arsenic(V) from spent ion exchange brine using a new class of starch-bridged magnetite nanoparticles. Water Res. **45**(5), 1961–1972 (2011)
11. Pakzadeh, B., Batista, J.R.: Surface complexation modeling of the removal of arsenic from ion-exchange waste brines with ferric chloride. J. Hazard. Mater. **188**(1–3), 399–407 (2011)
12. Zhang, G.S., Liu, F.D., Liu, H.J., et al.: Respective role of Fe and Mn oxide contents for arsenic sorption in iron and manganese binary oxide: an X-ray absorption spectroscopy investigation. Environ. Sci. Technol. **48**(17), 10316–10322 (2014)
13. Dou, X.M., Mohan, D., Pittman Jr., C.U.: Arsenate adsorption on three types of granular schwertmannite. Water Res. **47**(9), 2938–2948 (2013)
14. Na, L., Maohong, F., Johannes, V.L., et al.: Oxidation of As (III) by potassium permanganate. J. Environ. Sci. **19**(7), 783–786 (2007)
15. Sorlini, S., Gialdini, F.: Conventional oxidation treatments for the removal of arsenic with chlorine dioxide, hypochlorite, potassium permanganate and monochloramine. Water Res. **44**(19), 5653–5659 (2010)
16. Lee, Y.H., Um, I., Yoon, J.: Arsenic(III) oxidation by iron(VI) (ferrate) and subsequent removal of arsenic(V) by iron(III) coagulation. Environ. Sci. Technol. **37**(24), 5750–5756 (2003)
17. Molnár, L., Virčíkova, E., Lech, P.: Experimental study of As (III) oxidation by hydrogen peroxide. Hydrometallurgy **35**(1), 1–9 (1994)
18. Hug, S.J., Leupin, O.: Iron-catalyzed oxidation of arsenic(III) by oxygen and by hydrogen peroxide: pH-dependent formation of oxidants in the fenton reaction. Environ. Sci. Technol. **37**(12), 2734–2742 (2003)
19. Laat, J.D., Le, T.G.: Kinetics and modeling of the Fe(III)/$H_2O_2$ system in the presence of sulfate in acidic aqueous solutions. Environ. Sci. Technol. **39**(6), 1811–1818 (2005)
20. Sorlini, S., Gialdini, F., Stefan, M.: Arsenic oxidation by UV radiation combined with hydrogen peroxide. Water Sci. Technol. **61**(2), 339–344 (2010)
21. Bissen, M., Vieillard-Baron, M.-M., Schindelin, A.J., et al.: $TiO_2$-catalyzed photooxidation of arsenite to arsenate in aqueous samples. Chemosphere **44**(4), 751–757 (2001)

22. Klaning, U.K., Bielski, B.H.J., Sehested, K.: Arsenic(IV). A pulse-radiolysis study. Inorg. Chem. **28**(14), 2717–2724 (1989)
23. Rastogi, A., Al-Abed, S.R., Dionysiou, D.D.: Sulfate radical-based ferrous-peroxymonosulfate oxidative system for PCBs degradation in aqueous and sediment systems. Appl. Catal. B **85**(3–4), 171–179 (2009)
24. Gomathi Devi, L., Girish Kumar, S., Mohan Reddy, K., et al.: Photo degradation of methyl orange an azo dye by advanced fenton process using zero valent metallic iron: Influence of various reaction parameters and its degradation mechanism. J. Hazard. Mater. **164**(2–3), 459–467 (2009)
25. Yang, Y., Pignatello, J.J., Ma, J., Mitch, W.A.: Comparison of halide impacts on the efficiency of contaminant degradation by sulfate and hydroxyl radical-based advanced oxidation processes (AOPs). Environ. Sci. Technol. **48**(4), 2344–2351 (2014)
26. Hori, H., Hayakawa, E., Einaga, H., et al.: Decomposition of environmentally persistent perfluorooctanoic acid in water by photochemical approaches. Environ. Sci. Technol. **38** (22), 6118–6124 (2004)
27. Schröder, H.F., Meesters, R.J.W.: Stability of fluorinated surfactants in advanced oxidation processes–a follow up of degradation products using flow injection-mass spectrometry, liquid chromatography-mass spectrometry and liquid chromatography-multiple stage mass spectrometry. J. Chromatogr. A **1082**(1), 110–119 (2005)
28. Zhang, B.T., Zhang, Y., Teng, Y.G., et al.: Sulfate radical and its application in decontamination technologies. Crit. Rev. Environ. Sci. Technol. **45**(16), 1756–1800 (2015)
29. Lei, Z., Wei, Z., Jinfeng, Z., et al.: Ferrous-activated persulfate oxidation of arsenic (III) and diuron in aquatic system. J. Hazard. Mater. **263**(Part 2), 422–430 (2013)
30. Jing, X., Wei, D., Feng, W., et al.: Rapid catalytic oxidation of arsenite to arsenate in an iron (III)/sulfite system under visible light. Appl. Catal. B Environ. **186**, 55–61 (2016)
31. Neppolian, B., Doronila, A., Ashokkumar, M.: Sonochemical oxidation of arsenic (III) to arsenic (V) using potassium peroxydisulfate as an oxidizing agent. Water Res. **44**(12), 3687–3695 (2010)
32. Neppolian, B., Celik, E., Choi, H.: Photochemical oxidation of arsenic(III) to arsenic(V) using peroxydisulfate ions as an oxidizing agent. Environ. Sci. Technol. **42**(16), 6179–6184 (2008)
33. Woods, R., Kolthoff, I.M., Meehan, E.J.: Arsenic(IV) as an intermediate in the induced oxidation of arsenic(III) by the iron(II)- persulfate reaction and the photoreduction of iron (III). I. Absence of oxygen. J. Am. Chem. Soc. **85**(16), 2385–2390 (1963)
34. Woods, R., Kolthoff, I.M., Meehan, E.J.: Arsenic(IV) as an intermediate in the iron(III) and copper(II) catalyzed arsenic(III)-per sulfate reaction. Inorg. Chem. **4**(5), 697–704 (1965)
35. Takahashi, M., Chiba, K., Li, P.: Formation of hydroxyl radicals by collapsing ozone microbubbles under strongly acidic conditions. J. Phys. Chem. B **111**(39), 11443–11446 (2007)
36. Li, Y., Cai, X.J., Guo, J.W., et al.: UV-induced photoactive adsorption mechanism of arsenite by anatase $TiO_2$ with high surface hydroxyl group density. Colloids Surf. A **462**, 202–210 (2014)
37. Robins, R.G.: The solubility of metal arsenates. Metall. Trans. B **12**(1), 103–109 (1981)
38. Trigub, A.L., Tagirov, B.R., Kvashnina, K.O., et al.: X-ray spectroscopy study of the chemical state in “invisible” Au in synthetic minerals in the Fe-As-S system. Am. Miner. **102**(5), 1057–1065 (2017)
39. Gonzalez-Contreras, P., Weijma, J., Buisman, C.J.N.: Bioscorodite crystallization in an airlift reactor for arsenic removal. Cryst. Growth Des. **12**(5), 2699–2706 (2012)
40. Fujita, T., Taguchi, R., Abumiya, M., et al.: Novel atmospheric scorodite synthesis by oxidation of ferrous sulfate solution. Hydrometallurgy **90**(2–4), 92–102 (2008)
41. Paktunc, D., Dutrizac, J., Gertsman, V.: Synthesis and phase transformations involving scorodite, ferric arsenate and arsenical ferrihydrite: implications for arsenic mobility. Geochim. Cosmochim. Acta **72**(11), 2649–2672 (2008)

42. Dutrizac, J.E., Jambor, J.L.: The synthesis of crystalline scorodite, $FeAsO_4 \cdot 2H_2O$. Hydrometallurgy **19**(3), 377–384 (1988)
43. Yang, J.Q., Chai, L.Y., Yue, M.Q., et al.: Complexation of arsenate with ferric ion in aqueous solutions. RSC Adv. **5**(126), 103936–103942 (2015)
44. Sergeyeva, E.I., Khodakovskiy, I.L.: Physicochemical conditions of formation of native arsenic in hydrothermal deposits. Geochem. Int. 846–859 (1969)
45. Technical Note 270-7. Schumm, R.H., Wagman, D.D., Evans, W.H., et al.: Selected Values of Chemical Thermodynamic Properties. National Bureau of Standards (1973)
46. Wagman, D.D., Evans, W.H., Parker, V.B., et al.: Erratum: The NBS tables of chemical thermodynamic properties. J. Phys. Chem. Ref. Data **18**(4), 1807–1812 (1989)
47. Bard, A.J., Parsons, R., Jordan, J.: Standard Potentials in Aqueous Solution. CRC Press, New York (1985)
48. Shock, E.L., Helgeson, H.C.: Calculation of the thermodynamic and transport properties of aqueous species at high pressures and temperatures: correlation algorithms for ionic species and equation of state predictions to 5 kb and 1000 °C. Geochim. Cosmochim. Acta **52**(8), 2009–2036 (1988)
49. Nordstrom, D.K., Archer, D.G.: Arsenic Thermodynamic Data and Environmental Geochemistry. Arsenic in Ground Water, pp. 1–25. Springer, (2003)
50. Jin-qin, Y.A.N.G., Li-yuan, C.H.A.I., Qing-zhu, L.I., et al.: Redox behavior and chemical species of arsenic in acidic aqueous system. Trans. Nonferrous Species Arsen. Acidic Aqueous **27**(9), 2063–2072 (2017)
51. Long, H.: A Fundamental Study of the Acidic Pressure Oxidation of Orpiment and Pyrite at High Temperature. University of British Columbia, British Columbia (2000)
52. Masscheleyn, P.H., Delaune, R.D., Patrick, J.R.W.H.: Effect of redox potential and pH on arsenic speciation and solubility in a contaminated soil. Environ. Sci. Technol. **25**(8), 1414–1419 (1991)
53. Pieczaba, E., Sanak-rydlewska, S., Zieba, D.: Removal of arsenic from aqueous solutions by the method of precipitate flotation. Arch. Min. Sci. **50**(1), 131–142 (2005)
54. Yazdi, M.R.S., Darban, A.K.: Effect of arsenic speciation on remediation of arsenic-contaminated soils and waters. In: 15th International Conference on Heavy Metals in the Environment (ICHMET), Gdansh, pp. 492–495 (2010)
55. Yi, X.W.: An empirical estimation of standard entropy for some complex cations and the E-pH diagram of As-$H_2O$ system at elevated temperature. J. Kunming Univ. Sci. Technol. **3**, 58–73 (1982)
56. Debusschere, L., Demesmay, C., Rocca, J.L.: Arsenic speciation by coupling capillary zone electrophoresis with mass spectrometry. Chromatographia **51**(5–6), 262–268 (2000)
57. Gout, R., Pokrovski, G., Schott, J., et al.: Raman spectroscopic study of arsenic speciation in aqueous solutions up to 275 °C. J. Raman Spectrosc. **28**(9), 725–730 (1997)
58. Loehr, T.M., Plane, R.A.: Raman spectra and structures of arsenious acid and arsenites in aqueous solution. Inorg. Chem. **7**(9), 1708–1714 (1968)
59. Mähler, J., Persson, I., Herbert, R.B.: Hydration of arsenic oxyacid species. Dalton Trans. **42** (5), 1364–1377 (2013)
60. Bu, L.J., Gu, T., Ma, Y.X., et al.: Enhanced cathodic preconcentration of As(0) at Au and Pt electrodes for anodic stripping voltammetry analysis of As(III) and As(V). J. Phys. Chem. C **119**(21), 11400–11409 (2015)
61. Marini, L., Accornero, M.: Prediction of the thermodynamic properties of metal–arsenate and metal–arsenite aqueous complexes to high temperatures and pressures and some geological consequences. Environ. Geol. **52**(7), 1343–1363 (2007)
62. Pettine, M., Campanella, L., Millero, F.J.: Arsenite oxidation by $H_2O_2$ in aqueous solutions. Geochim. Cosmochim. Acta **63**(18), 2727–2735 (1999)
63. Vink, B.W.: Stability relations of antimony and arsenic compounds in the light of revised and extended Eh-pH diagrams. Chem. Geol. **130**(1–2), 21–30 (1996)

64. Marini, L., Accornero, M.: Erratum to: Prediction of the thermodynamic properties of metal–arsenate and metal–arsenite aqueous complexes to high temperatures and pressures and some geological consequences. Environ. Earth Sci. **59**(7), 1601–1606 (2010)
65. Knight, R.J., Sylva, R.N.: Spectrophotometric investigation of iron (III) hydrolysis in light and heavy water at 25 °C. J. Inorg. Nucl. Chem. **37**(3), 779–783 (1975)
66. Langmuir, D., Mahoney, J., Rowson, J.: Solubility products of amorphous ferric arsenate and crystalline scorodite ($FeAsO_4 \cdot 2H_2O$) and their application to arsenic behavior in buried mine tailings. Geochim. Cosmochim. Acta **70**(12), 2942–2956 (2006)
67. Robins, R.G.: The stability and solubility of ferric arsenate: an update. In: EPD Congress'90, pp. 93–104 (1990)
68. Whiting, K.S.: The Thermodynamics and Geochemistry of Arsenic with Application to Subsurface Waters at the Sharon Steel Superfund Site at Midvale, Utah. Colorado School of Mines, Midvale (1992)
69. Hug, S.J., Canonica, L., Wegelin, M., et al.: Solar oxidation and removal of arsenic at circumneutral pH in iron containing waters. Environ. Sci. Technol. **35**(10), 2114–2121 (2001)
70. Wang, K.L., Jia, Y.F.: Effects of temperature and pH on the transformation of ferric arsenate to scorodite in acidic solution. Adv. Mater. Res. **726–731**, 2165–2168 (2013)
71. Paktunc, D., Dutrizac, J., Gertsman, V.: Synthesis and phase transformations involving scorodite, ferric arsenate and arsenical ferrihydrite: implications for arsenic mobility. Geochim. Cosmochim. Acta **72**(11), 2649–2672 (2008)
72. Welham, N.J., Malatt, K.A., Vukcevic, S.: The effect of solution speciation on iron–sulfur–arsenic–chloride systems at 298 K. Hydrometallurgy **57**(3), 209–223 (2000)
73. Raposo, J.C., Olazabal, M.A., Madariaga, J.M.: Complexation and precipitation of arsenate and iron species in sodium perchlorate solutions at 25 °C. J. Solut. Chem. **35**(1), 79–94 (2006)
74. Khoe, G.H., Robins, R.G.: ChemInform abstract: the complexation of iron(III) with sulfate, phosphate, or arsenate ion in sodium nitrate medium at 25 °C. J. Chem. Soc. Dalton Trans. **8**, 2015–2021 (1988)
75. Chai, L., Yang, J., Zhang, N., et al.: Structure and spectroscopic study of aqueous Fe(III)-As (V) complexes using UV-Vis, XAS and DFT-TDDFT. Chemosphere **182**, 595–604 (2017)
76. Ikeda-Ohno, A., Hennig, C., Tsushima, S., et al.: Speciation and stuctural study of U(IV) and (VI) in perchloric and nitric acid solutions. Inorg. Chem. **48**(15), 7201–7210 (2009)
77. Stefansson, A., Lemke, K.H., Seward, T.M.: Iron(III) complexation in hydrothermal solutionse an experimental and theoretical study. In: 15th International Conference on the Properties of Water and Steam, Berlin (2008)
78. Chen, Z., Zhu, Y.G., Liu, W.J., et al.: Direct evidence showing the effect of root surface iron plaque on arsenite and arsenate uptake into rice (Oryza sativa) roots. New Phytol. **165**(1), 91–97 (2005)
79. Glastras, M.: The Precipitation of Arsenic from Aqueous Solutions. University of New South Wales, Sydney (1988)
80. Lee, J.S., Nriagu, J.O.: Stability constants for metal arsenates. Environ. Chem. **4**(2), 123–133 (2007)
81. Nordstrom, D.K., Parks, G.A.: Solubility and stability of scorodite, $FeAsO_4 \cdot 2H_2O$: discussion. Am. Miner. **72**(7–8), 849–851 (1987)
82. Galal-Gorchev, H., Stumm, W.: The reaction of ferric iron with ortho-phos-phate. J. Inorg. Nucl. Chem. **25**(5), 567–574 (1963)
83. Wilhelmy, R.B., Patel, R.C., Matijevic, E.: Thermodynamics and kinetics of aqueous ferric phosphate complex formation. Inorg. Chem. **24**(20), 3290–3297 (1985)
84. Harris, D., Loew, G.H., Komornicki, A.: Structure and relative spin-state ener-getics of $[Fe(H_2O)_6]^{3+}$: a comparison of UHF, møller-plesset, nonlocal DFT, and semiempircal INDO/S calculations. J. Phys. Chem. A **101**(21), 3959–3965 (1997)

85. Jarzecki, A.A., Anbar, A.D., Spiro, T.G.: DFT analysis of $[Fe(H_2O)_6]^{3+}$ and $[Fe(H_2O)_6]^{2+}$ structure and vibrations; implications for isotope fractionation. J. Phys. Chem. A **108**(14), 2726–2732 (2004)
86. Collins, R.N., Rosso, K.M., Rose, A.L., et al.: An in situ XAS study of ferric iron hydrolysis and precipitation in the presence of perchlorate, nitrate, chloride and sulfate. Geochim. Cosmochim. Acta **177**, 150–169 (2016)
87. Sherman, D.M., Randall, S.R.: Surface complexation of arsenic (V) to iron(III) (hydr)oxides: structural mechanism from ab initio molecular geometries and EXAFS spectroscopy. Geochim. Cosmochim. Acta **67**(22), 4223–4230 (2003)
88. Waychunas, G.A., Rea, B.A., Fuller, C.C.: Surface chemistry of ferri-hydrite: Part 1. EXAFS studies of the geometry of coprecipitated and adsorbed arsenate. Geochim. Cosmochim. Acta **57**(10), 2251–2269 (1993)
89. Kitahama, K., Kiriyama, R., Baba, Y.: Refinement of the crystal structure of scorodite. Acta Crystallogr. Sect. B **31**(1), 322–324 (1975)
90. Mikutta, C., Mandaliev, P.N., Kretzschmar, R.: New clues to the local atomic structure of short-range ordered ferric arsenate from extended X-ray absorption fine structure spectroscopy. Environ. Sci. Technol. **47**(22), 13201–13202 (2013)
91. Chen, N., Jiang, D.T., Cutler, J., et al.: Structural characterization of poorly-crystalline scorodite, iron(III) earsenate co-precipitates and uranium mill neutralized raffinate solids using X-ray absorption fine structure spectroscopy. Geochim. Cosmochim. Acta **73**(11), 3260–3276 (2009)
92. Mikutta, C., Michel, F.M., Mandaliev, P., et al.: Structure of Amorphous Ferric Arsenate from EXAFS Spectroscopy and Total X-ray Scattering. EGU General Assembly, Vienna (2013)
93. Guido, C.A., Cortona, P., Mennucci, B., et al.: On the metric of charge transfer molecular excitations: a simple chemical descriptor. J. Chem. Theory Comput. **9**(7), 3118–3125 (2013)
94. Goguel, R.: Direct spectrophotometric determination of sulfate in natural waters by formation of the ferric sulfate complex. Anal. Chem. **41**(8), 1034–1038 (1969)
95. Harharan, A., Sudhakar, C.H., Rao, B.V.: Studies on the solvent extraction of iron (III) with tri-iso-octylamine from aqueous mineral acid solutions. Orient. J. Chem. **28**(4), 1785–1790 (2012)
96. Sen, B., Mukherjee, M., Pal, S., et al.: A water soluble copper(II) complex as a $HSO_4^-$ ion selective turn-on fluorescent sensor applicable in living cell imaging. RSC Adv. **5**(62), 50532–50539 (2015)
97. Drahota, P., Filippi, M.: Secondary arsenic minerals in the environment: a review. Environ. Int. **35**(8), 1243–1255 (2009)
98. Parker, V.B., Khodakovskii, I.L.: Thermodynamic properties of the aqueous ions (2+ and 3+) of iron and the key compounds of iron. J. Phys. Chem. Ref. Data **24**(5), 1699 (1995)
99. Jia, Y.F., Xu, L.Y., Wang, X., et al.: Infrared spectroscopic and X-ray diffraction characterization of the nature of adsorbed arsenate on ferrihydrite. Geochim. Cosmochim. Acta **71**(7), 1643–1654 (2007)
100. Le Berre, J.F., Gauvin, R., Demopoulos, G.P.: Characterization of poorly-crystalline ferric arsenate precipitated from equimolar Fe(III)-As(V) solutions in the pH range 2 to 8. Metall. Mater. Trans. B **38**(5), 751–762 (2007)
101. Roque-Malherbe, R., Polanco-Estrella, R., Marquez-Linares, F.: Study of the interaction between silica surfaces and the carbon dioxide molecule. J. Phys. Chem. C **114**(41), 17773–17787 (2010)
102. Zhu, H.F., Tang, P.G., Feng, Y.J., et al.: Intercalation of IR absorber into layered double hydroxides: preparation, thermal stability and selective IR absorption. Mater. Res. Bull. **47**(3), 532–536 (2012)
103. Kim, C.R., Noh, T.H., Yoo, K.H., et al.: Anionic indicators on the surface of submicrospheres consisting of ionic palladium(II) complex. Bull. Korean Chem. Soc. **30**(12), 3057–3060 (2009)

104. Yoon, S.H., Lee, S., Kim, T.H., et al.: Oxidation of methylated arsenic species by UV/$S_2O_8^{2-}$. Chem. Eng. J. **173**(2), 290–295 (2011)
105. Chai, L., Yang, J., Liao, F., et al.: Kinetics and molecular mechanism of arsenite photochemical oxidation based on sulfate radical. Mol. Catal. **438**, 113–130 (2017)
106. Neppolian, B., Celik, E., Choi, H.: Photochemical oxidation of arsenic(III) to arsenic(V) using peroxydisulfate ions as an oxidizing agent. Environ. Sci. Technol. **42**(16), 6179–6184 (2008)
107. Zhou, L., Zheng, W., Ji, Y.F., et al.: Ferrous-activated persulfate oxidation of arsenic(III) and diuron in aquatic system. J. Hazard. Mater. **263**(Part 2), 422–430 (2013)
108. Davies, M.J., Gilbert, B.C., Stell, J.K., et al.: Nucleophilic substitution reactions of spin adducts. Implications for the correct identification of reaction intermediates by EPR/spin trapping. J. Chem. Soc. Perkin Trans. **2**(3), 333–335 (1992)
109. Mottley, C., Mason, R.P.: Sulfate anion free radical formation by the peroxidation of (Bi) sulfite and its reaction with hydroxyl radical scavengers. Arch. Biochem. Biophys. **267** (2), 681–689 (1988)
110. Wang, Z., Bush, R.T., Sullivan, L.A., et al.: Selective oxidation of arsenite by peroxymonosulfate with high utilization efficiency of oxidant. Environ. Sci. Technol. **48** (7), 3978–3985 (2014)
111. Buxton, G.V., Greenstock, C.L., Helman, W.P., et al.: Critical review of rate constants for reactions of hydrated electrons, hydrogen atoms and hydroxyl radicals ($OH/\cdot O^-$ in aqueous solution. J. Phys. Chem. Ref. Data **17**(2), 513–886 (1988)
112. Neta, P., Grodkowski, J., Ross, A.B.: Rate constants for reactions of aliphatic carbon-centered radicals in aqueous solution. J. Phys. Chem. Ref. Data **25**(3), 709–1050 (1996)
113. Klaening, U.K., Bielski, B.H.J., Sehested, K.: Arsenic(IV). A pulse-radiolysis study. Inorg. Chem. **28**(14), 2717–2724 (1989)
114. Yamazaki-Nishida, S., Kimura, M.: Kinetics of the oxidation reaction of arsenious acid by peroxodisulfate ion, induced by irradiation with visible light of aqueous solutions containing tris(2,2′-bipyridine) ruthenium(II) ion. Inorg. Chim. Acta **174**(2), 231–235 (1990)
115. Yoon, S.H., Lee, J.H., Oh, S.E., et al.: Photochemical oxidation of As(III) by vacuum-UV lamp irradiation. Water Res. **42**(13), 3455–3463 (2008)
116. Fan, Z.Y., Huang, J.L., Wang, P., et al.: Kinetics of aniline oxidation with chlorine dioxide. J. Environ. Sci. **16**(2), 238–241 (2004)
117. Gomes, A.C., Nunes, J.C., Simões, R.M.S.: Determination of fast ozone oxidation rate for textile dyes by using a continuous quench-flow system. J. Hazard. Mater. **178**(1–3), 57–65 (2010)
118. Iwai, M., Majima, H., Awakura, Y.: Oxidation of As(III) with dissolved molecular oxygen in alkaline solutions. Trans. Jpn. Inst. Met. **26**(7), 492–498 (1985)
119. Pettine, M., Campanella, L.G., Millero, F.J.: Arsenite oxidation by $H_2O_2$ in aqueous solutions. Geochim. Cosmochim. Acta **63**(18), 2727–2735 (1999)
120. Lau, T.K., Chu, W., Graham, N.J.D.: The aqueous degradation of butylated hydroxyanisole by UV/$S_2O_8^{2-}$: study of reaction mechanisms via dimerization and mineralization. Environ. Sci. Technol. **41**(2), 613–619 (2007)
121. Chai, L., Yue, M., Yang, J., et al.: Formation of tooeleite and the role of direct removal of As (III) from high-arsenic acid wastewater. J. Hazard. Mater. **320**, 620–627 (2016)
122. Nishimura, T., Robins, R.G.: Confirmation that tooeleite is a ferric arsenite sulfate hydrate, and is relevant to arsenic stabilization. Miner. Eng. **21**(4), 246–251 (2008)
123. McLeod, J., Paterson, A.H.J., Jones, J.R., et al.: Primary nucleation of alpha-lactose monohydrate: the effect of supersaturation and temperature. Int. Dairy J. **21**(7), 455–461 (2011)
124. Dang, S.V., Kawasaki, J., Abella, L.C., et al.: Removal of arsenic from simulated groundwater by adsorption using iron-modified rice husk carbon. J. Water Environ. Technol. **7**(2), 43–56 (2009)

125. Mercer, K.L., Tobiason, J.E.: Removal of arsenic from high ionic strength solutions: effects of ionic strength, pH, and preformed versus in situ formed HFO. Environ. Sci. Technol. **42** (10), 3797–3802 (2008)
126. Wang, Y.X., Duan, J.M., Liu, S.X., et al.: Removal of As(III) and As(V) by ferric salts coagulation—implications of particle size and zeta potential of precipitates. Sep. Purif. Technol. **135**, 64–71 (2014)
127. Li, X.F., Zhao, F.H., Deng, S.M.: The removal of arsenic(III) from acid mine drainage by mineral trap of tooeleite ($Fe_6(AsO_3)_4SO_4(OH)_4{\cdot}4H_2O$). In: An Interdisciplinary Response to Mine Water Challenges, pp. 671–674 (2014)
128. Swash, P.M., Monhemius, A.J.: Comparison of the solubilities of arsenic-bearing wastes from hydrometallurgical and pyrometallurgical processes. GDMB **83**, 141–152 (2000)
129. Paikaray, S., Göttlicher, J., Peiffer, S.: As(III) retention kinetics, equilibrium and redox stability on biosynthesized schwertmannite and its fate and control on schwertmannite stability on acidic (pH 3.0) aqueous exposure. Chemosphere **86**, 557–564 (2012)
130. Rahman, N., Haseen, U.: Development of polyacrylamide chromium oxide as a new sorbent for solid phase extraction of As(III) from food and environmental water samples. RSC Adv. **5**, 7311–7323 (2015)
131. Tokoro, C., Yatsugi, Y., Koga, H., et al.: Sorption mechanisms of arsenate during coprecipitation with ferrihydrite in aqueous solution. Environ. Sci. Technol **44**, 638–643 (2010)
132. Kaksonen, A.H., Riekkola-Vanhanen, M.L., Puhakka, J.A.: Optimization of metal sulfide precipitation in fluidized-bed treatment of acidic wastewater. Water Res. **37**(2), 255–266 (2003)
133. Mokone,Cas T.P., van Hille, R.P., Lewis, A.E.: Effect of solution chemistry on particle characteristics during metal sulfide precipitation. J. Colloid Interface Sci. **351**(1), 10–18 (2010)
134. Long, G., Peng, Y.J., Bradshaw, D.: Flotation separation of copper sulfides from arsenic minerals at Rosebery copper concentrator. Miner. Eng. **66–68**, 207–214 (2014)
135. Padilla, R., Rodriguez, G., Ruiz, M.C.: Copper and arsenic dissolution from chalcopyrite-enargite concentrate by sulfidation and pressure leaching in $H_2SO_4$–$O_2$. Hydrometallurgy **100**(3–4), 152–156 (2010)
136. Wang, T., Yang, W.C., Song, T.T., et al.: Cu doped $Fe_3O_4$ magnetic adsorbent for arsenic: synthesis, property, and sorption application. RSC Adv. **5**(62), 50011–50018 (2015)
137. Wang, Z.F., Cui, Z.J., Liu, L., et al.: Toxicological and biochemical responses of the earthworm eisenia fetida exposed to contaminated soil: effects of arsenic species. Chemosphere **154**, 161–170 (2016)
138. Yan, X., Li, Q.Z., Chai, L.Y., et al.: Formation of abiological granular sludge-A facile and bioinspired proposal for improving sludge settling performance during heavy metal wastewater treatment. Chemosphere **113**, 36–41 (2014)
139. Wang, T., Zhang, L.Y., Li, C.F., et al.: Synthesis of core-shell magnetic $Fe_3O_4$@poly (m-phenylenediamine) particles for chromium reduction and adsorption. Environ. Sci. Technol. **49**(9), 5654–5662 (2015)
140. Chai, L.Y., Wang, Q.W., Li, Q.Z., et al.: Enhanced removal of Hg(II) from acidic aqueous solution using thiol-functionalized biomass. Water Sci. Technol. **62**(9), 2157–2165 (2010)
141. Stalidis, G.A., Matis, K.A., Lazaridis, N.K.: Selective separation of Cu, Zn, and As from solution by flotation techniques. Sep. Sci. Technol. **24**(1–2), 97–109 (1989)
142. Alison Emslie Lewis: Review of metal sulfide precipitation. Hydrometallurgy **104**(2), 222–234 (2010)
143. Lian-hua, Z., Yu-lan, X.: Sulfide precipitation flotation for treatment of acidic mine waste water. Trans. Nonferrous Met. Soc. China **10**, 106–109 (2000)
144. Huisman, J.L., Schouten, G., Schultz, C.: Biologically produced sulfide for purification of process streams, effluent treatment and recovery of metals in the metal and mining industry. Hydrometallurgy **83**(1), 106–113 (2006)

145. Bhattacharyya, D., Jumawan Jr., A.B., Grieves, R.B.: Separation of toxic heavy metals by sulfide precipitation. Sep. Sci. Technol. **14**(5), 441–452 (1979)
146. Veeken, A.H.M., de Vries, S., van Der Mark, A., et al.: Selective precipitation of heavy metals as controlled by a sulfide-selective electrode. Sep. Sci. Technol. **38**(1), 1–19 (2003)
147. Jiang, G.M., Peng, B., Chai, L.Y., et al.: Cascade sulfidation and separation of copper and arsenic from acidic wastewater via gas–liquid reaction. Trans. Nonferrous Met. Soc. China **27**(4), 925–931 (2017)
148. Zheng, J.X., Ye, H.Q., Huang, N.D., et al.: Selective separation of Hg(II) and Cd(II) from aqueous solutions by complexation-ultrafiltration process. Chemosphere **76**(5), 706–710 (2009)
149. Yavuz, C.T., Mayo, J., Yu, W.W., et al.: Low-field magnetic separation of monodisperse $Fe_3O_4$ nanocrystals. Science **314**(5801), 964–967 (2006)
150. Zeng, H., Singh, A., Basak, S., et al.: Nanoscale size effects on uranium(VI) adsorption to hematite. Environ. Sci. Technol. **43**(5), 1373–1378 (2009)
151. Yean, S., Cong, L., Yavuz, C.T., et al.: Effect of magnetite particle size on adsorption and desorption of arsenite and arsenate. J. Mater. Res. **20**(12), 3255–3264 (2005)
152. Cotten, G.B., Eldredge, H.B.: Nanolevel magnetic separation model considering flow limitations. Sep. Sci. Technol. **37**(16), 3755–3779 (2002)
153. Kelland, D.R.: Magnetic separation of nanoparticles. IEEE Trans. Magn. **34**(4), 2123–2125 (1998)
154. Mou, F.Z., Guan, J.G., Ma, H., et al.: Magnetic iron oxide chestnutlike hierarchical nanostructures: preparation and their excellent arsenic removal capabilities. ACS Appl. Mater. Interfaces **4**(8), 3987–3993 (2012)
155. Ge, J.P., Huynh, T., Hu, Y.X., et al.: Hierarchical magnetite/silica nanoassemblies as magnetically recoverable catalyst–supports. Nano Lett. **8**(3), 931–934 (2008)
156. Wei, Z.H., Xing, R., Zhang, X., et al.: Facile template-free fabrication of hollow nestlike $\alpha$-$Fe_2O_3$ nanostructures for water treatment. ACS Appl. Mater. Interfaces **5**(3), 598–604 (2013)
157. Wang, T., Zhang, L., Li, C., et al.: Synthesis of core–shell magnetic $Fe_3O_4$@ poly (m-phenylenediamine) particles for chromium reduction and adsorption. Environ. Sci. Technol. **49**(9), 5654–5662 (2012)
158. Mou, F.Z., Guan, J.G., Xiao, Z.D., et al.: Solvent-mediated synthesis of magnetic $Fe_2O_3$ chestnut-like amorphous-core/$\gamma$-phase-shell hierarchical nanostructures with strong As(V) removal capability. J. Mater. Chem. **21**(14), 5414–5421 (2011)
159. Wang, P., Lo, I.M.C.: Synthesis of mesoporous magnetic $\gamma$-$Fe_2O_3$ and its application to Cr (VI) removal from contaminated water. Water Res. **43**(15), 3727–3734 (2009)
160. Wang, T., Zhang, L.Y., Wang, H.Y., et al.: Controllable synthesis of hierarchical porous $Fe_3O_4$ particles mediated by Poly(diallyldimethylammonium chloride) and their application in arsenic removal. ACS Appl. Mater. Interfaces **5**(23), 12449–12459 (2013)
161. Jia, B.P., Gao, L.: Morphological transformation of $Fe_3O_4$ spherical aggregates from solid to hollow and their self-assembly under an external magnetic field. J. Phys. Chem. C **112**(3), 666–671 (2008)
162. Fan, T., Pan, D., Zhang, H., et al.: Study on formation mechanism by monitoring the morphology and structure evolution of nearly monodispersed $Fe_3O_4$ submicroparticles with controlled particle sizes. Ind. Eng. Chem. Res. **50**(15), 9009–9018 (2011)
163. Liu, Z.H., Yang, X.J., Makita, Y., et al.: Preparation of a polycation-intercalated layered manganese oxide nanocomposite by a delamination/reassembling process. Chem. Mater. **14** (11), 4800–4806 (2002)
164. Liu, K.P., Zhang, J.J., Yang, G.H., et al.: Direct electrochemistry and electrocatalysis of hemoglobin based on poly(diallyldimethylammonium chloride) functionalized graphene sheets/room temperature ionic liquid composite film. Electrochem. Commun. **12**(3), 402–405 (2010)

165. Yu, X.Y., Luo, T., Jia, Y., et al.: Porous hierarchically micro-/nanostructured MgO: morphology control and their excellent performance in As(III) and As(V) removal. J. Phys. Chem. C **115**(45), 22242–22250 (2011)
166. Hang, C., Li, Q., Gao, S.A., et al.: As(III) and As(V) adsorption by hydrous zirconium oxide nanoparticles synthesized by a hydrothermal process followed with heat treatment. Ind. Eng. Chem. Res. **51**(1), 353–361 (2012)
167. Zhong, L.S., Hu, J.S., Liang, H.P., et al.: Self-assembled 3D flowerlike iron oxide nanostructures and their application in water treatment. Adv. Mater. **18**(18), 2426–2431 (2006)
168. Zhu, H., Hou, C., Li, Y.J., et al.: One-pot solvothermal synthesis of highly water-dispersible size-tunable functionalized magnetite nanocrystal clusters for lipase immobilization. Chem. Asian J. **8**(7), 1447–1454 (2013)
169. Cao, C.Y., Qu, J., Yan, W.S., et al.: Low-cost synthesis of flowerlike $\alpha$-$Fe_2O_3$ nanostructures for heavy metal ion removal: adsorption property and mechanism. Langmuir **28**(9), 4573–4579 (2012)
170. Kanel, S.R., Greneche, J.M., Choi, H.: Arsenic(V) removal from groundwater using nano scale zero-valent iron as a colloidal reactive barrier material. Environ. Sci. Technol. **40**(6), 2045–2050 (2006)
171. Nesbitt, H.W., Muir, I.J.: Oxidation states and speciation of secondary products on pyrite and arsenopyrite reacted with mine waste waters and air. Mineral. Petrol. **62**(1–2), 123–144 (1998)
172. Gomes, J.A.G., Daida, P., Kesmez, M., et al.: Arsenic removal by electrocoagulation using combined Al–Fe electrode system and characterization of products. J. Hazard. Mater. **139**(2), 220–231 (2007)
173. Chen, B., Zhu, Z.L., Ma, J., et al.: Surfactant assisted Ce–Fe mixed oxide decorated multiwalled carbon nanotubes and their arsenic adsorption performance. J. Mater. Chem. A **1**(37), 11355–11367 (2013)
174. Wielant, J., Hauffman, T., Blajiev, O., et al.: Influence of the iron oxide acid-base properties on the chemisorption of model epoxy compounds studied by XPS. J. Phys. Chem. C **111**(35), 13177–13184 (2007)
175. Ramos, M.A.V., Yan, W.L., Li, X.Q., et al.: Simultaneous oxidation and reduction of arsenic by zero-valent iron nanoparticles: understanding the significance of the core-shell structure. J. Phys. Chem. C **113**(33), 14591–14594 (2009)
176. Lim, S.F., Zheng, Y.M., Chen, J.P. Organic arsenic adsorption onto a magnetic sorbent. Langmuir **25**(9), 4973–4978 (2009)
177. Manning, B.A., Hunt, M.L., Amrhein, C., et al.: Arsenic (III) and arsenic (V) reactions with zerovalent iron corrosion products. Environ. Sci. Technol. **36**(24), 5455–5461 (2002)
178. Pena, M., Meng, X., Koratis, G.P., et al.: Adsorption mechanism of arsenic on nanocrystalline titanium dioxide. Environ. Sci. Technol. **40**(4), 1257–1262 (2006)
179. Sandoval, R., Cooper, A.M., Aymar, K., et al.: Removal of arsenic and methylene blue from water by granular activated carbon media impregnated with zirconium dioxide nanoparticles. J. Hazard. Mater. **193**, 296–303 (2011)
180. Xu, W.H., Wang, J., Wang, L., et al.: Enhanced arsenic removal from water by hierarchically porous $CeO_2$–$ZrO_2$ nanospheres: role of surface-and structure-dependent properties. J. Hazard. Mater. **260**, 498–507 (2013)
181. Lv, X.J., Yang, W.G., Quan, Z.W., et al.: Enhanced electron transport in Nb-Doped $TiO_2$ nanoparticles via pressure-induced phase transitions. J. Am. Chem. Soc. **136**(1), 419–426 (2014)
182. Li, H., Zhang, L.Z.: Oxygen vacancy induced selective silver deposition on the 001 facets of BiOCl single-crystalline nanosheets for enhanced Cr(VI) and sodium pentachlorophenate removal under visible light. Nanoscale **6**(14), 7805–7810 (2014)
183. Zhang, Y., Yang, M., Dou, X.M., et al.: Arsenate adsorption on an Fe-Ce bimetal oxide adsorbent: Role of surface properties. Environ. Sci. Technol. **39**(18), 7246–7253 (2005)

184. Warner, C.L., Chouyyok, W., Mackie, K.E., et al.: Manganese doping of magnetic iron oxide nanoparticles: tailoring surface reactivity for a regenerable heavy metal sorbent. Langmuir **28**(8), 3931–3937 (2012)
185. Wang, Y.J., Chen, D.G., Wang, Y.D., et al.: Tunable surface charge of ZnS: Cu nano-adsorbent induced the selective preconcentration of cationic dyes from wastewater. Nanoscale **4**(12), 3665–3668 (2012)
186. Neagu, D., Irvine, J.T.: Enhancing electronic conductivity in strontium titanates through correlated A and B-site doping. Chem. Mater. **23**(6), 1607–1617 (2011)
187. Norris, D.J., Efros, A.L., Erwin, S.C.: Doped nanocrystals. Science **319**(5871), 1776–1779 (2008)
188. Chen, Z., Pina, C.D., Falletta, E., et al.: A green route to conducting polyaniline by copper catalysis. J. Catal. **267**(2), 93–96 (2009)
189. Chai, L.Y., Wang, T., Zhang, L.Y., et al.: A Cu–m-phenylenediamine complex induced route to fabricate poly(m-phenylenediamine)/reduced graphene oxide hydrogel and its adsorption application. Carbon **81**, 748–757 (2015)
190. Deng, H., Li, X.L., Peng, Q., et al.: Monodisperse magnetic single-crystal ferrite microspheres. Angew. Chem. **117**(18), 2842–2845 (2005)
191. Yin, A.Y., Guo, X.Y., Dai, W.L., et al.: The nature of active copper species in Cu-HMS catalyst for hydrogenation of dimethyl oxalate to ethylene glycol: new insights on the synergetic effect between $Cu^0$ and $Cu^+$. J. Phys. Chem. C **113**(25), 11003–11013 (2009)
192. Derrouiche, S., Lauron-Pernot, H., Louis, C.: Synthesis and Treatment Parameters For Controlling Metal Particle Size And Composition in Cu/ZnO materials first evidence of $Cu_3Zn$ alloy formation. Chem. Mater. **24**(12), 2282–2291 (2012)
193. McFarland, E.W., Metiu, H.: Catalysis by doped oxides. Chem. Rev. **113**(6), 4391–4427 (2013)
194. Aldon, L., Kubiak, P., Picard, A., et al.: Size particle effects on lithium insertion into Sn-doped $TiO_2$ anatase. Chem. Mater. **18**(6), 1401–1406 (2006)
195. Deiana, C., Fois, E., Coluccia, S., et al.: Surface structure of $TiO_2$ P25 nanoparticles: infrared study of hydroxy groups on coordinative defect sites. J. Phys. Chem. C **114**(49), 21531–21538 (2010)
196. Chai, L.Y., Wang, Y.Y., Zhao, N., et al.: Sulfate-doped $Fe_3O_4/Al_2O_3$ nanoparticles as a novel adsorbent for fluoride removal from drinking water. Water Res. **47**(12), 4040–4049 (2013)
197. Mohapatra, M., Sahoo, S.K., Anand, S., et al.: Removal of As(V) by Cu(II)-, Ni(II)-, or Co (II)-doped goethite samples. J. Colloid Interface Sci. **298**(1), 6–12 (2006)
198. Yu, X.Y., Luo, T., Jia, Y., et al.: Porous hierarchically micro-/nanostructured MgO: morphology control and their excellent performance in As (III) and As (V) removal. J. Phys. Chem. C **115**(45), 22242–22250 (2011)
199. Xu, W.H., Wang, L., Wang, J., et al.: Superparamagnetic mesoporous ferrite nanocrystal clusters for efficient removal of arsenite from water. CrystEngComm **15**(39), 7895–7903 (2013)
200. Couture, R.M., Rose, J., Kumar, N., et al.: Sorption of arsenite, arsenate, and thioarsenates to iron oxides and iron sulfides: a kinetic and spectroscopic investigation. Environ. Sci. Technol. **47**(11), 5652–5659 (2013)
201. Can, M., Uzun, S.: Oxidizing effect of the $Cu(ClO_4)_2$ on chemical polymerization of aniline in anhydrous media. Asian J. Chem. **22**, 867–872 (2010)
202. Izumi, C.M.S., Constantino, V.R.L., Temperini, M.L.A.: Spectroscopic characterization of polyaniline formed by using copper (II) in homogeneous and MCM-41 molecular sieve media. J. Phys. Chem. B **109**(47), 22131–22140 (2005)
203. Mou, F.Z., Guan, J.G., Ma, H.R., et al.: Magnetic iron oxide chestnutlike hierarchical nanostructures: preparation and their excellent arsenic removal capabilities. ACS Appl. Mater. Interfaces **4**(8), 3987–3993 (2012)

204. Toulemon, D., Pichon, B.P., Cattoen, X., et al.: 2D assembly of non-interacting magnetic iron oxide nanoparticles via "click" chemistry. Chem. Commun. **47**(43), 11954–11956 (2011)
205. Feng, L.Y., Cao, M.H., Ma, X.Y., et al.: Superparamagnetic high-surface-area $Fe_3O_4$ nanoparticles as adsorbents for arsenic removal. J. Hazard. Mater. **217–218**, 439–446 (2012)
206. Saiz, J., Bringas, E., Ortiz, I.: Functionalized magnetic nanoparticles as new adsorption materials for arsenic removal from polluted waters. J. Chem. Technol. Biotechnol. **89**(6), 909–918 (2014)
207. Chen, B., Zhu, Z.L., Ma, J., et al.: One-pot, solid-phase synthesis of magnetic multiwalled carbon nanotube/iron oxide composites and their application in arsenic removal. J. Colloid Interface Sci. **434**, 9–17 (2014)
208. Li, H., Yu, S., Han, X.X.: Fabrication of CuO hierarchical flower-like structures with biomimetic superamphiphobic, self-cleaning and corrosion resistance properties. Chem. Eng. J. **283**, 1443–1454 (2016)
209. Meshram, S.P., Adhyapak, P.V., Mulik, U.P., et al.: Facile synthesis of CuO nanomorphs and their morphology dependent sunlight driven photocatalytic properties. Chem. Eng. J. **204–206**, 158–168 (2012)
210. Peng, B., Song, T., Wang, T., et al.: Facile synthesis of $Fe_3O_4$@$Cu(OH)_2$ composites and their arsenic adsorption application. Chem. Eng. J. **299**, 15–22 (2016)
211. Escudero, C., Fiol, N., Villaescusa, I., et al.: Arsenic removal by awaste metal (hydr)oxide entrapped into calcium alginate beads. J. Hazard. Mater. **164**(2–3), 533–541 (2009)
212. Li, Z.J., Deng, S.B., Yu, G., et al.: As(V) and As(III) removal from water by a Ce–Ti oxide adsorbent: behavior and mechanism. Chem. Eng. J. **161**(1–2), 106–113 (2010)
213. Yang, Y., Yang, M., Dou, X.M., et al.: Arsenate adsorption on an Fe-Ce bimetal oxide adsorbent: role of surface properties. Environ. Sci. Technol. **39**(18), 7246–7253 (2005)
214. Zhang, S.W., Li, J.X., Wen, T., et al.: Magnetic $Fe_3O_4$@NiO hierarchical structures: preparation and their excellent As(V) and Cr(VI) removal capabilities. RSC Adv. **3**(8), 2754–2764 (2013)
215. Wang, T., Zhang, L.Y., Li, C.F., et al.: Synthesis of core-shell magnetic $Fe_3O_4$@poly (m-phenylenediamine) particles for chromium reduction and adsorption. Environ. Sci. Technol. **49**(9), 5654–5662 (2015)
216. Xu, L.J., Wang, J.L.: Magnetic nanoscaled $Fe_3O_4$/$CeO_2$ composite as an efficient fenton-like heterogeneous catalyst for degradation of 4-chlorophenol. Environ. Sci. Technol. **46**(18), 10145–10153 (2012)
217. Gai, L.G., Li, Z.L., Hou, Y.H., et al.: Preparation of core–shell $Fe_3O_4$/$SiO_2$ microspheres as adsorbents for purification of DNA. J. Phys. D Appl. Phys. **43**, 445001 (2010)
218. Ma, Z.Y., Guan, Y.P., Liu, H.Z.: Synthesis and characterization of micron-sized monodisperse superparamagnetic polymer particles with amino groups. J. Polym. Sci. Part A Polym. Chem. **43**(15), 3433–3439 (2005)
219. Lu, C.H., Qi, L.M., Yang, J.H., et al.: Simple template-free solution route for the controlled synthesis of $Cu(OH)_2$ and CuO nanostructures. J. Phys. Chem. B **108**(46), 17825–17831 (2004)
220. Hua, R., Li, Z.K.: Sulfhydryl functionalized hydrogel with magnetism: synthesis, characterization, and adsorption behavior study for heavy metal removal. Chem. Eng. J. **249**, 189–200 (2014)
221. Martinson, C.A., Reddy, K.J.: Adsorption of arsenic(III) and arsenic(V) by cupric oxide nanoparticles. J. Colloid Interface Sci. **336**(2), 406–411 (2009)
222. Wang, X.L., Liu, Y.K., Zheng, J.T.: Removal of As(III) and As(V) from water by chitosan and chitosan derivative: a review. Environ. Sci. Pollut. Res. **23**(14), 13789–13801 (2016)
223. Zhang, G.S., Ren, Z.M., Zhang, X.W., et al.: Nanostructured iron(III)-copper(II) binary oxide: a novel adsorbent for enhanced arsenic removal from aqueous solutions. Water Res. **47**(12), 4022–4031 (2013)
224. Maeda, K., Domen, K.: New non-oxide photocatalysts designed for overall water splitting under visible light. J. Phys. Chem. C **111**(22), 7851–7861 (2007)

225. Huang, M.R., Ding, Y.B., Li, X.G.: Lead-ion potentiometric sensor based on electrically conducting microparticles of sulfonic phenylenediamine copolymer. Analyst **138**(13), 3820–3829 (2013)
226. Huang, M.R., Rao, X.W., Li, X.G., et al.: Lead ion-selective electrodes based on polyphenylenediamine as unique solid ionophores. Talanta **85**(3), 1575–1584 (2011)
227. Huang, J.Y., Li, S.H., Ge, M.Z., et al.: Robust superhydrophobic $TiO_2$@fabrics for UV shielding, self-cleaning and oil–water separation. J. Mater. Chem. A 3(6), 2825–2832 (2015)
228. Li, X.G., Ma, X.L., Sun, J., et al.: Powerful reactive sorption of silver(I) and mercury(II) onto poly(o-phenylenediamine) microparticles. Langmuir **25**(3), 1675–1684 (2009)
229. Wang, J.J., Jiang, J., Hu, B., et al.: Uniformly shaped poly(p-phenylenediamine) microparticles: shape-controlled synthesis and their potential application for the removal of lead ions from water. Adv. Func. Mater. **18**(7), 1105–1111 (2008)
230. Huang, M.R., Lu, H.J., Song, W.D., et al.: Dynamic reversible adsorption and desorption of lead ions through a packed column of poly(m-phenylenediamine) spheroids. Soft Mater. **8**(2), 149–163 (2010)
231. Zhang, L.Y., Wang, T., Wang, H.Y., et al.: Graphene@poly(m-phenylenediamine) hydrogel fabricated by a facile post-synthesis assembly strategy. Chem. Commun. **49**(85), 9974–9976 (2013)
232. Yu, W.T., Zhang, L.Y., Meng, Y., et al.: High conversion synthesis of functional poly (m-phenylenediamine) nanoparticles by Cu-OH-assisted method and its superior ability toward Ag+ adsorption. Synth. Met. **176**, 78–85 (2013)
233. Wang, H.Y., Chai, L.Y., Hu, A.J., et al.: Self-assembly microstructures of amphiphilic polyborate in aqueous solutions. Polymer **50**(13), 2976–2980 (2009)
234. Harris, J.K., Rose, G.D., Bruening, M.L.: Spontaneous generation of multilamellar vesicles from ethylene oxide/butylene oxide diblock copolymers. Langmuir **18**(14), 5337–5342 (2002)
235. Li, H.Q., Lai, Y.K., Huang, J.Y., et al.: Multifunctional wettability patterns prepared by laser processing on superhydrophobic $TiO_2$ nanostructured surfaces. J. Mater. Chem. B **3**(3), 342–347 (2015)
236. Dey, J., Kumar, S., Nath, S., et al.: Additive induced core and corona specific dehydration and ensuing growth and interaction of Pluronic F127 micelles. J. Colloid Interface Sci. **415**, 95–102 (2014)
237. Sang, P.L., Wang, Y.Y., Zhang, L.Y., et al.: Effective adsorption of sulfate ions with poly (m-phenylenediamine) in aqueous solution and its adsorption mechanism. Trans. Nonferrous Met. Soc. China **23**(1), 243–252 (2013)
238. Li, X.G., Wang, L.X., Jin, Y., et al.: Preparation and identification of a soluble copolymer from pyrrole and o-toluidine. J. Appl. Polym. Sci. **82**(2), 510–518 (2001)
239. Chai, L.Y., Zhang, L.Y., Wang, H.Y., et al.: An effective and scale-up self-assembly route to prepare the rigid and smooth oligo(o-phenylenediamine) microfibers in acidic solution by $NaClO_2$. Mater. Lett. **64**(21), 2302–2305 (2010)
240. Li, X.G., Huang, M.R., Duan, W.: Novel multifunctional polymers from aromatic diamines by oxidative polymerizations. Chem. Rev. **102**(9), 2925–3030 (2002)
241. Liu, M.L., Ye, M., Yang, Q., et al.: A new method for characterizing the growth and properties of polyaniline and poly(aniline-co-o-aminophenol) films with the combination of EQCM and in situ FTIR spectroelectrochemisty. Electrochim. Acta **52**(1), 342–352 (2006)
242. Huang, M.R., Li, X.G., Yang, Y.L.: Oxidative polymerization of o-phenylenediamine and pyrimidylamine. Polym. Degrad. Stab. **71**(1), 31–38 (2000)
243. Yu, W.T., Zhang, L.Y., Wang, H.Y., et al.: Adsorption of Cr(VI) using synthetic poly (m-phenylenediamine). J. Hazard. Mater. **260**, 789–795 (2013)
244. Zhang, L.Y., Chai, L.Y., Liu, J., et al.: pH manipulation: a facile method for lowering oxidation state and keeping good yield of poly(m-phenylenediamine) and its powerful $Ag^+$ adsorption ability. Langmuir **27**(22), 13729–13738 (2011)

245. Losito, I., Malitesta, C., De Bari, I., et al.: X-ray photoelectron spectroscopy characterization of poly(2,3-diaminophenazine) films electrosynthesised on platinum. Thin Solid Films **473** (1), 104–113 (2005)
246. Frost, R.L., Xi, Y.F., Wood, B.J.: Thermogravimetric analysis, PXRD, EDX and XPS study of chrysocolla (Cu, Al)$_2$H$_2$Si$_2$O$_5$(OH)$_4$·nH$_2$O-structural implications. Thermochim. Acta **545**, 157–162 (2012)
247. Pelissier, B., Beaurain, A., Fontaine, H., et al.: Investigations on HCl contaminated Cu 200 mm wafers using parallel angle resolved XPS. Micro Microelectron. Eng. **36**(5–6), 1013–1016 (2009)
248. Ayad, M.M., Amer, W.A., Stejskal, J.: Effect of iodine solutions on polyaniline films. Thin Solid Films **517**(21), 5969–5973 (2009)
249. Han, J., Dai, J., Guo, R.: Highly efficient adsorbents of poly(o-phenylenediamine) solid and hollow sub-microspheres towards lead ions: a comparative study. J. Colloid Interface Sci. **356**(2), 749–756 (2011)
250. Stejskal, J., Trchová, M., Brožová, L., et al.: Reduction of silver nitrate by polyaniline nanotubes to produce silver-polyaniline composites. Chem. Pap. **63**(1), 77–83 (2009)
251. Izumi, C.M.S., Brito, H.F., Ferreira, A.M.D.C., et al.: Spectroscopic investigation of the interactions between emeraldine base polyaniline and Eu(III) ions. Synth. Met. **159**(5–6), 377–384 (2009)
252. Huang, M.R., Huang, S.J., Li, X.G.: Facile synthesis of polysulfoaminoanthraquinone nanosorbents for rapid removal and ultrasensitive fluorescent detection of heavy metal ions phys. J. Phys. Chem. C **115**(13), 5301–5315 (2011)

# Chapter 4
# Arsenic Pollution Control Technologies for Arsenic-Bearing Solid Wastes

**Xiao-Bo Min, Li-Yuan Chai, Yan-Jie Liang and Yong Ke**

Arsenic-bearing solid wastes from industrial processes are classified as hazardous wastes due to their high leachable content of arsenic and other toxic metal ions. The current treatments of arsenic-bearing solid wastes originated from nonferrous metals smelter, taking the wastewater treatment sludge and arsenic-bearing anode slime for example, are mainly both solidification and secondary utilization.

Solidification is to reduce the toxicity and the mobility of harmful components by combining with the different solidification materials. As a result, the arsenic-bearing solid wastes become more stable in the physical and chemical properties, and the impact on environment is diminished. Solidification/stabilization processes such as cementitious solidification [1], polymeric encapsulation [2], vitrification [3], sulfidation [4] and crystallization [5, 6], are widely applied to reduce the mobility of arsenic and other heavy metals for environmental protection. Portland cement, hydrated lime and fly ash are among the most popularly used materials for stabilizing arsenic in solid wastes, sludges and soils [7–9]. In those methods, cement solidification is used practically because of being more economic and efficient, furthermore the final product can be regarded as building materials.

X.-B. Min · L.-Y. Chai · Y.-J. Liang (✉) · Y. Ke
School of Metallurgy and Environment, Central South University, Changsha, Hunan, China
e-mail: liangyanjie2015@csu.edu.cn

X.-B. Min
e-mail: mxb@csu.edu.cn

L.-Y. Chai
e-mail: lychai@csu.edu.cn

Y. Ke
e-mail: keyong000ke@csu.edu.cn

L.-Y. Chai
Chinese National Engineering Research Center for Control and Treatment of Heavy Metal Pollution (CNERC-CTHMP), Changsha, Hunan, China

 
L.-Y. Chai (ed.), *Arsenic Pollution Control in Nonferrous Metallurgy*,
https://doi.org/10.1007/978-981-13-6721-2_4

Secondary utilization is the recovery of valuable components. Arsenic-bearing solid wastes are the secondary resources, from which we can recover arsenic and many other heavy metals. With regards to recovery methods, one is redox roasting, arsenic in the solid wastes is recovered in the form of white arsenic, the other is hydrometallurgy treatment such as acid or alkali leaching and salt leaching, arsenic is separated from the solid wastes [10]. The former has the characteristics of large treatment capacity, being suitable for the solid wastes with more than 10% arsenic in the total content, however, it has disadvantages of secondary pollution and poor labor conditions and the lower removal rate. The latter has the characteristic of low-cost, no secondary pollution, efficiency and high removal rate, but the operation is more complicate. Recovery and utilization of valuable metals from these solid wastes are important for saving metal resources and for protecting the environment.

## 4.1 Arsenic Stabilization Technologies

### *4.1.1 Lime-Based Stabilization of High Alkaline Arsenic-Bearing Sludges*

Calcium arsenate crystallization is one of the most efficient techniques for lowering arsenic concentration of alkaline arsenic-bearing sludges. Bothe and Brown [11] detected the formation of calcium arsenate complex species $Ca_4(OH)_2(AsO_4)_2{\cdot}4H_2O$, $Ca_5(AsO_4)_3OH$ and $Ca_3(AsO_4)_2{\cdot}3^2/_3H_2O$ and determined the minimum equilibrium arsenic concentrations of $Ca_4(OH)_2(AsO_4)_2{\cdot}4H_2O$ (0.01 mg $L^{-1}$) and $Ca_5(AsO_4)_3OH$ (0.5 mg $L^{-1}$). Donahue and Hendry [12] confirmed that the sludge from neutralization of arsenic-bearing tailings wastewater would likely precipitate $Ca_4(OH)_2(AsO_4)_2{\cdot}4H_2O$ type calcium arsenate minerals under high Ca/As ratios. Camacho et al. [13] used lime as a stabilizing agent on arsenic desorption from ferric water treatment residual and indicated that lime reduced the leached arsenic; however, the method was not stable in the long term, so cement addition was necessary.

High alkaline arsenic-bearing sludges (HAABS) generated in antimony smelters with pH values of 11–13 and arsenic contents of 15–30% are the typical arsenic-bearing solid wastes, especially in Hunan province, China [14]. They were generated in the treatment process of highly concentrated arsenic wastewater by the lime-ferrate process. All samples were homogenized after drying at 60 °C for 24 h and sieved below 0.15 mm. High contents of As, Ca, Na, S, and Fe are present in the HAABS (Table 4.1). Arsenic leaching concentration (ALC) is 4628 mg $L^{-1}$ which is far above the regulation limit (5 mg $L^{-1}$).

To overcome the limitations of traditional methods on parameter optimization and interactive influences, multivariate statistical techniques of the Response Surface Methodology (RSM) were used to improve the efficiency of such process

**Table 4.1** Chemical composition of HAABS analyzed by ICP-AES (wt%)

| As | Ca | Na | S | Fe | Sb | Si | Al | Mg | K |
|---|---|---|---|---|---|---|---|---|---|
| 19.8 | 19.1 | 7.7 | 2.1 | 1.6 | 0.609 | 0.557 | 0.346 | 0.114 | 0.097 |

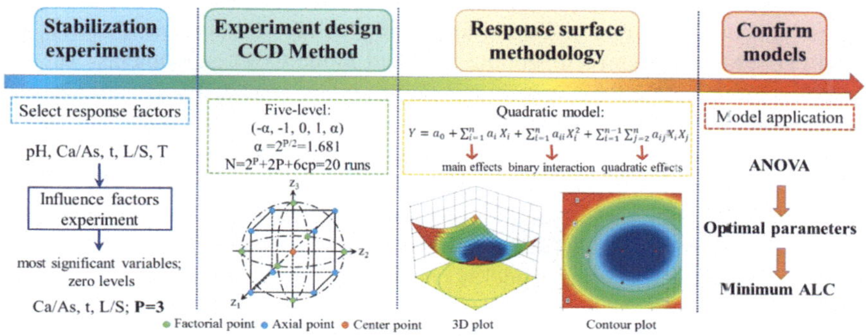


**Fig. 4.1** Design of experiments for RSM. Reprinted from Ref. [16] Copyright 2017, with permission from Taylor & Francis

[15]. In this section, the application of a Central Composite Design (CCD) based on the RSM was presented for modeling and optimizing the stabilization of HAABS using calcium as a stabilizer based on calcium arsenate crystallization [16]. The objective is to investigate the relationship between ALC and the stabilization parameters, to develop a prediction model, to study the interactive effects of the process parameters and to obtain the optimal process parameters and conditions.

The optimization using the RSM approach was divided into five stages (Fig. 4.1): (1) Selection of the independent variables and possible responses, (2) Selection of the experimental design strategy, (3) Execution of the experiments and fitting the models, (4) Verification of the models via Analysis of Variance (ANOVA), and (5) Determination of the optimal conditions.

The first and most important step in whole procedure was to select the most important variables and their ranges. Five critical factors such as pH value, Ca/As mole ratio (Ca/As), reaction time (t), liquid/solid ratio (L/S) and temperature (T) were evaluated to reduce the number of experiments using CCD. The pH value of 12.5 was near that of the HAABS with calcium hydroxide addition and the temperature of 25 °C was approximately room temperature. This methodology is helpful for predicting important parameters in the stabilization of arsenic-bearing sludges from the metallurgical industry.

The effects of the influence factors on the stabilization of HAABS is that the influence of the initial pH value on the ALC is very significant. The ALC decreases with the initial pH value increasing, and a concentration of 4.48 mg $L^{-1}$ is achieved around the initial pH value of 12.5. The ALC decreases rapidly at first and then slows down with an increasing Ca/As mole ratio from 2.5 to 4.5; the optimal Ca/As mole ratio is 4.0 when the ALC is 1.18 mg $L^{-1}$. The ALC first decreases and then increases as the reaction time increases from 1 h to 8 h. There is an obvious impact point in the curve at 3 h, being the optimal reaction time. The ALC decreases with an increasing liquid/solid ratio. When it reaches 10, the ALC decreases to 1.60 mg $L^{-1}$, which is much lower than the regulation limited value of 5 mg $L^{-1}$. Temperature has little effect on the process of stabilization, which is a spontaneous process at room temperature. Taking all the impact factors into consideration, the initial pH value and room temperature are fixed in the CCD experiments. The Ca/As mole ratio, reaction time, and liquid/solid ratio are considered as variables. All variables are controlled at five levels; the zero level of the Ca/As mole ratio is 4.0, that of the reaction time is 3 h, and that of the liquid/solid ratio is 10.

The 20 runs of experiments were conducted from Design Expert Software (Stat Ease, USA). By applying a multiple regression analysis to the experimental data, the empirical relationships between the responses and independent variables were obtained. The accuracy of the model is checked by the correlation coefficient ($R^2$) between the experimental and predicted values of the response variable. Figure 4.2 shows the relationship between the actual and predicted values of ALC of the stabilized HAABS. The actual data are the original measurements of the arsenic concentration in the leachate. On the other hand, the predicted values are generated by the model. A high value of $R^2$ (0.9322) for ALC indicates a high dependence and correlation between the actual and predicted values of the response. The $R^2$ value indicates good agreement between the calculated and observed results within the range of experiment.

**(1) Combined effects of variables**

The 3D response surface plots [17] based on quadratic models are generated to investigate the effects of three parameters on the stabilization of HAABS and are shown in Fig. 4.3a–c. Figure 4.3a shows the 3D response surface and contour plots

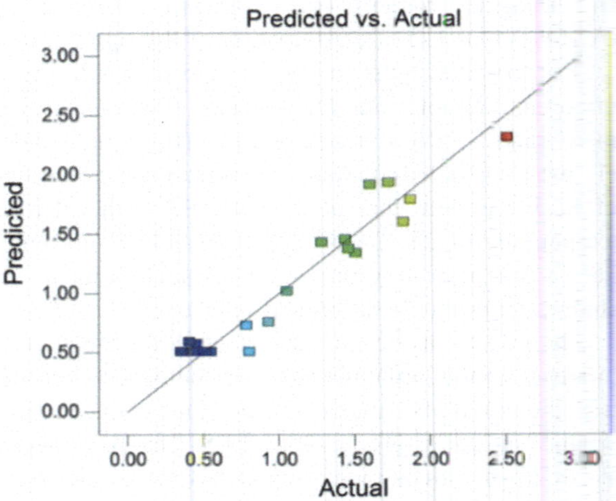


**Fig. 4.2** The actual and predicted response plots of the ALC. Reprinted from Ref. [16] Copyright 2017, with permission from Taylor & Francis

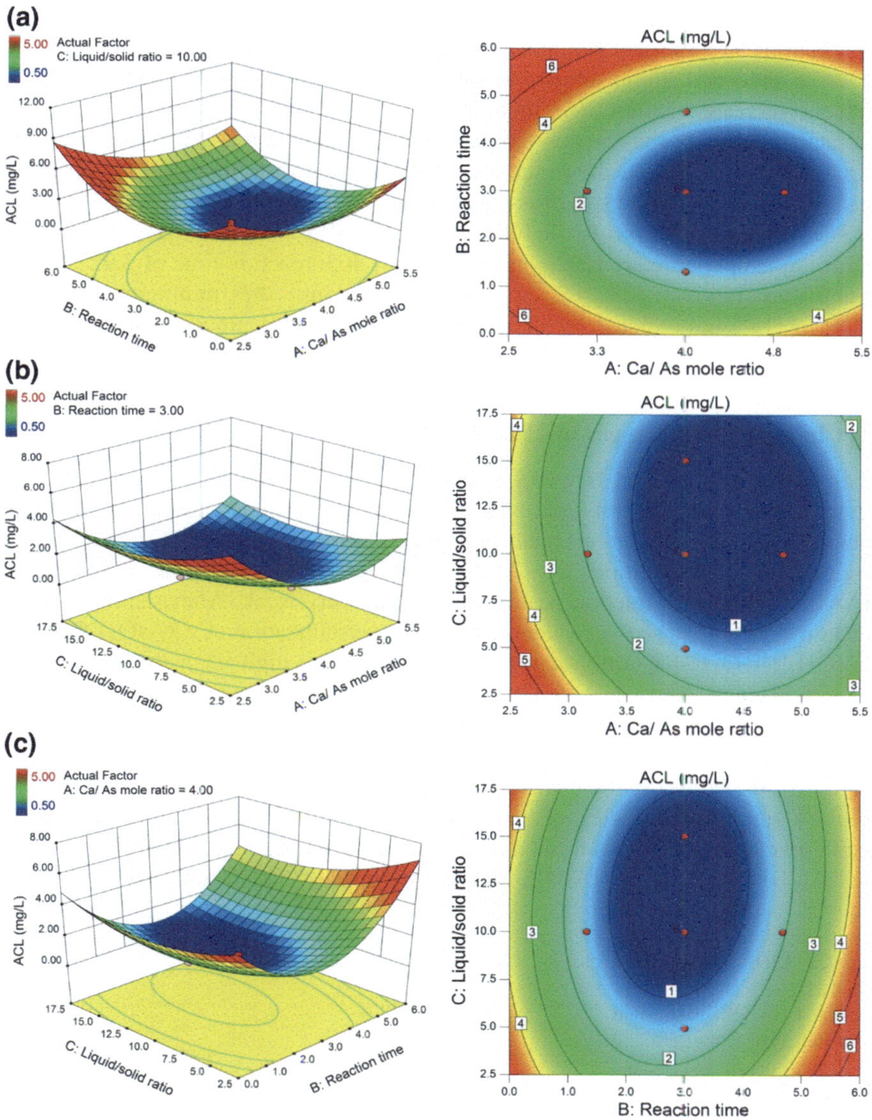


**Fig. 4.3** Response surface and contour plots for the combined effects of (**a**) the Ca/As mole ratio and reaction time, (**b**) the Ca/As mole ratio and liquid/solid ratio, and (**c**) the reaction time and liquid/solid ratio on the ALC of stabilized HAABS by TCLP. Reprinted from Ref. [16] Copyright 2017, with permission from Taylor & Francis

of the Ca/As mole ratio and reaction time on ALC of the stabilized HAABS. Ca/As mole ratio is an important parameter. Figure 4.3b is the combined effects of the Ca/As mole ratio and liquid/solid ratio for a constant reaction time. ALC decreases with the increase of the Ca/As mole ratio or liquid/solid ratio. Figure 4.3c is the interactive effects of the reaction time and liquid/solid ratio on ALC of the HAABS. ALC first decreases and then increases with reaction time increasing, whereas it decreases with the liquid/solid ratio increasing.

The optimal conditions for stabilization of HAABS using $Ca(OH)_2$ at initial pH and room temperature are Ca/As mole ratio of 4.463, reaction time of 2.927 h and liquid/solid ratio of 11.715. At these conditions, the minimum ALC of 0.312 mg $L^{-1}$ is predicted by the model. The optimized process parameters are validated with the experimental result of 0.306 mg $L^{-1}$, which is close to the predicted value.

**(2) Characteristic changes of HAABS**

Main phase in the initial HAABS is $CaCO_3$ (Fig. 4.4), the calcium arsenate precipitate formed because of lime addition is not stable under the attack of atmospheric $CO_2$. The calcium arsenate precipitate decomposes to $CaCO_3$ and releases arsenic easily. In addition, $Ca_3(AsO_4)_2$ and $Na_2SO_4$ also appear in the initial HAABS, but their diffraction peak intensities are very weak. After stabilization by calcium addition based on calcium arsenate crystallization, $Ca_5(AsO_4)_3OH$ and $Ca_4(OH)_2(AsO_4)_2{\cdot}4H_2O$ diffraction peaks appear with high intensity. Amorphous calcium arsenate converts into a crystal structure calcium arsenate which is more stable than the amorphous state calcium arsenate.

The initial particles of the HAABS have spherical and irregular bulk shapes (Fig. 4.5), the largest particle was approximately 10 μm, and the smallest one was below 1 μm. The particles decrease and loosen after stabilization and many needle-like crystal particles formed at the surface of the stabilized HAABS. The $d_{50}$ and $d_{90}$ of the initial HAABS are approximately 18.30 μm and 47.26 μm,

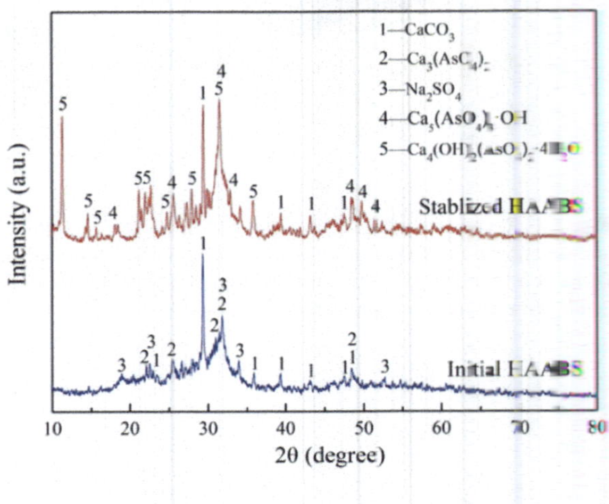


**Fig. 4.4** XRD patterns of the initial HAABS and the stabilized HAABS. Reprinted from Ref. [16] Copyright 2017, with permission from Taylor & Francis

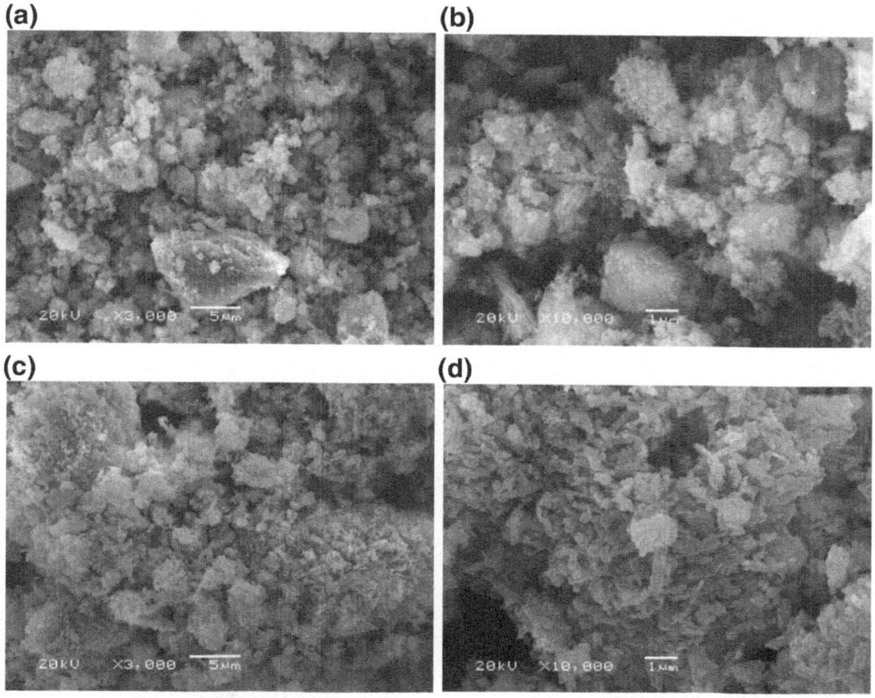


**Fig. 4.5** SEM images of the initial HAABS (**a**, **b**) and the stabilized HAABS (**c**, **d**). Reprinted from Ref. [16] Copyright 2017, with permission from Taylor & Francis

respectively, whereas the $d_{50}$ and $d_{90}$ of the stabilized HAABS are 8.09 μm and 26.14 μm, respectively, which are much smaller than those of the initial HAABS (Fig. 4.6). The stabilized HAABS contains small particles, which agrees with the results shown in Fig. 4.5. Combined with the XRD analysis in Fig. 4.4, the crystal $Ca_4(OH)_2(AsO_4)_2{\cdot}4H_2O$ and $Ca_5(AsO_4)_3OH$ formation results in a smaller particle size distribution of the stabilized HAABS.

### (3) Environmental risk or bioavailability of arsenic

The initial HAABS has a significantly high arsenic content, with an acid soluble fraction of 61.66%, therefore it easily releases to groundwater in a weak acidic environment (Fig. 4.7). The acid soluble fraction of arsenic decreases from 61.66 to 0.72% after the stabilization process. The stabilized HAABS is very stable in acid rain conditions. The number of reducible fractions (F-2) from the European Community Bureau of Reference (BCR) method is roughly equivalent to the content of Fe–Mn phase, and these two fractions (F-1 and F-2) are classified into the direct effect phases. The contents of the reducible form of arsenic are 23.16% in the initial HAABS and 60.67% in the stabilized HAABS. Most acid soluble fractions transformed into reducible fractions after stabilization. The percentage of

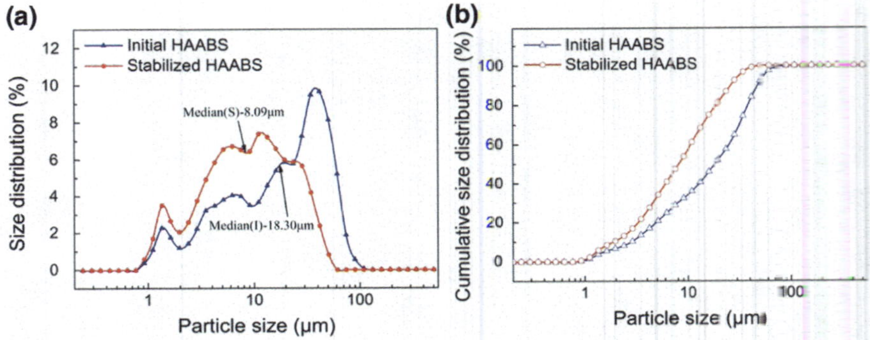


**Fig. 4.6** Particle size distribution of the initial HAABS and the stabilized HAABS: (**a**) size distribution and (**b**) cumulative size distribution. Reprinted from Ref. [16] Copyright 2017, with permission from Taylor & Francis

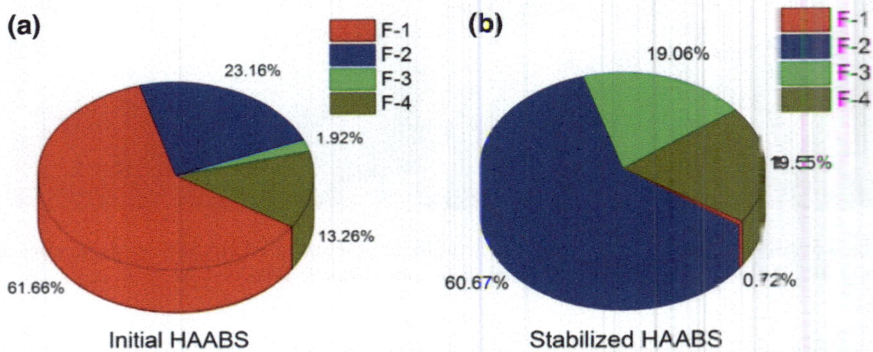


**Fig. 4.7** Chemical fraction of arsenic species in the HAABS by the BCR sequential extraction procedure: F-1 acid soluble fractions, F-2 reducible fractions, F-3 oxidizable fractions, and F-4 residual fraction. Reprinted from Ref. [16] Copyright 2017, with permission from Taylor & Francis

direct effect phases in the initial and the stabilized HAABS are 84.82% and 79.73%, respectively. The stabilized HAABS has high environmental activity in the long term, even though their leaching toxicity is lower than the Toxicity Characteristic Leaching Procedure (TCLP) regulation. The amount of oxidizable fractions (F-3) from the BCR method is roughly equivalent to the sum of organic matter and sulfide combination. F-3 is identified as a potential fraction due to the metals liberated or transformed into F-1 and F-2 under oxidizing conditions. This form in the stabilized HAABS increases the potential environmental risk. The residual fraction (F-4), known as the stable fraction, contains mainly primary and secondary minerals that hold metals within their crystal structure. Only 6.29% is added in this fraction after stabilization. The stabilized HAABS are stable in weak acidic environments by the TCLP, but they still pose an environmental risk.

The morphology transformation of calcium arsenate is the main reason for stabilizing HAABS. The crystals $Ca_5(AsO_4)_3OH$ and $Ca_4(OH)_2(AsO_4)_2{\cdot}4H_2O$ form by calcium treatment. However, although the stabilized HAABS is very stable in a weak acidic environment for short time contact, a significant amount of arsenic in the reducible and oxidizable forms change into an acid soluble form to cause potential environmental pollution for a long time. The further solidification technology should be used after stabilization of arsenic-bearing sludges to reduce the long-term environmental risk further.

### 4.1.2 *Stabilization of Arsenic Sludge with Mechanochemically Modified Zero Valent Iron*

Zero-valent iron (ZVI) for remediation of arsenic pollution has gained considerable interests. It can achieve spontaneous adsorption and coprecipitation of arsenic with iron(II) and iron(III) oxides/hydroxides, which form in situ during ZVI oxidation (corrosion) [18, 19]. However, the use of ZVI may be characterized by specific disadvantages in arsenic sludge management, such as the progressive reduction of the ZVI stabilization effect due to the low activity and poor mixing efficiency in solid state reaction system. Mechanochemical method based on the high energy mechanical force has been identified as a promising option to improve the materials' mixing efficiency, as well as the reaction activity [20]. Mechanochemical modified ZVI was obtained from co-ground commercial iron powder and manganese dioxide through mechanochemical method to improve the reaction activity of ZVI. $MnO_2$ was selected as arsenic remediation promoter, as it exhibits excellent oxidizability and arsenic absorbability [21].

The arsenic sludge sample was collected from an antimony smelter of China, generated after the batch disposal process of arsenic-containing wastewater by air oxidation through ferrous sulfate method. The arsenic containing sludge was acid (pH 4.8), with a moisture content around 85%. The main elements of the sludge are As (33.0%), Fe (23.1%), and Na (6.2%), respectively (Table 4.2). XRD analysis indicates that arsenic in the sludge presents in amorphous ferric arsenate. TCLP results show that its leaching concentration of arsenic is 78.6 mg $L^{-1}$, far beyond the limit threshold (As < 5 mg $L^{-1}$). Commercially available iron powder with the particle size from 150 to 800 μm, is used as ZVI sources: (Fe > 99%, Mn < 0.35%, Si < 0.1%, C < 0.03%, S < 0.02%, P < 0.02%).

**Table 4.2** The elemental composition and leaching concentration of arsenic sludge

| Element | As | Fe | Mn | Sb | K | Na |
|---|---|---|---|---|---|---|
| Composition (wt%) | 33.1 | 23.1 | 0.8 | 0.1 | 0.1 | 6.2 |
| Leaching concentration (mg $L^{-1}$) | 78.6 | 18.2 | – | – | 8.8 | 489.5 |

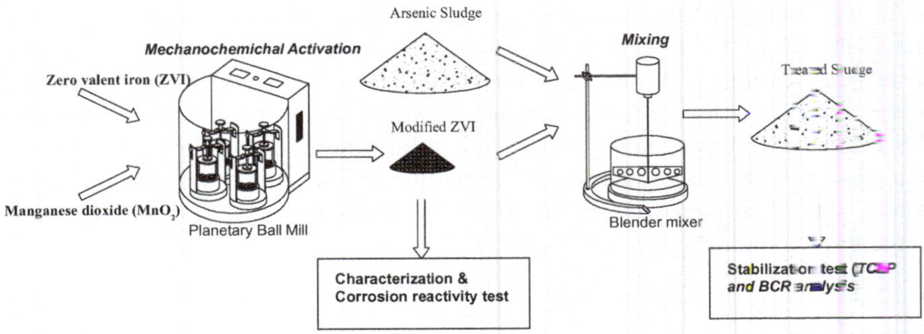


**Fig. 4.8** Experimental schematic. Reprinted from Ref. [21] Copyright 2017, with permission from Elsevier

A schematic diagram of the stabilization process is shown in Fig. 4.8. The modified ZVI was prepared in a planetary ball mill reactor. First, 50.00 g of ZVI and $MnO_2$ mixture, whose initial $MnO_2$: Fe molar ratio varies from 0:1 to 0.6:1 was loaded into a stainless pot (500 mL inner volume). Then, 500.0 g of stainless steel balls between 5 and 10 mm in diameter were placed into the pot. The ball mill reactor was operated at 500 rpm for different time durations (from 0 to 4 h). After grinding, the modified ZVI were collected from the pot and certain amounts of it with dry weight ratio ranging from 0 to 15%, were respectively mixed with raw sludge in a blender mixer for 60 min. Thereafter, the different varieties of the treated sludge were collected and finally dried at 80 °C overnight in a vacuum oven for stabilization test. The corrosion reactivity test for modified ZVI was performed in simulated acid solution and raw sludge TCLP leachate with commercial ZVI was taken as a control sample. The simulated acid solution was prepared by sulfuric acid with a pH value of 4.8, being similar to the sludge pH value. After mixed with these solutions for 12 h under solid to liquid ratio of 1:10, ZVI samples were carefully filtered and dried in vacuum drying oven for further characterization.

### (1) Stabilization performance of the modified ZVI

The arsenic stabilization effect of the modified ZVI relies on the coefficient effects of $MnO_2$ and milling process. The arsenic concentration in TCLP leachate decreases rapidly from 78.60 to 8.45 mg $L^{-1}$ when 10% of ZVI ($MnO_2$: Fe = 0:1) is added, and then decreases dramatically from 8.45 to 0.63 mg $L^{-1}$ with molar ratio of $MnO_2$: Fe increasing from 0:1 to 0.2:1 (Fig. 4.9a, illustrating that the addition of $MnO_2$, as a role of promoter for ZVI corrosion, can enhance stabilization of arsenic efficiently.

Increasing the ball-milling time significantly strengthens the specific energy demanded for grinding [22]. The mechanochemical activation duration is deemed to be a very important factor on the arsenic stabilization of ZVI (Fig. 4.9b). Intensive mechanical stressing, caused by an excessive ball-milling duration, improves the

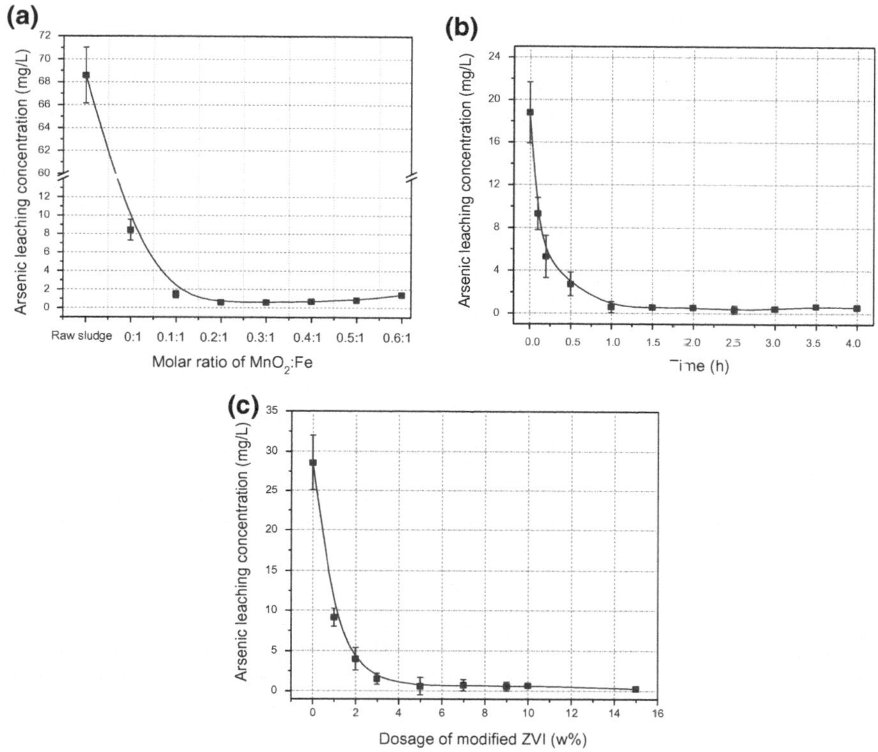


**Fig. 4.9** Arsenic stabilization performance of modified ZVI. (**a**) Impact of $MnO_2$:Fe molar ratio (1 h, dosage of modified ZVI:10%); (**b**) impact of ball-milling duration ($MnO_2$:Fe = 0.2:1, dosage of Fe-$MnO_2$:10%); (**c**) impact of modified ZVI dosage ($MnO_2$:Fe = 0.2:1, 2 h) Reprinted from Ref. [21] Copyright 2017, with permission from Elsevier

ZVI absorption of arsenic. But if energy consumption is taken into consideration, 2 h is the optimal milling duration.

When the dosage of ZVI varies from 0 to 5% (by weight), the arsenic concentration decreases from 78.60 to 0.64 mg $L^{-1}$ (Fig. 4.9c) and then changes slightly in arsenic concentration with further dosage increase from 5 to 15%. The optimum modified ZVI dosage is 5%. As for this highly arsenic-bearing sludge, the dosage application of modified ZVI is much less than other stabilization agents [23, 24], and it can bring about great advantages in low-cost and less-volume disposal of the arsenic sludge.

**(2) ZVI mechanical activation procedure**

The improved arsenic stabilization performance is attributed to the possible mechanochemical reaction between ZVI and $MnO_2$. Generally, an intensive mechanochemical stress can easily induce reactions in a variety of metal oxide and

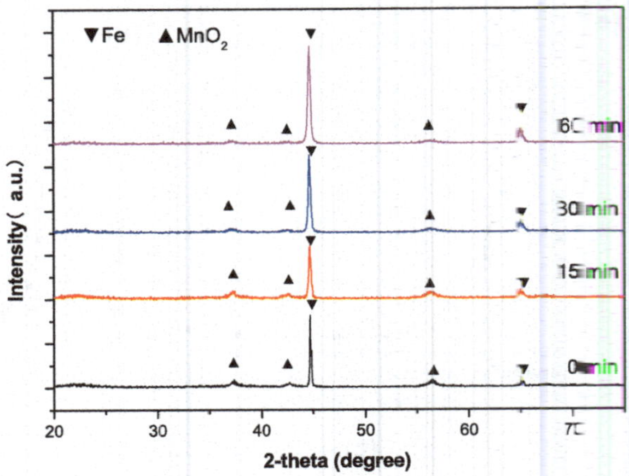


**Fig. 4.10** XRD patterns of modified ZVI during mechanochemical activation process. Reprinted from Ref. [21] Copyright 2017, with permission from Elsevier

reactive metals, and most of these reactions have been classified as mechanically induced self-propagating reactions (MSR) [25–28]. The MSR, which is highly exothermic and can quickly propagate through the entire mixture, may make an adverse impact on the activation of ZVI. It is caused by the high exothermicity which leads to high reactivity small particles agglomerated and melted to form bulk particles with low reactivity [29]. However, MSR were not observed in this procedure. The peaks corresponding to Fe and $MnO_2$ are predominant and no other phases are detected in different ball milling duration (Fig. 4.10). The result is attributed to their thermodynamic property and the following equation can be used as a necessary criterion for MSR.

$$-\Delta H^0_{f298}\Big/C_{298} = \Delta H/C > 2000\,\text{K} \tag{4.1}$$

$-\Delta H^0_{f298}\Big/C_{298}$, written simply as ΔH/C, is the enthalpy of reaction (ΔH) divided by the room temperature heat capacity (C) for reaction. As for the MSR, ΔH/C is an effective thermodynamic criterion over Gibbs free energy changes ($\Delta G_{298}$), and ΔH/C > 2000 K can be used as a necessary criterion.

Table 4.3 shows the thermodynamic calculations results of Fe and $MnO_2$ reactions. ΔH/C of all reactions are lower than the critical value of 2000 K, demonstrating that the MSR of Fe and $MnO_2$ cannot occur in theory.

However, the redox reactions of Fe and $MnO_2$ may be produced in certain degree due to the negative Gibbs free energy changes ($\Delta G_{298}$) as shown in Table 4.3. Thus, the surface compositions of the selected milled samples were determined by XPS, and Gaussian-Lorentzian resolving was performed to analyze the component of the layer [30]. Fe oxide peak from the 15 min milled sample is barely visible (Fig. 4.11a), however, after 60 min milling, significantly greater proportion of iron oxide was observed in the surface region (Fig. 4.11b). It suggests

**Table 4.3** Thermodynamic calculations results of Fe and $MnO_2$ reactions

| No. | Reactions | $\Delta G_{298}$ (kJ $mol^{-1}$) | $\Delta H/C$ (K) |
|---|---|---|---|
| 1 | $2Fe + 3MnO_2 = Fe_2O_3 + 3MnO$ | −102.32 | 1746.8 |
| 2 | $2Fe + 6MnO_2 = Fe_2O_3 + 3Mn_2O_3$ | −138.88 | 1418.8 |
| 3 | $2Fe + 4.5MnO_2 = Fe_2O_3 + 1.5Mn_3O_4$ | −134.67 | 1765.3 |
| 4 | $Fe + MnO_2 = FeO + MnO$ | −33.79 | 1386.9 |
| 5 | $Fe + 2MnO_2 = FeO + Mn_2O_3$ | −45.99 | 1227.8 |
| 6 | $2Fe + 3MnO_2 = 2FeO + Mn_3O_4$ | −89.16 | 1484.5 |
| 7 | $3Fe + 4MnO_2 = Fe_3O_4 + 4MnO$ | −142.23 | 1828.1 |
| 8 | $3Fe + 8MnO_2 = Fe_3O_4 + 4Mn_2O_3$ | −190.98 | 1464.4 |
| 9 | $3Fe + 6MnO_2 = Fe_3O_4 + 2Mn_3O_4$ | −185.36 | 1826.2 |

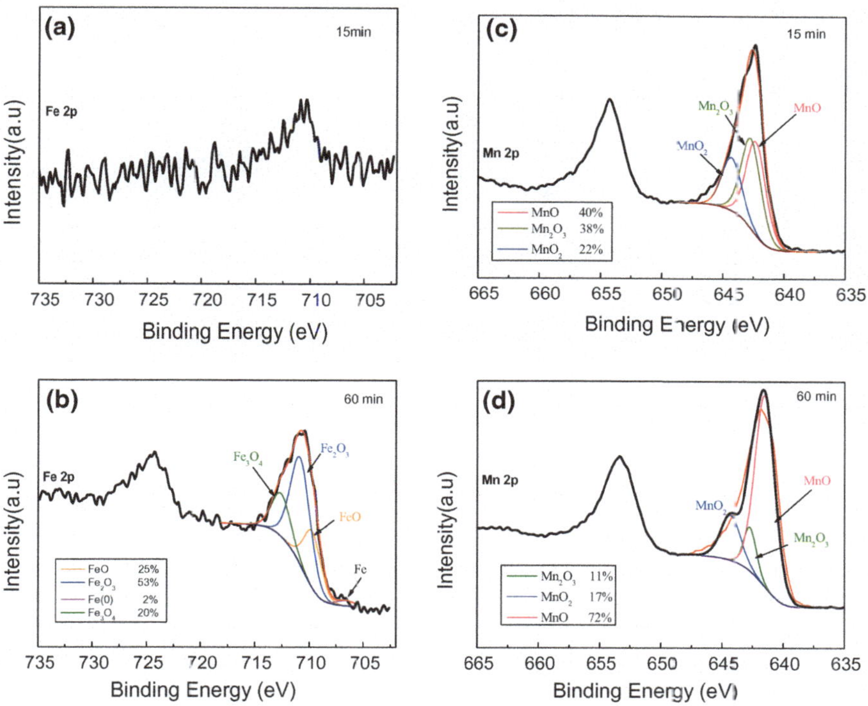


**Fig. 4.11** XPS analysis of modified ZVI. Fe 2p spectra of 15 min (**a**) and 60 min (**b**) milling; Mn 2p spectra of 15 min (**c**) and 60 min (**d**) milling. Reprinted from Ref. [21] Copyright 2017, with permission from Elsevier

that the iron components are composed by 2% of Fe, 20% of $Fe_3O_4$, 25% of FeO and 53% of $Fe_2O_3$. Mn 2p spectrum can be deconvoluted into three constituents corresponding bonds such as $MnO_2$ at 644.3 eV, $Mn_2O_3$ at 642.7 eV and MnO at 642.3 eV, respectively. After 60 min milling, the content of $MnO_2$ (72% in content)

**Fig. 4.12** SEM images of starting materials (**a**, **b**); and modified ZVI (**c**, **d**). Reprinted from Ref. [21] Copyright 2017, with permission from Elsevier

predominates the Mn speciation on the surface of manganese dioxide, followed by $MnO_2$ (17%) and $Mn_2O_3$ (11%) (Fig. 4.11c, d). However, XPS spectrum only confirmed the redox reactions occurring on the modified ZVI surface, the inner part of which is still composed by Fe(0), since the redox reactions cannot propagate to the entire ZVI or $MnO_2$ due to the low $\Delta H/C$ value. Therefore, the above analysis reveals that the modified ZVI consists of an elemental iron core surrounded by a shell of Fe–Mn binary oxides, which is formed during the mechanochemical reaction.

The morphology and structure of modified ZVI of various milling time were examined by SEM. The initial materials are bulk Fe particles with more than 100 μm in size (Fig. 4.12a) and $MnO_2$ particles with 0.2–1.0 μm in diameter, gather on the smooth surface of Fe particle (Fig. 4.12b). After grinding for [illegible] h, some small iron plates and chips can be seen in Fig. 4.12c, since they are formed from bulk iron particle with the repeated fracture and laminating during ball-powder collisions, with an accompanying increase in the surface roughness. In closer

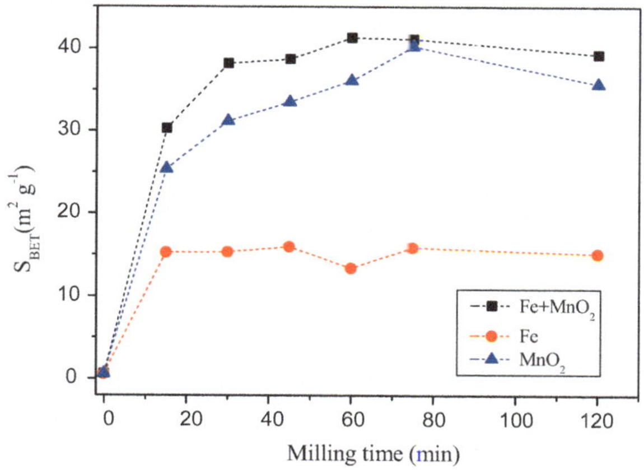


**Fig. 4.13** Variation in specific surface area of Fe, $MnO_2$ and Fe & $MnO_2$ with milling time. Reprinted from Ref. [21] Copyright 2017, with permission from Elsevier

inspection (Fig. 4.12d), an outer layer on the iron particles can be observed and is actually made up from a porous coating, which is composed by $Fe_2O_3$ as suggested by XPS in Fig. 4.11b. At the same time, dispersive fine $MnO_2$, whose particle size has been reduced from 0.1 to 0.5 μm, is embed in the $Fe_2O_3$ coating. It can be deduced that the combination of $MnO_2$ and ZVI is relatively tight due to the rough and porous superficial structure, implying that the $MnO_2$ can be attached closely to ZVI and plays a role of a steadily oxidizer for ZVI corrosion after mixing with sludge.

The layer of binary oxides leads to an increasing in specific surface areas ($S_{BET}$) (Fig. 4.13). The $S_{BET}$ of ZVI, $MnO_2$, and ZVI and $MnO_2$ mixture (ZVI + $MnO_2$, 0.2:1 of $MnO_2$: Fe molar ratio), are respectively measured after different milling times. The initial materials (ZVI, $MnO_2$ and ZVI + $MnO_2$) have similar $S_{BET}$, from 0.58 $m^2\ g^{-1}$, 0.51 $m^2\ g^{-1}$ to 0.60 $m^2\ g^{-1}$, respectively. After 120 min of milling, $S_{BET}$ of these samples increases to 15.14 $m^2\ g^{-1}$, 35.78 $m^2\ g^{-1}$ and 39.33 $m^2\ g^{-1}$, respectively. It should be noticed that, the $S_{BET}$ of milled ZVI + $MnO_2$, namely modified ZVI, is much higher than milled ZVI or milled $MnO_2$. The phenomena support the previous statement that the shell of oxides layer, which presents distinguished morphological and structural properties, is formed on the modified ZVI surface.

**(3) Modified ZVI Corrosion and arsenic stabilization**

The modified ZVI exhibits high specific area and tight combination with oxidizer ($MnO_2$), both of which are generally regarded as desirable properties for its corrosion. We tested the corrosion reactivity of the modified ZVI in the simulated acid leachate (pH = 4.8, As = 0 mg $L^{-1}$) and the raw sludge TCLP leachate (pH = 4.8, As = 78.6 mg $L^{-1}$), respectively. Figure 4.14 shows the XRD pattern of the corrosion product of the ordinary ZVI and the modified ZVI, as well as the modified ZVI after contacting with sludge leachate. Maghemite ($\gamma$-$Fe_2O_3$), a common crystalline phase often found as a result of the iron corrosion process [31], is the

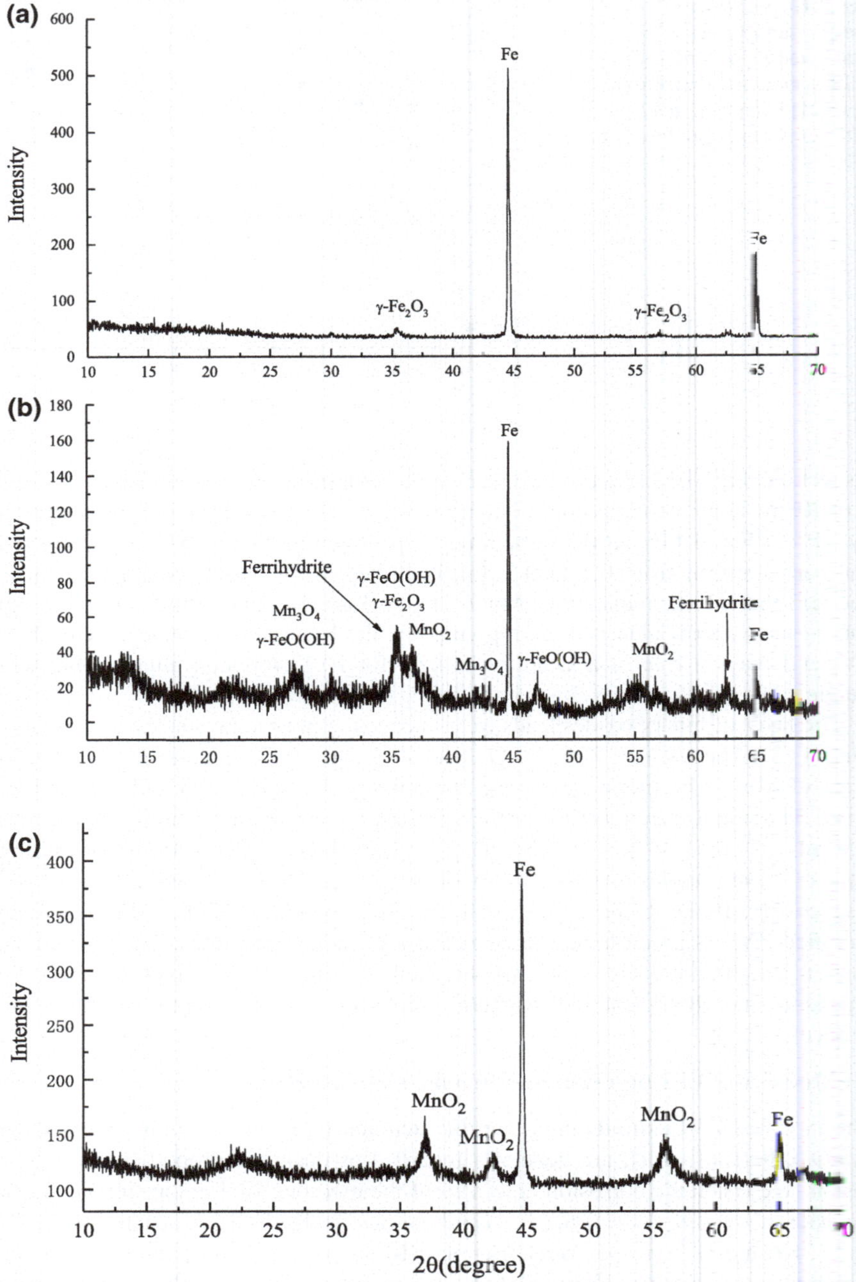


**Fig. 4.14** XRD patterns of ordinary ZVI corrosion product (**a**), modified ZVI corrosion product (**b**) and modified ZVI after contacting with sludge leachate (**c**). Reprinted from Ref. [21] Copyright 2017, with permission from Elsevier

predominant corrosion specie of ordinary ZVI corrosion products (Fig. 4.14a). In terms of modified ZVI (Fig. 4.14b), it gives lower Fe peak intensities in comparison with Fig. 4.14a, illustrating that more intensive corrosion occurs in modified ZVI. Moreover, a mass of widened diffraction peaks (Fig. 4.14b), corresponding to the extremely small size and amorphous structure of particles imply that the corrosion products possibly are composed by various amorphous iron (hydr)oxides, such as ferrihydrite, whose relative peaks are in agreement with natural ferrihydrite [32]. After contacting with arsenic leachate (Fig. 4.14c), the peaks corresponding to Fe and $MnO_2$ are predominant and the widened diffraction peaks disappear, demonstrating that the amorphous corrosion products present high reactivity to arsenic ions.

SEM support that mechanochemical activation can enhance corrosion extent, as well as the arsenic stabilization performance. After corrosion, the ordinary ZVI consists of spherical particles in the size range about 100 μm, covered with a trivial layer of corrosion products (Fig. 4.15a). The closer inspection (Fig. 4.15b) reveals that the corrosion products are composed by a huge amount of chopstick-shaped crystal, which has been certified as maghemite ($\gamma$-$Fe_2O_3$) by XRD analysis. The morphology of modified ZVI corrosion products are distinguishable from that of ordinary ZVI and the average size of particles decrease to 50 μm accompanied with the significant formation of flocculent deposits (Fig. 4.15c). Actually, as can be seen in high magnification of the coated layer (Fig. 4.15d), the deposits are composed by vast nanosized spherical aggregate, which is similar to the ferrihydrite microstructure in previous research [33]. In the present of arsenic, the surface of ZVI becomes smooth and compact (Fig. 4.15e). In closer inspection (Fig. 4.15f), needle-like aggregative ferric arsenate and sticks like lepidocrocite and/or goethite (200–300 nm length) are formed on the surface of the particles, accompanied with $MnO_2$ and a small quantity of unreacted ferrihydrite.

It is evident that the modified ZVI presents an enhanced corrosion extent, as well as a distinguishing corrosion product. The improved corrosion extent is caused by the increase of $S_{BET}$ and the porous Fe–Mn binary oxide coating, which allows the corrosion/oxidation reaction to occur efficiently in the inner part of ZVI. In addition, the amorphous corrosion products, formed on the surface of ZVI, possess a much higher arsenic absorbability than crystalline iron (hydr)oxides [34–36]. Generally speaking, the amorphous corrosion products are unstable iron-based intermediates, therefore, and it can be easily converted to hematite, maghemite or goethite in common corrosion process [37]. Amorphous iron (hydr)oxides, such as ferrihydrite, predominate the ultimate corrosion products instead of the crystallize ones. It is attributed to the present of Mn(II), which can form the Mn-ferrihydrite coprecipitate during the corrosion process, Mn-ferrihydrite has a different structure with iron (hydr)oxides crystal, thus it can prevent the crystallization of iron (hydr)oxides [38].

Three-staged BCR sequential extraction is adopt to assess the environmental activity and potential ecological risks. Generally, the arsenic in acid soluble and reducible fraction is classified as direct effect phases for environmental availability and ecological risk because they are presented as a loosely bound phase or thermodynamically unstable phase respectively, which are liable to release into the

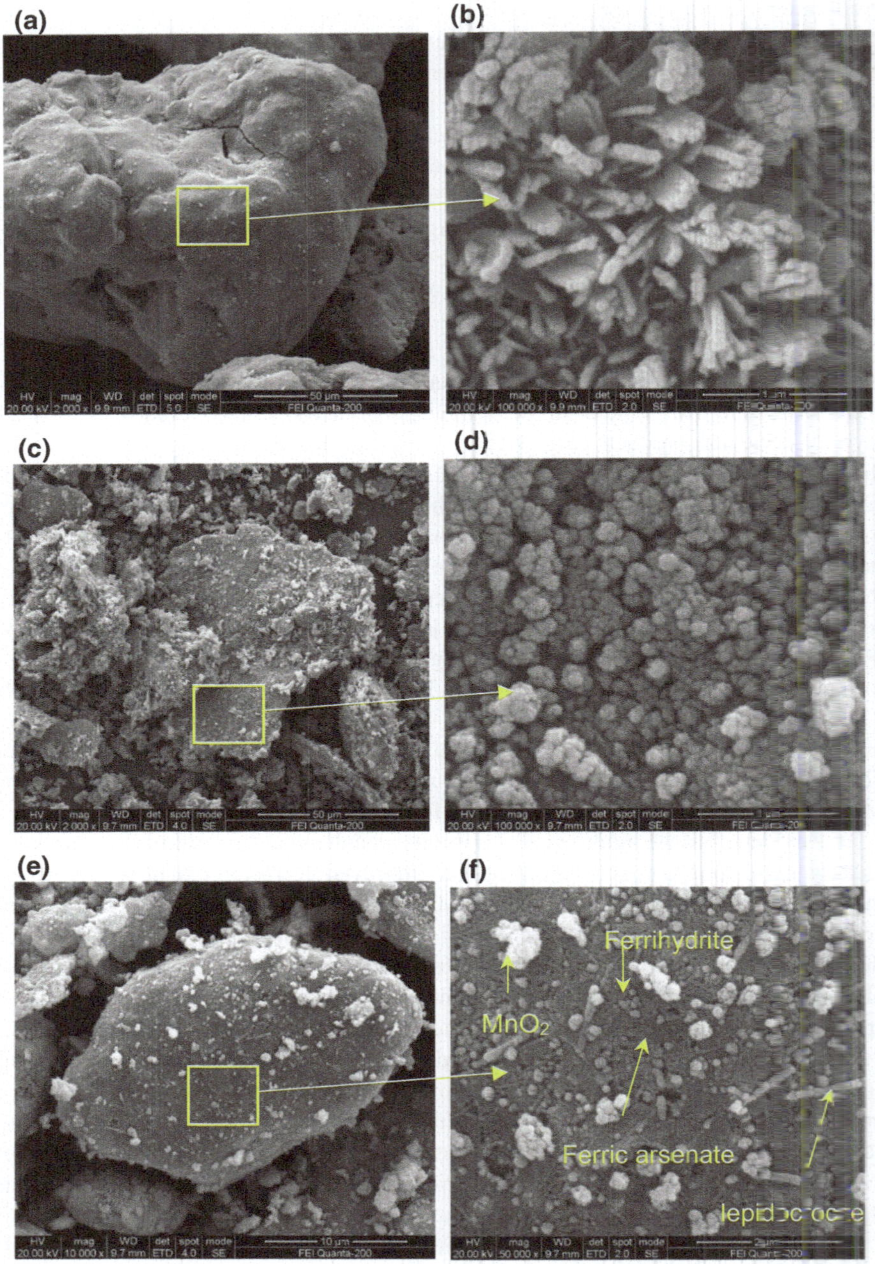


**Fig. 4.15** SEM images of ordinary ZVI corrosion product (**a**, **b**); modified ZVI corrosion product (**c**, **d**) and modified ZVI after contacting with sludge leachate (**e**, **f**). Reprinted from Ref. [21] Copyright 2017, with permission from Elsevier

environment. Meanwhile, arsenic associated with the oxidizable fraction is identified as a potential effect fraction, because it can be liberated or transformed into acid soluble and reducible fraction under oxidizing conditions. Only the residual fraction is believed as a stable fraction, because it contains mainly primary and secondary minerals, which may retain metal elements within their crystal structure. After the stabilization treatment, the acid soluble and reducible soluble fractions dramatically decrease from 0.4% and 1.0% to 0.06% and 0.31%, respectively, indicating that direct toxicity effect fractions are reduced [39, 40]. In terms of residual fraction, it may rise or be caused by the presence of $MnO_2$ on the contrary and has a slight descent from 1.20 to 1.11%. This phenomenon is attributed to the effective formation ferric arsenate from the unstable phase, which can be proved by the increase of residual fraction from 97.40 to 98.52%. Thus, an important conclusion is that the chemical species of arsenic in sludge are significantly transformed to residual fraction by the stabilization of mechanochemical activated ZVI and $MnO_2$, resulted in a restrained environmental availability.

## 4.2 Arsenic Solidification Technologies

### *4.2.1 Solidification/Stabilization of Arsenic and Heavy Metals with Minimal Cement Clinker*

During the pyro-metallurgical processing of copper, most of the arsenic containing in copper concentrates flows into smelting slag or volatilizes into the exhaust gas [41]. It was reported that production of one ton of refined copper generates approximately 2.2–3 tons of copper slag [42]. In generally, such copper slag is subjected to recovery copper through flotation, which results in the generation of a large amount of flotation wastes, so-called flotation waste of copper slag (FWCS). As for the arsenic-containing exhaust gas, it is commonly washed with the diluted acid, accompanied by the generation of large quantities of acid wastewater. Generally, the acid wastewater is treated by lime-ferrate precipitation before discharged, with the generation of arsenic-containing gypsum sludge (GS) [43]. In addition, neutralization sludge (NS) also produces abundantly due to the treatment of comprehensive wastewater from copper smelter with lime precipitation. Due to the stringent environmental regulations and declining storage space, the treatment of FWCS, GS and NS are currently of great urgency for copper smelter. Particularly, co-treatment technologies of these wastes need to be developed.

Recently, the management of arsenic and heavy metals containing waste has become a major public concern. Numerous treatment methods, including vitrification, extraction, and stabilization/solidification (S/S), have been developed to prevent arsenic pollution [16, 21, 44–50]. Up to now, S/S with binders, such as cement, pozzolanic matter and geopolymer, is the most widely used method for the disposal of arsenic and other heavy metals-containing waste [51–56]. However, it will result in a large compatibilization ratio and high cost when FWCS, GS and NS are treated with cement separately. Our previous studies showed that co-treatment of waste containing toxic components can reduce the compatibilization ratio and cost [46, 57].

It had been proved that the FWCS can be used in the production of cement clinker since more than 59% of $Fe_2O_3$ (mainly in the form of fayalite ($Fe_2SiO_4$) and magnetite ($Fe_3O_4$)), $SiO_2$ and small amounts of $Al_2O_3$ are contained in FWCS [58]. When it is applied in clinker production, it can act as both iron adjusting and mineralizing component. Calcium sulfate and its hydrate in NS and GS can be used as retarder and activator for cement production. The FWCS and NS can replace one part of cement during the S/S of arsenic and heavy metals in GS, which can greatly reduce the cost of S/S. Therefore, we proposed co-treatment of FWCS, NS and GS for solidification/stabilization of arsenic and heavy metals with minimal cement clinker. The FWCS, NS and cement clinker were used to synthesize the binder. The preparation parameters (FWCS dosage, NS dosage, mill time for ball mill mixing, and water-to-binder ratio) of the synthesized binder were optimized. Then, the binder was used for the S/S of GS. The co-disposal effectiveness was evaluated by testing the unconfined compressive strength (UCS) and heavy metal leachability.

The chemical composition and mineral composition of the FWCS, NS, GS and cement clinker (Table 4.4 and Fig. 4.16) showed that the FWCS contained high amount of Fe, O and Si elements, with a total content of more than 88%. Fayalite ($Fe_2SiO_4$), magnetite ($Fe_3O_4$) and quartz ($SiO_2$) peaks were the major mineral phase in FWCS (Fig. 4.16). The NS contained high amount of O, Ca and S elements, with a total content of 98%. The crystalline phase in NS is calcium sulfate dihydrate ($CaSO_4 \cdot 2H_2O$). The GS contained high amount of Ca, O, S and As elements. The major heavy metals was As (11.96%), Zn (1.23%) and Cu (0.55%). Therefore, GS was regarded as the main research object which should be solidified or stabilized. The calcium sulfate dihydrate ($CaSO_4 \cdot 2H_2O$), calcite ($CaCO_3$) and pharmacolite ($CaHAsO_4 \cdot 2H_2O$) are the major mineral phase of GS.

For FWCS, the leaching concentration of Pb reached 9.15 mg $L^{-1}$, above the limit 5 mg $L^{-1}$. The concentrations of As and Cu were 4.11 mg $L^{-1}$ and 80.16 mg $L^{-1}$, respectively, being closed to the limits. For NS, the leaching concentrations of As, Pb, Cd and Zn were far below the limits. For GS, the leaching concentration of As was 356.3 mg $L^{-1}$, which far exceeded the limit allowed. The leaching concentration of other heavy metals is extremely low.

**Table 4.4** Main elemental compositions of the materials (wt%)

| Elemental | Fe | O | Si | Al | Ca | S | Zn | As | Pb | Cr | Cu | Cd |
|---|---|---|---|---|---|---|---|---|---|---|---|---|
| FWCS | 44.52 | 23.89 | 20.05 | 2.50 | 2.90 | 0.52 | 1.95 | 0.33 | 0.34 | 0.04 | 0.49 | ND |
| GS | 1.00 | 33.91 | 1.24 | 0.65 | 33.45 | 11.63 | 1.23 | 11.96 | 0.21 | ND | 0.55 | [illegible] |
| NS | 0.17 | 40.57 | 0.75 | 0.26 | 35.65 | 22.21 | 0.02 | 0.02 | 0.005 | 0.003 | ND | ND |
| Cement clinker | 2.66 | 28.02 | 11.30 | 2.59 | 51.27 | 0.30 | ND | ND | ND | ND | ND | ND |

Notice: *ND* not detected

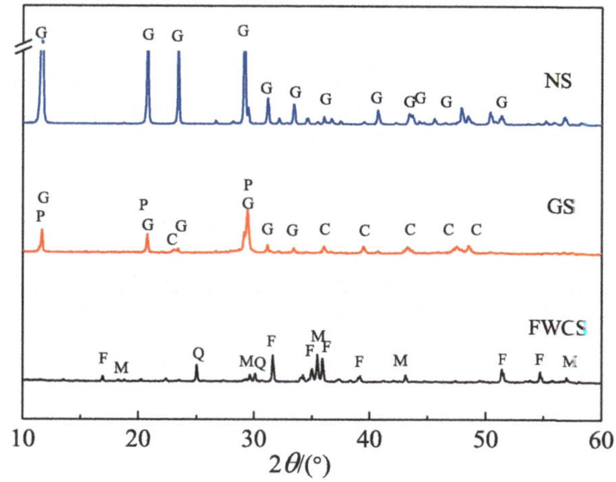


**Fig. 4.16** XRD pattern of raw materials. Label: symbol in figure-mineral (PDF number from ICDD). M: Magnetite-$Fe_3O_4$ (89-2355), F: Fayalite-$Fe_2SiO_4$ (70-1861), Q: Quartz-$SiO_2$ (80-2147), G: Gypsum-$CaSO_4{\cdot}2H_2O$ (33-0311), C: Calcite-$CaCO_3$ (85-0849), P Pharmacolite-$CaHAsO_4{\cdot}2H_2O$ (74-0602). Reprinted from Ref. [58], Copyright 2017, Springer-Verlag GmbH Germany, part of Springer Nature

### (1) Preparation of the binder with FWCS, NS and cement clinker

The effects of FWCS dosage, NS dosage, mill time, water-to-binder ratio on unconfined compressive strength (UCS) of the binder was investigated (Fig. 4.17). The UCS decreased with the increase in the dosage of FWCS. When the dosage of FWCS was more than 40%, the UCS decreased significantly. Since FWCS have a low pozzolanic activity, excessive addition will result in a decrease in the UCS of matrices. The UCS of the binder can reach 19.82 MPa after curing 28 days when the FWCS dosage was 40%. Comprehensive consideration of UCS and the FWCS dosage, 40% is selected as the optimum dosage of FWCS.

The UCS varied slightly with the increase in the dosage of NS, indicating that the dosage of NS has little effect on UCS. When the NS dosage changed from 2 to 8%, the UCS matrix increased slightly. Thereafter, the UCS declined when NS dosage varied from 8 to 12%. The UCS of the matrix is the best with a dosage of NS of 8%. However, when the dosage of NS is 10%, the UCS of the matrix is close to that of the dosage of 8%. Comprehensive consideration of the UCS and NS dosage, 10% is selected as the optimum dosage of NS.

The UCS increases and then decreases with the mill time increasing from 0 to 2.5 h. Mechanical milling can activate potential activity of volcanic materials [59]. This is the reason why UCS increased after milling. However, excessive mill time will result in an exceedingly fine material size, which will be detrimental to the increase of UCS. Although the UCS of the binder prepared with mill time of 1.5 h was a bit lower than that of the binder prepared with mill time of 2 h, from the view of energy saving, 1.5 h was determined to be the optimal mill time.

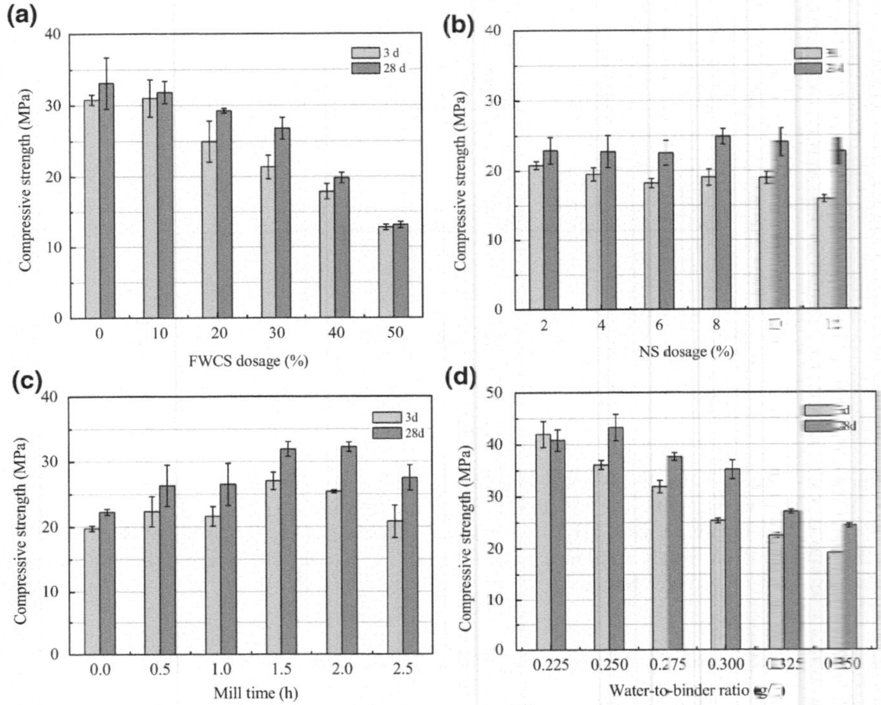


**Fig. 4.17** Effects of FWCS dosage, NS dosage, mill time, and water-to-binder ratio on the UCS of the binder. Reprinted from Ref. [58] Copyright 2017, Springer-Verlag GmbH Germany, part of Springer Nature

The UCS was inversely proportional to the water-to-binder ratio, which was similar to the Abrams' law [60]. The maximum UCS of the binder reached 43.24 MPa after curing 28 days when water-to-binder ratio was 0.25.

The optimal parameters for the synthesis of binder were the FWCS dosage of 40%, NS dosage of 10%, cement clinker dosage of 50%, mill time of 1.5 h, and water-to-binder ratio of 0.25. On such an optimal condition, the UCS of the binder reached 43.24 MPa after hydration of 28 days.

**(2) Solidification/stabilization of GS with the binder**

The UCS decreased with the increase of GS dosage (Fig. 4.18). When the ratio of binder-to-GS reached 5:5, the UCS of matrix was 11.06 MPa after hydration of 28 days, which reached the required UCS level of MU10 brick in China.

The concentrations of main heavy metals in leachate and pH value of leachate for matrices A–E after hydration of 28 days were listed in Table 4.5. Compared with the raw FWCS, NS and GS, Cu, Zn, Cd and Ni were all not detected in the leachate for all matrices. Only Pb and As were detected in leachate. The leaching

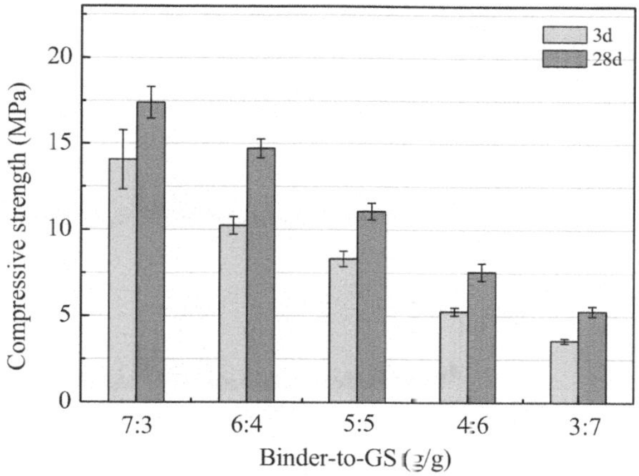


**Fig. 4.18** UCS of the matrices with various ratios of binder-to-GS. Reprinted from Ref. [58] Copyright 2017, Springer-Verlag GmbH Germany, part of Springer Nature

**Table 4.5** The results of CSLT for matrices cured 28 days (mg $L^{-1}$)

| Element | pH | As | Pb | Cu | Zn | Cd | Ni |
|---|---|---|---|---|---|---|---|
| CSLT limits | | $\leq 5$ | $\leq 5$ | $\leq 100$ | $\leq 100$ | $\leq 1$ | $\leq 5$ |
| Matrix A (7:3) | 12.36 | 0.78 | 0.66 | ND | ND | ND | ND |
| Matrix B (6:4) | 12.34 | 1.03 | 1.05 | ND | ND | ND | ND |
| Matrix C (5:5) | 12.35 | 1.36 | 1.01 | ND | ND | ND | ND |
| Matrix D (4:6) | 12.13 | 2.53 | 0.77 | ND | ND | ND | ND |
| Matrix E (3:7) | 11.43 | 26.84 | 0.71 | ND | ND | ND | ND |

concentration of Pb for matrices A–E was around 1 mg $L^{-1}$ and was below the China Standard Leaching Test (CSLT) limit allowed. The arsenic concentration in leachate for matrix A–D was also below the limit. However, arsenic concentration for the matrix E was an exception, which was more than 5 times of the limit. Therefore, it can be concluded that the binder has a strong S/S capacity for As and other heavy metals.

The pH value of leachate for the matrices A–E varied from 11.4 to 12.4. The pH value has a significant influence on the arsenic concentration in $Ca^{2-}$-rich aqueous medium. The arsenic concentration reduces with the pH value increasing in leachate. The pH value also has a remarkable effect on the speciation of heavy metals in aqueous medium. Seco et al. [61] had pointed out that Pb may deposit as a metal lead hydroxide, lead sulfate or re-dissolved as plumbite with the variation of pH value. Therefore, the change in the concentration of heavy metals in leachates is related not only to the composition of the matrices but also to the pH value of leachate.

The matrix prepared with binder-to-GS ratio of 5:5 is characterized with XRD (Fig. 4.19). The major minerals phases of the binder were fayalite, magnetite, quartz, ettringite, portlandite, calcium silicate hydrate and calcium silicate. Fayalite

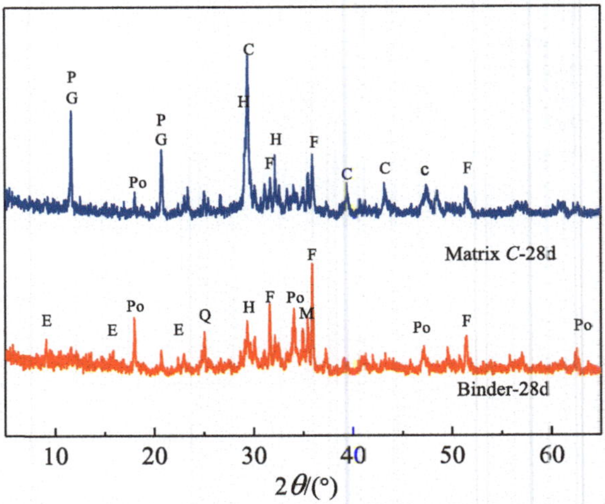


**Fig. 4.19** XRD patterns of binder and matrix *C* hydration of 28 days. Label: symbol in figure-mineral (PDF number from ICDD). M: Magnetite-$Fe_3O_4$ (89-2355), F: Fayalite-$Fe_2SiO_4$ (70-1861), Q: Quartz-$SiO_2$ (80-2147), G: Gypsum-$CaSO_4{\cdot}2H_2O$ (33-0311), C: Calcite-$CaCO_3$ (85-0849), P: Pharmacolite-$CaHAsO_4{\cdot}2H_2O$ (74-0602). H: Calcium silicate hydrate-$Ca_5(SiO_4)_2(OH)_2$ (84-0148). Po: Portlandite-$Ca(OH)_2$ (72-0156), E: Ettringite-$3CaO{\cdot}Al_2O_3{\cdot}3CaSO_4{\cdot}32H_2O$ (41-1451), Cs: Calcium silicate-$Ca_3SiO_5$ (49-0442). Reprinted from Ref. [58] Copyright 2017, Springer-Verlag GmbH Germany, part of Springer Nature

and magnetite were the original minerals phases of the FWCS. The ettringite, portlandite, calcium silicate hydrate was generated during the hydration of binder via the reactions shown in Eqs. 4.2–4.6. The strength of the synthesized binder mainly depends on the calcium silicate hydrate, and the strength increased with the content of calcium silicate hydrate. In addition, ettringite has a gelling property and can improve the strength of the hydrated binder. The hydration products calcium silicate hydrate and ettringite in the binder increased day by day until the hydration was completed, therefore, the strength of the binder increased continuously with the curing time.

$$3CaO{\cdot}SiO_2 + H_2O \rightarrow CaO{\cdot}SiO_2{\cdot}H_2O + Ca(OH)_2 \tag{4.2}$$

$$2CaO{\cdot}SiO_2 + H_2O \rightarrow CaO{\cdot}SiO_2{\cdot}H_2O + Ca(OH)_2 \tag{4.3}$$

$$Ca(OH)_2 + SiO_2(\text{amorphous}) \rightarrow CaO{\cdot}SiO_2{\cdot}H_2O \tag{4.4}$$

$$3CaO{\cdot}Al_2O_3 + H_2O \rightarrow 3CaO{\cdot}Al_2O_3{\cdot}6H_2O \tag{4.5}$$

$$3CaO{\cdot}Al_2O_3 + CaSO_4{\cdot}2H_2O + H_2O \rightarrow 3CaO{\cdot}Al_2O_3{\cdot}3CaSO_4{\cdot}32H_2O\ (\text{ettringite}) \tag{4.6}$$

The portlandite, gypsum, calcite, calcium silicate hydrate, fayalite and pharmacolite were the major mineral composition of Matrix *C*. Gypsum, calcite, pharmacolite derived from the GS. It can be seen that the phase of arsenic did not change during the S/S process. The physical encapsulation played an important role in the solidification/stabilization of arsenic compounds. Based on the previous studies [62–64], calcium silicate hydrate has a strong ability to adsorb As(III) and As(V). Moreover, the $SO_4^{2-}$ in ettringite can be substituted in some degree by $AsO_4^{3-}$ [65]. In addition, $AsO_4^{3-}$ can be converted into $Ca_3(AsO_4)\cdot nH_2O$ and $CaHAsO_4\cdot nH_2O$ through the reactions given by Eqs. 4.7–4.8 in pH value ranges of 11–14 [66]. It is supposed that the reduction of As concentration in CLST leachates might be attributed to these reasons.

$$Ca(OH)_2 + AsO_4^{3-} + nH_2O \rightarrow Ca_3(AsO_4)_2\cdot nH_2O \quad (4.7)$$

$$Ca(OH)_2 + HAsO_4^{2-} + nH_2O \rightarrow CaHAsO_4\cdot nH_2O \quad (4.8)$$

### 4.2.2 Co-treatment of Gypsum Sludge and Pb/Zn Smelting Slag

The nonferrous smelting process generates a large amount of industrial hazardous waste, with an annual production of 32 million tons, and it is difficult to use the waste because of its complex composition. Stabilization technology has been used to transform potentially hazardous solid waste into solids that are less hazardous than raw waste or are nonhazardous [67]. Due to its technological, economic, and environmental benefits, the use of industrial wastes as a substitute for cement in solidifying materials is common practice [68].

Wastewater treatment sludge from the primary lead-zinc smelter is characterized as hazardous waste which requires treatment prior to disposal due to its significant arsenic and heavy metals contents. This section presents a method for the stabilization of arsenic sludge with slag-based binder composed of smelting slag, cement clinker and limestone. According to the Unconfined Compressive Strength (UCS) and optimal experiments, the binder was determined to be prepared with mass ratio of m (smelting slag): m (cement clinker): m (gypsum sludge): m (limestone) of 70:13:12:5. When the binder was mixed with arsenic sludge at mass ratio of 1:1 and at 25 °C for 28 d, the UCS reached 9.30 MPa. Moreover, TCLP test shows that arsenic concentration in leachate was always less than 5 $mg\ L^{-1}$, which is a safe level, and does not contribute to recontamination of the environment.

The arsenic sludge, gypsum sludge and smelting slag used were collected from a Pb/Zn smelter in south China, and the composition of them was listed in Table 4.6. The observed mineral phases in the gypsum sludge were dihydrate calcium sulfate ($CaSO_4\cdot 2H_2O$) and calcium sulfate hemihydrate ($CaSO_4\cdot 1/2H_2O$) (Fig. 4.20a). The arsenic sludge is composed of dihydrate calcium sulfate ($CaSO_4\cdot 2H_2O$) and

**Table 4.6** Compositions of the materials (wt%)

| Element composition (%) | Ca | S | As | Mg | Si | Cd | Fe | Zn | Pb |
|---|---|---|---|---|---|---|---|---|---|
| Gypsum sludge | 40.62 | 8.51 | 0.47 | 2.48 | 2.14 | 0.84 | 0.43 | 0.05 | 0.04 |
| Arsenic sludge | 30.69 | 15.43 | 6.81 | 2.60 | 2.38 | 0.48 | 0.61 | 11.56 | – |
| Composition (%) | CaO | $SiO_2$ | $Al_2O_3$ | $Fe_2O_3$ | MgO | $SO_3$ | Loss | | |
| Smelting slag | 15.96 | 34.15 | 17.49 | 22.13 | 2.51 | 1.0 | 0.50 | | |
| Cement clinker | 63.72 | 22.42 | 4.67 | 3.89 | 0.64 | 0.5 | 2.20 | | |

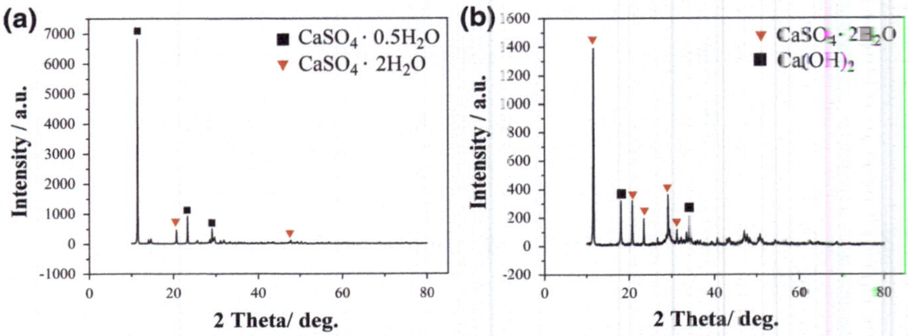


**Fig. 4.20** XRD patterns of the gypsum sludge (**a**) and the arsenic sludge (**b**). Reprinted from Ref. [46] Copyright 2016, with permission from Elsevier

calcium hydroxide ($Ca(OH)_2$) (Fig. 4.20b), which is generated during the treatment of waste acid using a lime-neutralization process [69]. Smelting slag was produced from a blast furnace of Pb/Zn smelting after being quenched with water.

### (1) Preparation of the binder

Smelting slag, cement clinker, gypsum sludge and limestone was used to prepare the binder. The mixtures were consolidated in cubic molds (20 mm × 20 mm × 20 mm). After curing for 3 and 7 d, the compressive strength (UCS) and leaching concentration of As for the binder was tested to optimize the preparation parameters. As the proportion of gypsum sludge increased, the UCS (3 d) of the binder changed slightly, whereas the USC of 7 d increased (Fig. 4.21). Although the leaching toxicity of the binder increased somewhat with the increase of gypsum sludge dosage, they remained at a low level in the whole gypsum sludge dosage range. In order to handle more gypsum sludge, the optimum proportion of gypsum sludge was chosen to be 12%.

Limestone contains nucleation sites of calcium hydroxide crystals at early hydration ages that can accelerate the hydration of cement clinker particles, especially tricalcium silicate ($C_3S$). The compressive strengths of the binder with

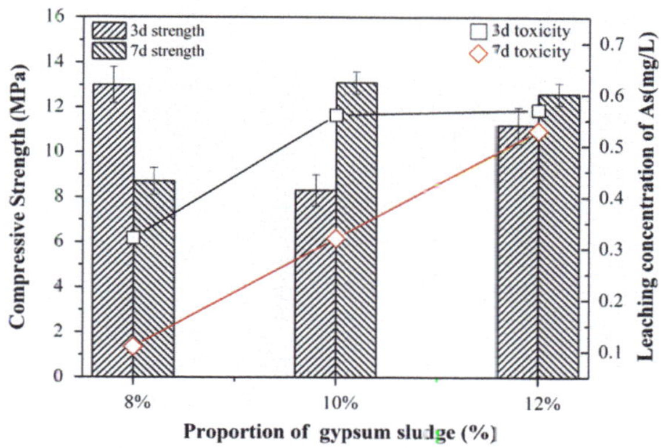


**Fig. 4.21** Effect of gypsum sludge proportion on the compressive strength and leaching concentration of As. Reprinted from Ref. [46] Copyright 2016, with permission from Elsevier

**Table 4.7** The UCS of binders under different conditions

| No. | Smelting slag (%) | Cement clinker (%) | Gypsum sludge (%) | Limestone (%) | Time (h) | 3 days | 7 days |
|---|---|---|---|---|---|---|---|
| 1 | 55 | 28 | 12 | 5 | 3.5 | 19.6 | 27.0 |
| 2 | 60 | 23 | 12 | 5 | 3.5 | 14.0 | 21.3 |
| 3 | 65 | 18 | 12 | 5 | 3.5 | 11.3 | 17.0 |
| 4 | 70 | 13 | 12 | 5 | 3.5 | 10.6 | 15.3 |
| 5 | 75 | 8 | 12 | 5 | 3.5 | 3.4 | 6.9 |

limestone dosage of 3, 5 and 8% reached 22.6, 26.9 and 21.3 MPa. Thus, the optimum proportion of limestone was 5%.

The principal hydration products of smelting slag are similar to the compound produced by the hydration of calcium silicates in Portland cement. Smelting slag also reduces the permeability and ionic diffusion of heavy metals in concrete. However, the replacement of cement clinker by smelting slag usually results in lower early strength and longer setting times, which limits the application of slag at large scales [70]. At present, mechanical activation is often regarded as necessary pre-treatment processes to activate raw materials and increase the amount of dissolved aluminosilicate [71]. The results show that the grain size is between 1 and 40 μm after 3.5 h of ball milling. When the samples were ground for more time, the fine samples agglomerates and the grain sizes of the products decreased, which affected the strength of the structure. Under the condition of ball milling of 3.5 h, the USC of the binder reached 15.3 MPa when the mass ratios of smelting slag: cement clinker: gypsum sludge: limestone was 70:13:12:5 (Table 4.7).

**(2) Solidification of arsenic sludge with the binder**

The effect of mass ratio of binder to arsenic sludge on the UCS and leaching concentration of As was investigated (Figs. 4.22 and 4.23). When the mass ratio of

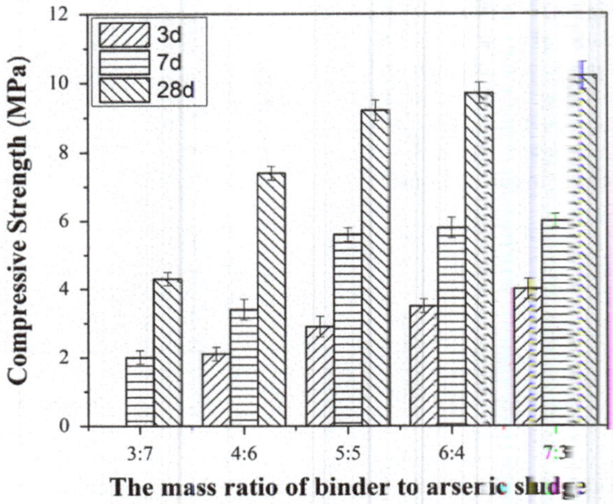


**Fig. 4.22** UCS of the solidified samples with various mass ratio of binder to arsenic sludge. Reprinted from Ref. [46] Copyright 2016, with permission from Elsevier

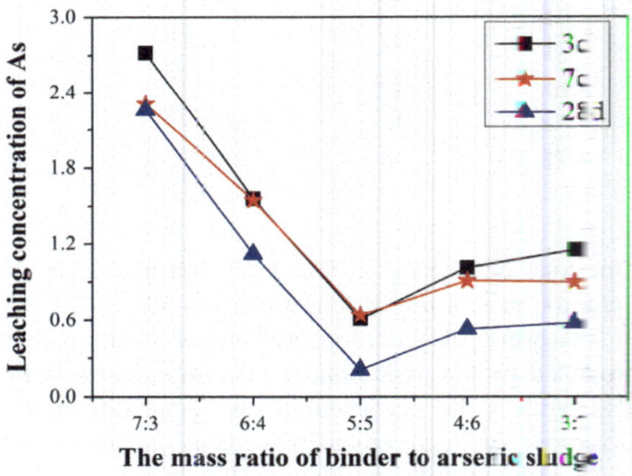


**Fig. 4.23** Leaching concentration of As with variable mass ratio of binder to arsenic sludge using CSLT and TCLP. Reprinted from Ref. [46] Copyright 2016, with permission from Elsevier

binder to arsenic sludge was larger than 5:5, the UCS hardly changed (Fig. 4.22). Hence, the mass ratio of binder to arsenic sludge was selected to be 5:5.

To verify the environmental stability of the matrices, the China Standard Leaching Test (CSLT) were introduced. Prior to the solidification process, we measured the leaching toxicity of gypsum sludge and arsenic sludge with CSLT method. The leaching concentrations for all heavy metals in gypsum sludge are below the limits of CSLT, whereas the leaching concentration of As in arsenic sludge reached 8.6 mg $L^{-1}$ which excessed the limit (5 mg $L^{-1}$ for As). After solidification treatment with the binder, the leachabilities of arsenic and heavy metals reduced dramatically. The reduction of the As leachability at a ratio of 5:5 was much greater than the reduction of the As leachability at the other mass ratios (Fig. 4.23). Similar trends were observed for that of both Pb and Zn (not shown here). The toxicity concentrations by the CSLT are 0.15, 0.01 and 0.05 mg $L^{-1}$ for As, Pb and Zn, respectively, all of which were always below the limits. In this case, the matrices appear safe and have low environmental risk.

### 4.2.3 Utilization of Red Mud and Arsenic Sludge for the Synthesis of a Red Mud-Based Cementitious Material

In general, cementitious materials use Ordinary Portland Cement (OPC) and industrial solid waste as raw materials, under water and other aqueous conditions through chemical reaction to form ettringite and C–S–H [72]. Calcium sulfoaluminate (ettringite) is a major hydration product of OPC and exhibits good early strength. Aluminate and calcium sulfate are the main component of synthetic calcium sulfoaluminate [73]. Red mud is a high alkalinity aluminosilicate raw material [74]. Smelting slag can be considering as a high-iron aluminum raw material. Calcium sulfate is the main components in the arsenic sludge. Red mud and arsenic sludge can be seen as synthetic raw materials of calcium sulfoaluminate. Hence, this section focuses on the use of red mud, arsenic sludge and lime to synthesize red mud-based cementitious material (RCM) for the solidification of arsenic sludge [75].

The red mud was obtained from a Bayer plant in southern China. Al, Si and Fe are the main elements composition of red mud. Hematite ($Fe_2O_3$), katoite ($Ca_3Al_2(SiO_4)(OH)_8$), aluminium silicate ($Al_{0.5}Si_{0.75}O_{2.25}$) and yugawaralite ($CaAl_2Si_6O_{16}(H_2O)_4$) are the main compounds in the red mud (Fig. 4.24). Therefore, red mud can be considering as a high alkalinity aluminosilicate raw

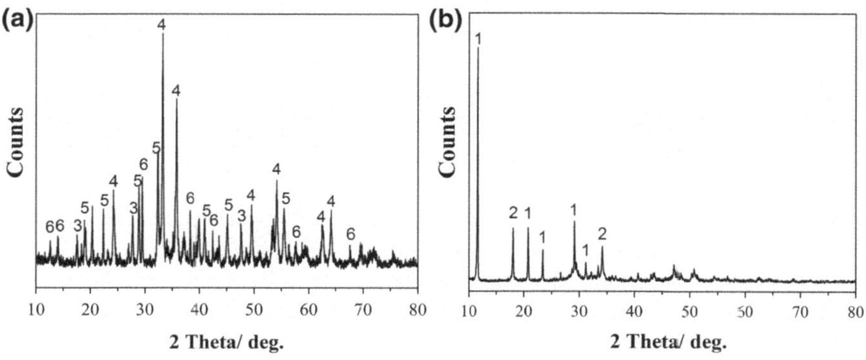


**Fig. 4.24** XRD patterns of the red mud (**a**) and the arsenic sludge (**b**) 1. $CaSO_4 \cdot 2H_2O$ (PDF# 74-1433); 2. $Ca(OH)_2$ (44-1481); 3. $Al_{0.5}Si_{0.75}O_{2.25}$ (37-1460); 4. $Fe_2O_3$ (79-1741); 5. $Ca_3Al_2(SiO_4)(OH)_8$ (38-0368); 6. $CaAl_2Si_6O_{16}(H_2O)_4$ (70-1921). Reprinted from Ref. [75] Copyright 2016, with permission from Elsevier

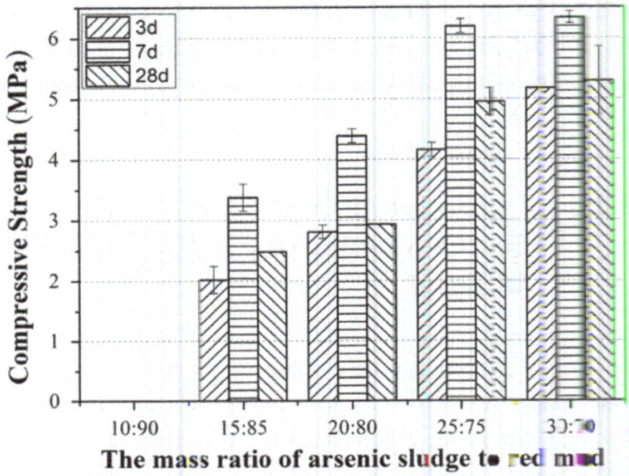


**Fig. 4.25** The UCS with variations of the red mud content. Reprinted from Ref. [75] Copyright 2016, with permission from Elsevier

material. The arsenic sludge contains high amount of arsenic, and the moisture content reach approximately 50.2%. Calcium sulfate dihydrate ($CaSO_4 \cdot 2H_2O$) and calcium hydroxide ($Ca(OH)_2$) are the main components in the arsenic sludge.

**(1) UCS of the red mud-based cementitious material**

The result represents the UCS of the binder (without the addition of lime) with various proportions of red mud and arsenic sludge (Fig. 4.25). The mass ratio of m (arsenic sludge): m (red mud) was varying from 10:90 to 30:70. Red mud does not have self-hardening properties. Therefore, with the decrease of the red mud the UCS of the binder increased. With the increase of the arsenic sludge from 10 to 25%, the UCS of the binder increased. It is possible that arsenic sludge reacts with red mud to form hydration products [76]. Gypsum and hydration products are conducive to the formation of ettringite in the presence of calcium hydroxide, which can improve the early compressive strength of the binder. Thus, these hydration products exhibit good early strength. With the increase of arsenic sludge from 25 to 30%, the pH of the reaction conditions is found to decrease (the pH from 9.62 to 8.98), which is not conductive to the hydration reaction. Therefore, compressive strength remains constant as the mass ratios of arsenic sludge increases. Taking the UCS of the RCM into account, the optimum mass ratio of m (arsenic sludge): m (red mud) is 25:75.

With the increase of the lime from 0 to 8%, the UCS of the RCM increased and then remained constant (Table 4.8). In general, the formation of hydration products required alkaline conditions. The lime reacts readily with water to form calcium hydroxide. The above reactions consume a lot of water to improve the system alkalinity (the pH from 9.17 to 10.52). The higher alkalinity is conducive to solvation of the aluminosilicate oxide. The calcium hydroxide reacts with gypsum and dissolved aluminosilicate oxide to form amorphous ettringite ($C_3A \cdot 3C\bar{S} \cdot 32H$) and calcium silicate hydrate (C–S–H) phases [77]. The dissolution concentration of Al and Si for RCM is 14.62 and 5.48 mg $L^{-1}$, respectively. However, the dissolution

**Table 4.8** The USC of RCM under different conditions

| No. | Binder[a] (%) | Lime (%) | 3 days (MPa) | 7 days (MPa) | 28 days (MPa) | pH |
|---|---|---|---|---|---|---|
| 1 | 100 | 0 | 4.16 | 6.19 | 4.95 | 9.17 |
| 2 | 98 | 2 | 4.28 | 7.29 | 8.[illegible]5 | 9.72 |
| 3 | 96 | 4 | 4.75 | 9.56 | 10.04 | 10.09 |
| 4 | 94 | 6 | 5.23 | 10.71 | 12.05 | 10.23 |
| 5 | 92 | 8 | 5.30 | 10.64 | 11.97 | 10.52 |

[a]The binder is composed of red mud and arsenic sludge with a mass ratic of 75 25

concentration of Al and Si for red mud is only 6.62 and 2.43 mg $L^{-1}$, respectively. Therefore, with the increase of the lime, the UCS of samples is found to increase. In addition, the calcium hydroxide will react with carbon dioxide to form limestone, which can improve sample packing and block capillary pores. Limestone contains nucleation sites of calcium hydroxide crystals at early hydration stages that can accelerate the hydration of amorphous ettringite. Taking the effective technology and performance into account, the optimum mass ratio of binder to lime was 94:6.

### (2) Leaching characteristics

CSLT has been used to identify whether general solid wastes and recycled products conform to hazardous waste standards. The concentration of As in leachate for arsenic sludge is higher than in the other samples and risked exceeding the As concentrations in the standards (Table 4.9). Therefore, in this study, the leaching concentration of As is used as an indicator of immobilization performance. We found that the leaching concentration of As in the RCM (0.65 mg $L^{-1}$) is significantly lower than in the arsenic sludge (8.6 mg $L^{-1}$). This means that the chemical speciation transformation occurs in the RCM and some of the heavy metal is absorbed by the ettringite and calcium sulfoaluminoferrite structure [5, 78]. At the same time, lime will react with arsenic to form complex As–Ca compounds. The leachable level of heavy metals from RCM is less than the limits set for hazardous waste.

### (3) Hydration characteristics of RCM

The XRD patterns of red mud, binder and RCM are shown in Fig. 4.26. The XRD patterns of Fig. 4.26b, c show that $CaCO_3$ is a major chemical constituent. This

**Table 4.9** Metal concentrations in the CSLT leachates at 28 days and CSLT limits ($\leq$ means under detection limit)

| | As (mg $L^{-1}$) | Pb (mg $L^{-1}$) | Zn (mg $L^{-1}$) | Cu (mg $L^{-1}$) | Cd (mg $L^{-1}$) | Cr (mg $L^{-1}$) |
|---|---|---|---|---|---|---|
| Arsenic sludge | 8.60 | 0.20 | 0.05 | 0.03 | 0.01 | 0.10 |
| Red mud | 0.25 | 0.01 | 0.10 | $\leq$0.01 | $\leq$0.01 | $\leq$0.01 |
| RCM | 0.65 | 0.05 | 0.01 | $\leq$0.01 | $\leq$0.01 | $\leq$0.01 |
| CSLT | 5.00 | 5.00 | 100 | 100 | 1.00 | 15.0 |

agrees well with the above analysis that limestone indeed present in the RCM. The peaks of $CaHAsO_4(H_2O)_3$ can be observed in Fig. 4.26b, c. Because $HAsO_4^{2-}$ and $AsO_4^{3-}$ are the two arsenic species present in solution under the alkaline conditions. The $HAsO_4^{2-}$ and $AsO_4^{3-}$ react with the calcium to form $CaHAsO_4(H_2O)_3$, which are stable arsenic compounds [79]. In addition, with the addition of lime, the peaks of aluminite and andradite disappear, and the peaks of $Ca(OH)_2$, ettringite and C–S–H are observed in Fig. 4.26c. It means that the aluminite and andradite in RCM are formed in a hydration reaction. It is evident from Fig. 4.26c that the broad and amorphous hump from 25° to 38° (2θ) is the characteristic peak of amorphous gels, which may be attributed to Si–OH and Si–O–Si gel structures with C–S–H character [80, 81]. The intensities of C–S–H peaks in RCM are higher than that in the binder (Fig. 4.26b). It means that RCM has more ettringite and $CaCO_3$. This also explains why the compressive strength of RCM is higher than that of the binder. At the same time, compared with raw arsenic sludge (Fig. 4.26b), the peaks of gypsum disappear, and the peaks of ettringite and C–S–H are observed in Fig. 4.26c. Therefore, the following reactions may occur:

$$3C_3A + 3(CaSO_4{\cdot}2H_2O) + 26H_2O \rightarrow 3CaO{\cdot}Al_2O_3{\cdot}3CaSO_4{\cdot}32H_2O \quad (4.9)$$

$$C_3A + 3(CaSO_4{\cdot}2H_2O) + 2Ca(OH)_2 + 24H_2O \rightarrow 3CaO{\cdot}Al_2O_3{\cdot}3CaSO_4{\cdot}32H_2O \quad (4.10)$$

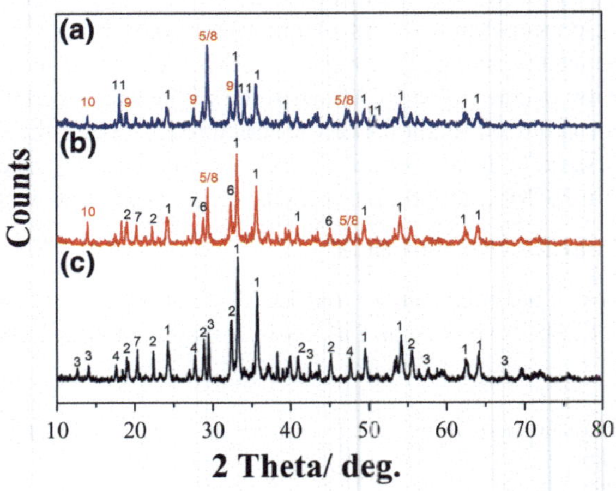


**Fig. 4.26** XRD diagrams of the red mud (**a**), binder (**b**) and RCM (**c**). 1. $Fe_2O_3$ (PDF# 79-1741); 2. $Ca_3Al_2(SiO_4)(OH)_8$ (38-0368); 3. $CaAl_2Si_6O_{16}(H_2O)_4$ (70-1921); 4. $Al_{0.5}Si_{0.75}O_{2.25}$ (37-1460); 5. $CaCO_3$ (72-1937); 6. Andradite (87-1971); 7. Aluminite (70-1103); 8. C–S–H (08-0148); 9. Ettringite (41-1451); 10. $CaHAsO_4(H_2O)_3$ (70-1280); 11. $Ca(OH)_2$ (84-1263). Reprinted from Ref. [75] Copyright 2016, with permission from Elsevier

$$3C_3A{\cdot}CaSO_4 + 8CaSO_4 + 6CaO + 96H_2O \rightarrow 3(3CaO{\cdot}Al_2O_3{\cdot}3CaSO_4{\cdot}32H_2O) \quad (4.11)$$

The experimental results also show that the intensities of hematite ($Fe_2O_3$) peaks in RCM is lower than that in red mud (Fig. 4.26a). With the results of the FTIR and SEM analyses, we speculate that some of the aluminium in sulfoaluminate is replaced by iron to form aluminoferrite phases ($Fe^{3+} = 0.675$ Å and $Al^{3+} = 0.535$ Å). The above chemical reaction will form new complex compounds of calcium sulfoaluminate-calcium sulfoferrite which could be regarded as complex compounds of calcium sulfoaluminoferrite [82]. The calcium sulfoaluminoferrite has many advantages over other OPC, such as good compressive strength. The experimental results show that the red mud and arsenic sludge can be utilized as RCM.

The infrared spectra of red mud, binder and RCM are shown in Fig. 4.27. The band at 1631 $cm^{-1}$ corresponds to H–O–H bending vibration of the water molecules from the C–S–H [83]. The bands at 1433 and 712 $cm^{-1}$ in the spectra of the RCM are attributed to the O–C–O asymmetric stretching of carbonates in the materials. The IR analysis of RCM is in agreement with the results of XRD. The bands at 1112 $cm^{-1}$ in the spectra of the red mud, binder and RCM are attributed to the Si–O–Si asymmetric stretching in tetrahedral [84].

The bands at 997 $cm^{-1}$ in the red mud are assigned to the asymmetric stretching vibration of Si–O, but the bands at 997 $cm^{-1}$ in the binder disappear. This means that the added arsenic sludge reacts with red mud. In addition, the addition of lime will improve the alkalinity of the system. The higher alkalinity is conducive to the dissolution of silicon dioxide. The band at 997 $cm^{-1}$ in RCM corresponds to a bending vibration of Si–O–Si for C–S–H. Unlike with raw arsenic sludge, the bands at 873 $cm^{-1}$ corresponding to Al(Fe)–O in the RCM are observed. This means that some of the Al–OH in ettringite is replaced by Fe–OH to form the calcium sulfoaluminoferrtite phase [82]. The above chemical reaction implies that the ettringite and C–S–H are the major ingredients of RCM. Ettringite and C–S–H are also the

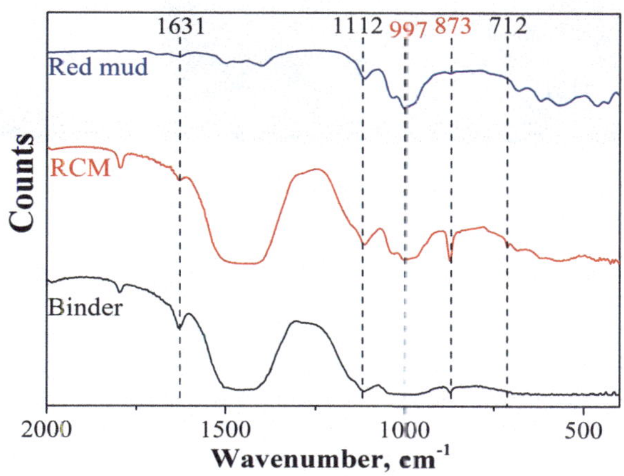


**Fig. 4.27** FTIR patterns of red mud, binder and RCM. Reprinted from Ref. [75] Copyright 2016, with permission from Elsevier

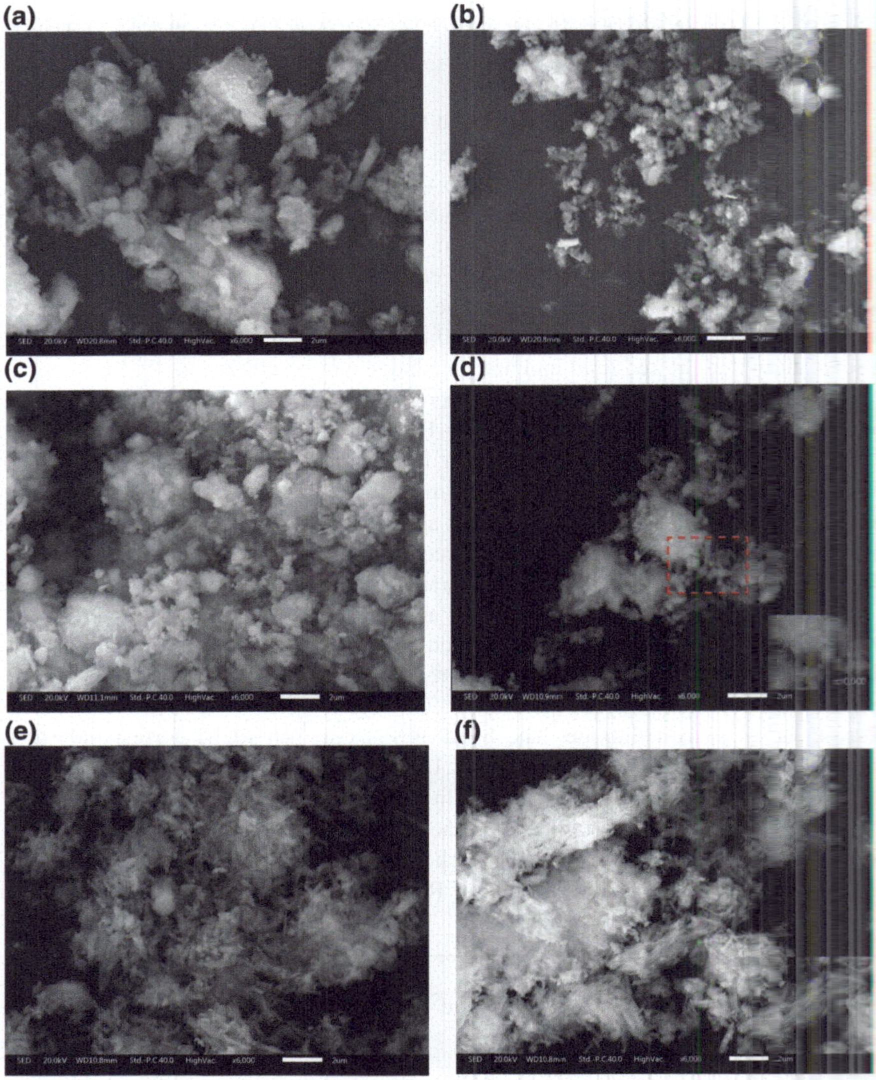

**Fig. 4.28** SEM images of arsenic sludge (**a**) and red mud (**b**); binder (**c**) and RCM (**d**); ettringite (**e**) and ettringite containing $Fe_2O_3$ (**f**). Reprinted from Ref. [75] Copyright 2016, with permission from Elsevier

major ingredients of OPC. Therefore, we conclude that the red mud could be considered as a raw material for RCM.

The microstructure of the arsenic sludge and red mud are presented in Fig. 4.28a and b, respectively. Columnar particles are a major component in Fig. 4.28a, because calcium sulfate dihydrate ($CaSO_4 \cdot 2H_2O$) is the main component of arsenic sludge. Figure 4.28b shows the SEM of red mud, the particle size of which is approximately 0.5 μm. We also found that a large number of unreacted particles are merely cohering as a new amorphous phase (Fig. 4.28c). In addition, by comparing Fig. 4.28c with Fig. 4.28d, it seems reasonable to expect that RCM may develop a higher UCS than the binder.

To study the characteristics of RCM, the ettringite and ettringite containing $Fe_2O_3$ are synthesized from $Al_2(SO_4)_3 \cdot 18H_2O$, CaO and $Fe_2O_3$. The ettringite and ettringite containing $Fe_2O_3$ are synthesized according to the study of Goetz-Neunhoeffer et al. [85]. The microstructures of the ettringite and ettringite containing $Fe_2O_3$ are presented in Fig. 4.28e and f, respectively. Columnar particles are a major component in Fig. 4.28e, because ettringite is the main component in Fig. 4.28e. Unlike Fig. 4.28e, needle-shaped particles are a major component in Fig. 4.28f. By the analyses of ettringite and ettringite containing $Fe_2O_3$ with XRD and FTIR, we speculate that some of the Al–OH in ettringite is replaced by Fe–OH to form calcium ferrite phases. The structural transformation means that red mud and arsenic sludge could be considered as cementitious raw materials for RCM. At the same time, the transformation of structure improves the structural stability of RCM, and improves the solidification effect.

Figure 4.29a shows that most of the As, Cr and Cd in arsenic sludge exist as the weak acid soluble fraction and the reducible fraction, implying that the heavy metals in arsenic sludge can dissolve out in combination with acid rain, thereby causing severe environmental problems. In Fig. 4.29b, most of the As and Cr in RCM with weak acid soluble fraction and reducible fraction has been transformed into the oxidizable and residual fractions. It is inferred that As and hydration products display another interaction behavior [86], and surface complexes is the predominant form of As, such as ettringite and various double-layered hydroxides of calcium alumino/ferric hydrates [45]. It is possible that the Cr in red mud is mainly distributed in the residual fractions. Therefore, most of Cr in the RCM with weak acid soluble fraction and reducible fraction is transformed into the oxidizable and residual fractions. The chemical speciation transformations of As and Cr

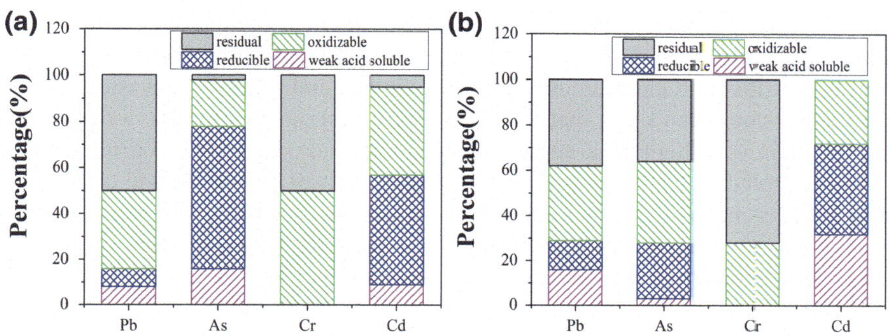


**Fig. 4.29** Chemical speciations of Pb, As, Cr, Cd in arsenic sludge (**a**), RCM (**b**). Reprinted from Ref. [75] Copyright 2016, with permission from Elsevier

indicate the outstanding encapsulation effect of the RCM. Thus, the RCM has reliable heavy metal immobilization ability.

Figure 4.29a shows that most of the Pb and Cd in arsenic sludge exist as the weak acid soluble fraction and the reducible fraction. However, the chemical speciation analyses of Pb and Cd are almost identical in RCM. $2PbCO_3{\cdot}Pb(OH)_2$ and $CdPbO_3$ are the main compounds of Pb and Cd in OPC. In arsenic sludge and red mud, they may also exist in the same form. Therefore, the chemical speciation analyses of Pb and Cd are almost identical.

Slag-based cementitious material and RCM were effective for controlling the release of As from arsenic sludge. The reaction of arsenic and calcium from the hydraulic binder can form calcium arsenates, which are stable arsenic compounds. Under the alkaline conditions observed in the solidified samples, the two As species present in solution are $HAsO_4^{2-}$ and $AsO_4^{3-}$. These anions react with the calcium from the pore water according to reactions (4.12) and (4.13):

$$HAsO_4^{2-} + Ca^{2+} + nH_2O \rightarrow CaHAsO_4{\cdot}nH_2O \tag{4.12}$$

$$2AsO_4^{3-} + 3Ca^{2+} + nH_2O \rightarrow Ca_3(AsO_4)_2{\cdot}nH_2O \tag{4.13}$$

The formation of $Ca_3(AsO_4)_2$ was regarded as a key process for As immobilization in this section [66]. The interaction of As with C–S–H exhibit another behavior, and As mainly forms surface complexes, such as ettringite, monosulfate, and various double-layered hydroxides of calcium alumino/ferric hydrates [87]. Ettringite and arsenate analog of ettringite were frequently identified in successful cement based solidification forms of arsenic wastes [88].

## 4.3 Arsenic Vitrification Technologies

### *4.3.1 Arsenic Oxides Vitrification by Iron Phosphate Glass*

Solidification and stabilization by cementation have been frequently used to treat solid waste worldwide. However, this technology increases the waste volume. By contrast, vitrification can immobilize heavy metals, and reduce the waste volume. A previous study also states that the leaching concentration of metals in crystalline slags is relatively high in comparison to amorphous slags [89]. Vitrification is widely accepted as the safest process for treating hazardous wastes and converting them into leach-resistant materials. Its major advantage is that the glassy matrix possesses a good chemical stability and can incorporate waste with a complex chemical composition into its vitreous structure, but it has a very significant drawback in the high energy consumption required by the melting processes.

Iron phosphate glasses have received considerable attention in the past decade for vitrification owing to their high chemical durability and low melting temperatures. Additionally, iron phosphate glasses can tolerate several waste components with high concentrations which lowers the volume and thus, the cost of the end waste product. This rare combination of properties suggests that these glasses may be suitable for the immobilization of certain toxic and radioactive wastes. It has been reported [90–92] that a 40 mol% $Fe_2O_3$–60 mol% $P_2O_5$ glass has the highest chemical durability among the various compositions of the iron phosphate glasses. In recent years, there has been an enormous amount of research on improving the physical properties and the chemical durability of 40 mol% $Fe_2O_3$–60 mol% $P_2O_5$ glasses by introducing the glass modifiers and formers such as monovalent (Li, Na or K), divalent (Mg, Ca, Ba, and Pb) oxides and boron oxide [93]. It has been demonstrated that small additions of $B_2O_3$ to iron phosphate glass compositions can substantially improve their thermal stability which may be useful in improving the performance of iron phosphate glasses for applications such as waste immobilization or sealing glasses.

$As_2O_3$ has been reported to be a strong network former and it is expected that $AsO_3$ groups form a single arsenic–phosphorus–oxygen framework in PbO–$P_2O_5$–$As_2O_3$ glass system and may strengthen its structure. In view of the excellent properties of modified iron phosphate melts and characteristics of arsenic, a series of modified iron phosphate glass with a basic composition of 20CaO–8$B_2O_3$–24$Fe_2O_3$–48$P_2O_5$ + x$As_2O_3$ (x = 0, 1, 5, 10, 15 mol%) are prepared using analytical grade > 99% purity $As_2O_3$, CaO, $H_3BO_3$, $Fe_2O_3$, $NH_4H_2PO_4$ as materials. The mixture is homogenized by grinding in a mortar and heated in ceramic crucibles for 2 h at 220 °C to expel ammonia and water which result from decomposition of the ammonium dihydrogen phosphate, and then heated to 1000 °C (heating rate is 10 °C $min^{-1}$) and melted at 1000 °C for 1 h. The melts are then quenched to room temperature in the air without annealing in order to produce vitreous samples [94].

**(1) Vitrification of arsenic-rich residue using iron phosphate glass**

The entire sample shows a regular and homogeneous surface with shiny, metallic luster on the external surface, which becomes more noticeable with increasing concentration of $As_2O_3$. No cracks are observed at the surface of the investigated glasses. XRD pattern of the studied glasses have shown that the compositions containing up to 15 mol% of $As_2O_3$ form glasses and they do not present any crystalline phase (Fig. 4.30). XRD has provided no evidence of arsenic partitioning in any crystallized samples, which indicate great compatibility of arsenic with the phosphate network.

The measured compositions of the studied glasses are shown in Table 4.10. The comparable differences between nominal compositions and the analyzed are mainly attributed to the volatilization of $As_2O_3$. And greater volatilization rates of $As_2O_3$ are observed as the $As_2O_3$ content increase.

TEM images of the glasses with x = 0 (Fig. 4.31a) and x = 15 (Fig. 4.31b), shows a homogeneous glass at the nanometer scale. And no crystalline particles are observed in other glasses (x = 1, 5, 10 mol%). The addition of 15 mol% $As_2O_3$ to the base glass results in a similarly featureless TEM micrograph, revealing a homogeneous glass free from phase separation at the sub-micron scale. Therefore,

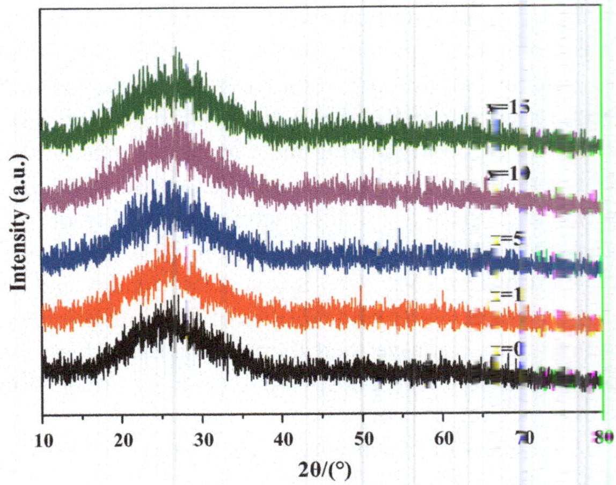


**Fig. 4.30** XRD patterns for 20CaO–8$B_2O_3$–24$Fe_2O_3$–48$P_2O_5$ + x$As_2O_3$ glass samples. Reprinted from Ref. [94] Copyright 2015, with permission from Elsevier

**Table 4.10** Batch composition (mol%) of modified iron phosphate glasses containing arsenic

| Glass | 0As | 1As | 5As | 10As | 15As |
|---|---|---|---|---|---|
| $P_2O_5$ (analyzed) | 48 (47.25) | 47.52 (46.23) | 45.6 (46.54) | 43.2 (46.78) | 40.8 (46.57) |
| $Fe_2O_3$ (analyzed) | 24 (25.81) | 23.76 (26.30) | 22.8 (26.24) | 21.6 (26.15) | 20.4 (26.[illegible]7) |
| $B_2O_3$ (analyzed) | 8 (7.25) | 7.92 (7.35) | 7.6 (7.32) | 7.2 (7.30) | [illegible] (7.3[illegible]) |
| CaO (analyzed) | 20 (19.68) | 19.8 (20.12) | 19 (19.90) | 18 (19.78) | 17 (19.8[illegible]) |
| $As_2O_3$ (analyzed) | 0 (0.00) | 1 (1.00) | 5 (4.26) | 10 (6.66) | 15 (7.53) |

$As_2O_3$ is confirmed to be well incorporated into the structural network of the parent base glass.

**(2) Structural properties of the modified iron phosphate glass for the immobilization of arsenic**

The spectra for the glasses containing up to 7.53 mol% of $As_2O_3$, 0As–15As, have similar spectral features (Fig. 4.32). Low-scattering cross-section of borate units in the Raman spectra and their low content in the studied glasses allow us to neglect here their effect on the Raman spectra. The Raman spectra can be deconvoluted using Gaussian lines to the component bands for each one to study the origin of this characteristic Raman symmetry. A number of 10–11 peaks have been obtained by deconvolution in the 100–1500 $cm^{-1}$ wavenumber region (Fig. 4.33). The band centers and relative area of the deconvoluted peaks, their assignments of the suitable bands are given in Table 4.11.

The specific deconvoluted spectrum for x = 0 glass is given in Fig. 4.33a. The band around 220 $cm^{-1}$ is due to a P–O–P bending mode. A low shoulder appears at 276 $cm^{-1}$ and can be likely related to the bending $Q^0$ with Fe as modifier. The band at ~340 $cm^{-1}$ has been assigned to bending vibrations of phosphate polyhedral. The bending and torsional vibrations occurs 576 $cm^{-1}$ and can be assigned to the

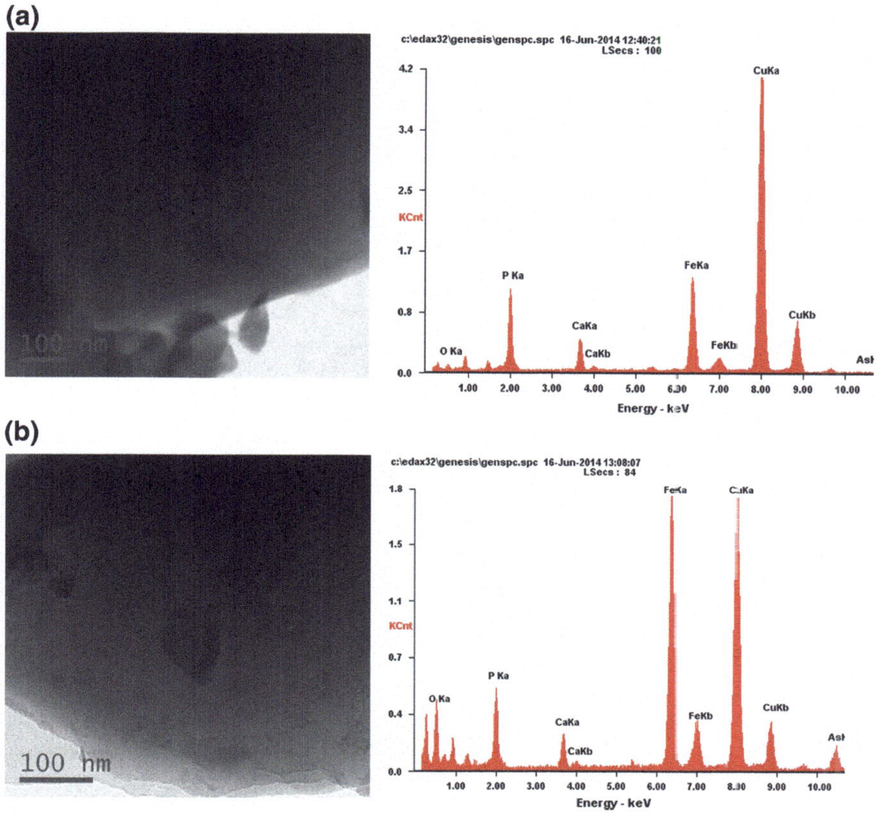


**Fig. 4.31** TEM and EDS micrograph of $20CaO–8B_2O_3–24Fe_2O_3–48P_2O_5 + xAs_2O_3$ glasses for **a** x = 0 mol%, **b** x = 15 mol%. Reprinted from Ref. [94] Copyright 2015, with permission from Elsevier

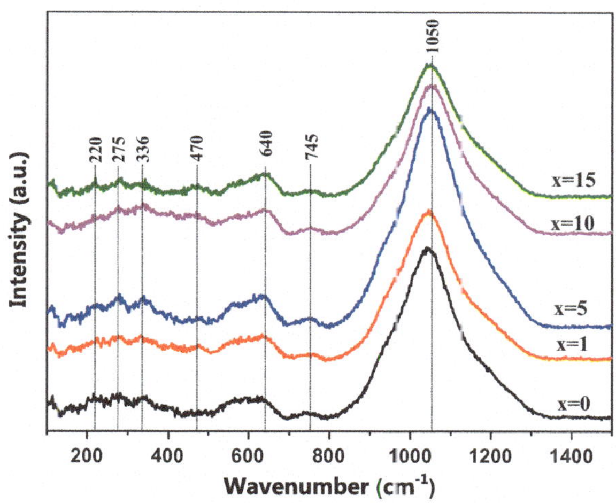


**Fig. 4.32** Raman spectra for $20CaO–8B_2O_3–24Fe_2O_3–48P_2O_5 + xAs_2O_3$ glass samples. Reprinted from Ref. [94] Copyright 2015, with permission from Elsevier

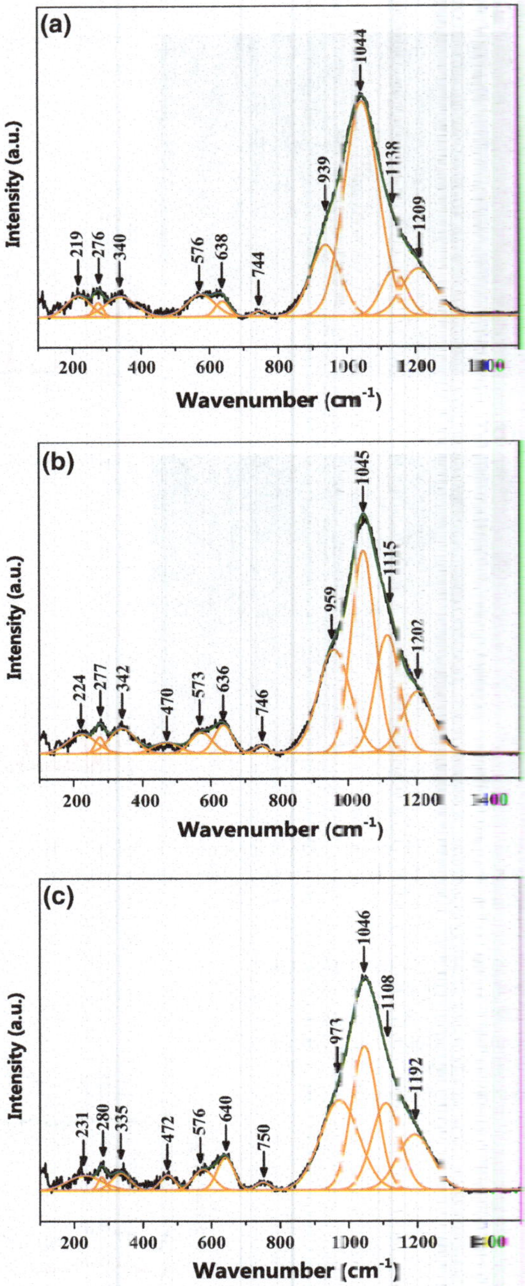


**Fig. 4.33** Deconvoluted Raman spectra of $20CaO–8B_2O_3–24Fe_2O_3–48P_2O_5 + xAs_2O_3$ glasses for **a** x = 0 mol%, **b** x = 5 mol %, **c** x = 15 mol%. Reprinted from Ref. [94] Copyright 2015, with permission from Elsevier

**Table 4.11** The assignments and relative areas of different vibrational bands from Raman spectra of $20CaO–8B_2O_3–24Fe_2O_3–48P_2O_5 + xAs_2O_3$ glasses

| Frequency regions ($cm^{-1}$) | Assignments | Relative areas dependence of vibrational bands versus $As_2O_3$ concentrations (x mol%) | | | | |
|---|---|---|---|---|---|---|
| | | 0 | 1 | 5 | 10 | 15 |
| 219–241 | Network bend | 3.50 | 2.35 | 3.46 | 5.93 | 3.66 |
| 275–280 | $Q^0$ bend with Fe as modifier | 0.90 | 1.80 | 0.95 | 0.66 | 0.75 |
| 335–348 | phosphate polyhedral bend | 4.23 | 4.74 | 4.06 | 5.50 | 2.25 |
| 470–472 | O–P–O bend ($Q^0$) and As–O–As stretch | – | 3.75 | 1.66 | 4.76 | 1.58 |
| 570–582 | Cation oxygen polyhedral and $(P_2O_7)^{4-}$ groups | 4.28 | 3.20 | 2.87 | 4.09 | 3.06 |
| 636–645 | $(P–O–P)_{asym}$ stretch ($Q^2$) | 1.49 | 4.16 | 3.21 | 2.60 | 3.96 |
| 744–754 | $(P–O–P)_{sym}$ stretch ($Q^1$) | 0.40 | 0.68 | 0.56 | 0.43 | 0.79 |
| 939–973 | $(PO_4)_{asym}$ stretch ($Q^0$) | 14.98 | 21.32 | 21.54 | 19.81 | 25.27 |
| 1043–1048 | $(O–P–O)_{sym}$ stretch ($Q^1$) | 51.13 | 30.99 | 31.26 | 27.73 | 26.91 |
| 1108–1138 | $(P–O)_{sym}$ stretch ($Q^2$) | 8.03 | 15.76 | 18.36 | 15.85 | 17.94 |
| 1192–1209 | $(O–P–O)_{asym}$ stretch ($Q^2$) | 11.07 | 11.25 | 12.06 | 12.63 | 13.82 |

overlapping vibrations involving iron oxygen polyhedral and $(P_2O_7)^{4-}$ groups which are characteristic for a structure dominated by $Q^1$ tetrahedrons. Another band due to the symmetric stretching vibrations of P–O–P bridge in $Q^1$ units is located at 744 $cm^{-1}$. Moreover, the strong and broad band at 1044 $cm^{-1}$ is related to the symmetric stretching mode of O–P–O non-bridging bond in $Q^1$ groups, which indicates that $Q^1$ units are a completely major component of the structure for the arsenic-free modified iron phosphate glass. The shoulder at 939 $cm^{-1}$ suggests the presence of $Q^0$ phosphate units. The band around 640 $cm^{-1}$ corresponds to the asymmetric stretching of the bridging oxygen of the metaphosphate group. 1138 $cm^{-1}$ and 1209 $cm^{-1}$ are attributed to symmetric stretching motions of the P–O and asymmetric stretching motions of the O–P–O non-bridging oxygen, respectively, in the $Q^2$ groups. The presence of a certain amount of orthophosphate ($Q^0$) and metaphosphate ($Q^2$) units along with the pyrophosphate ($Q^1$) units in these glasses results from the equilibrium between the pyrophosphate units and their disproportionation products in the melts according to the following equation:

$$2Q^1 \longleftrightarrow Q^0 + Q^2.$$

As examples represent for the nominal molar composition $0 \leq x \leq 15\%$ range the deconvolution in Gaussian bands of the spectrum for the x = 5% and x = 15% glasses, are given in Fig. 4.33b and Fig. 4.33c, respectively. Changes in the Raman spectra of $20CaO–8B_2O_3–24Fe_2O_3–48P_2O_5 + xAs_2O_3$ glasses (Fig. 4.32) with increasing $As_2O_3$ content unambiguously manifest the incorporation of $As_2O_3$ into the structural network of the parent modified iron phosphate glass. Structural changes in the phosphate network are reflected especially in the high-frequency part of the Raman spectra, whereas vibrations of structural units containing As–O bonds can be seen in the lower frequency region. Additions of $As_2O_3$ to the base glass result in appearing of a new peak at ~470 cm$^{-1}$. The specific peak is assigned to the As–O–As stretch, and appears to be merged with vibration of O–P–O bending vibrations of $Q^0$ units which exhibits a common meta centre about 470 cm$^{-1}$. As a consequence, we may expect linkages of the type P–O–As and the band in this region may be due to the vibrations of these bonds. Moreover, adding $As_2O_3$ into the iron phosphate glass resulting in a decreased intensity of the 1044 cm$^{-1}$ band and an increased intensity of the ~1138 and ~1209 cm$^{-1}$, indicates conversion of more $Q^1$ into $Q^2$ units in the glass network with increased concentration of $As_2O_3$ in the glass matrix, which means a less depolymerisation of phosphate chains. According to Krishna et al. [95] $As_2O_3$ is a strong network former with corner sharing $AsO_3$ pyramidal units; normal bond lengths of As–O lie between 1.72–1.81 Å and O–As–O, As–O–As bond angles lie in the ranges 90°–103° and 123°–135° respectively. So it can be assumed that the As takes part network-forming positions with $AsO_3$ pyramids structural units and interacts with $PO_2$ units. Thus, P–O–As linkages/bridges are formed, which connect the chains together. On the other hand, the increase of glass network former results in less non-bridging oxygens from the breaking of long phosphate chains by network modifier, CaO. In other words, the decreasing degree of depolymerisation of the phosphate glass with the addition of $As_2O_3$ can be ascribed by the creation of As–O–$Ca^{2+}$ at the expense of modifier cations.

IR spectra of $20CaO–8B_2O_3–24Fe_2O_3–48P_2O_5$ + $xAs_2O_3$ glasses (Fig. 4.34) show that all the glasses have similar spectra features. All the spectra show a strong broad absorption band near 522 cm$^{-1}$ which can be composed of two component bands near 480–500 cm$^{-1}$ and 540–550 cm$^{-1}$ assigned to deformation modes of

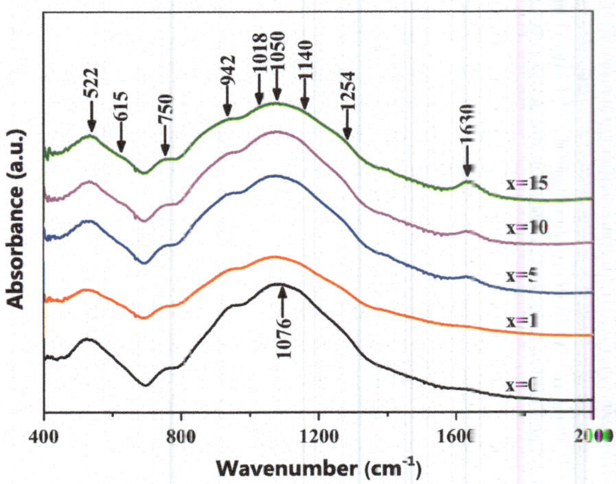


**Fig. 4.34** FTIR spectra for $20CaO–8B_2O_3–24Fe_2O_3–48P_2O_5$ + $xAs_2O_3$ glass samples. Reprinted from Ref. [94] Copyright 2015, with permission from Elsevier

$PO_4^{3-}$ groups ($Q^0$) and the bending mode of O–P–O in the $Q^-$ units respectively. Low shoulders appear at ~615 $cm^{-1}$ and ~750 $cm^{-1}$ can be attributed to stretching band of the Fe–O–P and symmetric stretching vibration of P–O–P rings ($Q^2$ species) respectively. The band observed at ~942 $cm^{-1}$ is assigned to $BO_4$ tetrahedral groups. The overlapping bands observed in the 840–950 $cm^{-1}$ have contributions from both phosphate and borate groups which indicate the existence of P–O–B in the structure of glass. The largest intensity for the IR lines is at 1000–1200 $cm^{-1}$ corresponding to stretching vibration of P–O–P linkages on $Q^1$ units and $Q^2$ units, which indicate a large number of pyrophosphate and metaphosphate groups in the glass network. A small absorption band appears in the spectra of all glasses around 1630 $cm^{-1}$. Such a band may be related to P–O–H or water, which may be brought by air moisture during the preparation of KBr pellets for IR measurements.

When the $As_2O_3$ concentration in the glass matrix has increased, the bands around 522, 615, 750 and 942 $cm^{-1}$ remain unaffected. However, absorption band of symmetric stretching vibration of $PO_4^{3-}$ tetrahedra in $Q^0$ units at ~1076 $cm^{-1}$ for the glass with x = 0 shift to lower frequencies (~1050 $cm^{-1}$) when $As_2O_3$ increases. The specific peak near 1050 $cm^{-1}$ is assigned to the symmetric stretching vibrations of $AsO_3$ structural units, and appears to be merged with vibration of P–$O^-$ groups (chain terminator) which exhibits a common meta center about 1050 $cm^{-1}$. As a consequence, we may expect linkages of the type P–O–As and the band in this region may be due to the vibrations of these bonds, which may also account for the observed broadness of this band. Moreover, a change that the relative intensities of the ~1018 $cm^{-1}$ band assigned to the symmetric stretching vibration of $(PO_3)^{2-}$ ($Q^1$ species) decreases and the one at 1140 $cm^{-1}$ due to symmetric stretching modes ($PO_2$) of metaphosphate chains ($Q^2$ speices) increases can be observed for the glasses, which indicate conversion from $Q^1$ units to $Q^2$ units in the glass network.

**(3) Structural modifications and property changes of iron phosphate glass for the immobilization of arsenic**

The glass transition temperature ($T_g$), first crystallization peak ($T_r$) and the value of ($T_r - T_g$) which is indicative of the thermal stability of glass, is analyzed by DTA (Differential Thermal Analysis). As shown in Table 4.12, with the addition of arsenic, the glass transition temperature ($T_g$) decreases and the value of ($T_r - T_g$) varies from 112.6 to 128.4 °C irregularly. However, ($T_r - T_g$) values for samples containing $As_2O_3$ are higher than that for the one without $As_2O_3$ which shows studied glasses substantially greater thermal stability than the As-free glass.

**Table 4.12** DTA parameters (°C) for 20CaO–8$B_2O_3$–24$Fe_2O_3$–48$P_2O_5$ + x$As_2O_3$ glass samples

| Glass | x = 0 | x = 1 | x = 5 | x = 10 | x = 15 |
|---|---|---|---|---|---|
| $T_g$ | 587 | 580 | 580 | 569 | 559 |
| $T_r$ | 700 | 708 | 699 | 686 | 677 |
| $T_r - T_g$ | 113 | 128 | 119 | 117 | 118 |

A systematic approach has been made to immobilize arsenic into modified iron phosphate glass. All the samples formed homogeneous glasses and no crystalline phases which manifest the incorporation of $As_2O_3$ into the parent phosphate glass. For the glasses with the As-free the $Q^1$ groups are the predominant structural units, indicated by Raman. The structural features of infrared spectra agree well with that of the Raman data, which indicates that with increase of $As_2O_3$ in glass compositions as a glass network leads to the conversion from $Q^1$ to $Q^2$. Arsenic takes part network-forming positions with $AsO_3$ pyramids structural units and incorporated into the phosphate glass network by formation of P–O–As bonds. The ($T_c - T_g$) values for samples containing $As_2O_3$ are higher than that for the As-free one, which indicates that the thermal stability is improved with the addition of $As_2O_3$.

### 4.3.2 *Arsenic-Rich Slag Vitrification by Iron Phosphate Glass*

The most common option to treat arsenic residue lies in encapsulating the contaminated material through solidification/stabilization techniques and disposing of the treated wastes in secure landfills [96], which can convert hazardous wastes into chemically stable solids [97]. Unfortunately, insufficient immobilization may lead to leaching of arsenic contaminating soil and underground water [98]. Therefore, new methods must be proposed to reduce or eliminate the secondary pollutant [99, 100]. Among these methods, vitrification has been an increasingly attractive approach to treat hazardous and toxic materials. A relatively inert glass is produced by vitrification at temperatures above 750 °C and the hazardous waste can be incorporated into the glass structure [101]. Moreover, such technology has the capacity to reduce volume of solid waste and completely stabilize the hazardous substances [99]. Hence, vitrification has the potential to reduce the leaching ability of hazardous constituents from waste and avoid potential environmental risk [101, 102].

Iron phosphate glasses have been received increasing attention as a host matrix for the vitrification of radioactive waste [103], and toxic wastes [93]. In our previous research [101, 104], iron phosphate glass was used to immobilize arsenic trioxide. The addition of $As_2O_3$ into the modified iron phosphate glass leads to the conversion of $Q^1$–$Q^2$ and arsenic is incorporated into the phosphate network in the form of P–O–As bond. Additionally, glass modifier (antimony oxide) was added into iron phosphate glass to strengthen glass structure containing arsenic and the proper quantities of glass modifier can enhance the immobilization of arsenic into iron phosphate glass. However, the vitrification of $As_2O_3$ had been extensively studied and the application of this approach on real arsenic residue is limited.

In this section, a typical arsenic-rich residue (ARR), which mainly contains of $Na_3AsO_4$, was collected to be vitrified by iron phosphate-based glass, of which the based glass composition is $4B_2O_3$–5CaO–$33Fe_2O_3$–$58P_2O_5$ (mol%). According to

**Table 4.13** Average chemical compositions of the ARR (wt%)

| $As_2O_5$ | $Na_2O$ | $SO_2$ | $SiO_2$ | $K_2O$ | $Fe_2O_3$ | $P_2O_5$ | CaO | Other oxides |
|---|---|---|---|---|---|---|---|---|
| 39.17 | 39.35 | 7.8 | 7.5 | 2.4 | 0.45 | 0.43 | 0.72 | 2.18 |

**Table 4.14** The compositions of the glasses containing different content of ARR (g)

| Specimen name | $H_3BO_3$ | CaO | $Fe_2O_3$ | $NH_4H_2PO_4$ | ARR | ARR (wt%)[a] |
|---|---|---|---|---|---|---|
| S1 | 2.46 | 1.68 | 11.52 | 33.12 | 0 | 0.00 |
| S2 | 2.46 | 1.68 | 11.52 | 33.12 | 5 | 9.29 |
| S3 | 2.46 | 1.68 | 11.52 | 33.12 | 10 | 17.01 |
| S4 | 2.46 | 1.68 | 11.52 | 33.12 | 15 | 23.52 |
| S5 | 2.46 | 1.68 | 11.52 | 33.12 | 20 | 29.07 |
| S6 | 2.46 | 1.68 | 11.52 | 33.12 | 25 | 33.88 |
| S7 | 2.46 | 1.68 | 11.52 | 33.12 | 30 | 38.08 |

[a]The mass fraction of ARR in the glass composition

the previous studies [101, 104], $Na_3AsO_4$ (the chemical formula can be seen as $3Na_2O{\cdot}As_2O_5$) can be seen as the combination of glass former and glass modifier, which means that ARR could be vitrified by iron phosphate glass.

The compositions of the ARR collected from an antimony smelting enterprise in Hunan province, China, were listed in Table 4.13. The main constituents of the ARR are $As_2O_5$, $Na_2O$, $SiO_2$, and $K_2O$ which could act as glass former and glass modifier in the glass [105] which can contribute to the formation of the glass to host arsenic.

$H_3BO_3$, CaO, $Fe_2O_3$, and $NH_4H_2PO_4$ (purity >99%) were used as sintering materials. The mixture of sintering materials and ARR was homogenized by grinding in a mortar with various mass ratio (Table 4.14). Then, the sample were heated to 1050 °C with the heating rate of 10 °C $min^{-1}$ in corundum crucible and melted at 1050 °C for 1 h. Subsequently, the vitrification products were quenched to room temperature in the air without annealing in order to produce vitreous samples.

The arsenic immobilization ratio ($\varphi$), which was measured by energy dispersive X-ray (EDX), was defined as the following formulation.

$$\varphi = \frac{\gamma \times m}{m_0} \times 100\% \quad (4.14)$$

where $\varphi$ (%) is the arsenic immobilization ratio; m (g) is the mass of the glass; γ (%) is the mass fraction of arsenic in the glass, which was calculated by EDX results; $m_0$ (g) is the mass of arsenic in the starting materials.

Main phase compositions of the ARR as detected by XRD are $Na_4As_2O_7$, $Na_3AsO_3$, $NaAsO_3$, $NaAsO_2$, and $Na_3As$ (Fig. 4.35a). These arsenic compounds have great solubility in water, resulting in great environmental hazard. For the sample S1, the XRD pattern shows the diffraction peaks obviously, corresponded to

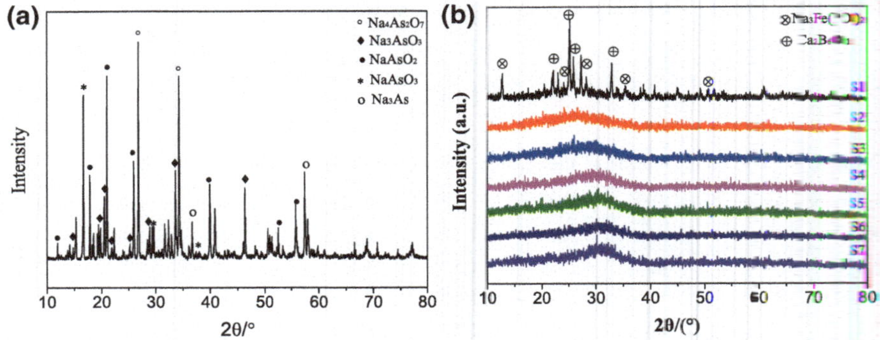


**Fig. 4.35** XRD patterns of the ARR (**a**) and products of vitrification (**b**). Reprinted from Ref. [102] Copyright 2017, with permission from the society of glass technology

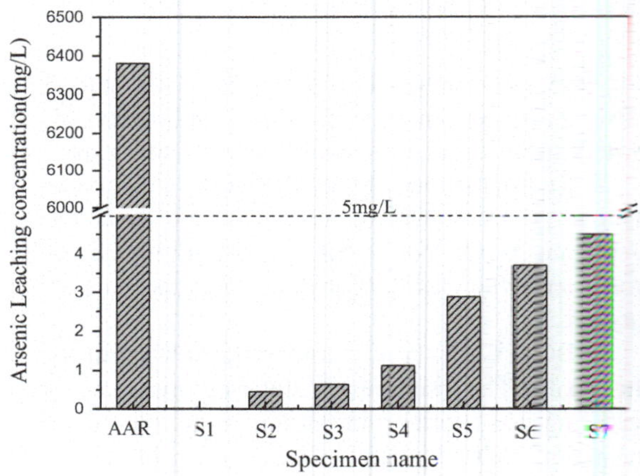


**Fig. 4.36** Arsenic leaching concentration of ARR and glasses. Reprinted from Ref. [102] Copyright 2017, with permission from the society of glass technology

$Na_3Fe(PO_4)_2$ and $Ca_2B_6O_{11}$ (Fig. 4.35b). The crystal peaks disappear with the addition of ARR into the specimen (S2–S7) and a broad hump around $2\theta = 26°–32°$ is observed, which indicates that As and Na from the ARR participate in the formation of glass.

The arsenic leaching concentration of the ARR is 6380 mg $L^{-1}$ (Fig. 4.36), far exceeding the USEPA limitation (<5 mg $L^{-1}$). As revealed in XRD patterns, the arsenic in ARR, which presents in the form of arsenates, has high solubility in acid extraction solution.

With different content of ARR being vitrified into the glass, the arsenic leaching concentration presents a sharp reduction and drops below 5 mg $L^{-1}$. However with the increase addition of ARR, the arsenic leaching concentrations increases from 0.46 mg $L^{-1}$ (S2) to as higher as 4.45 mg $L^{-1}$ (S7). Based on the fact that all specimens show low arsenic leaching concentrations in the TCLP experiments, it is reasonable to speculate that the arsenic in the ARR immobilized into the glass

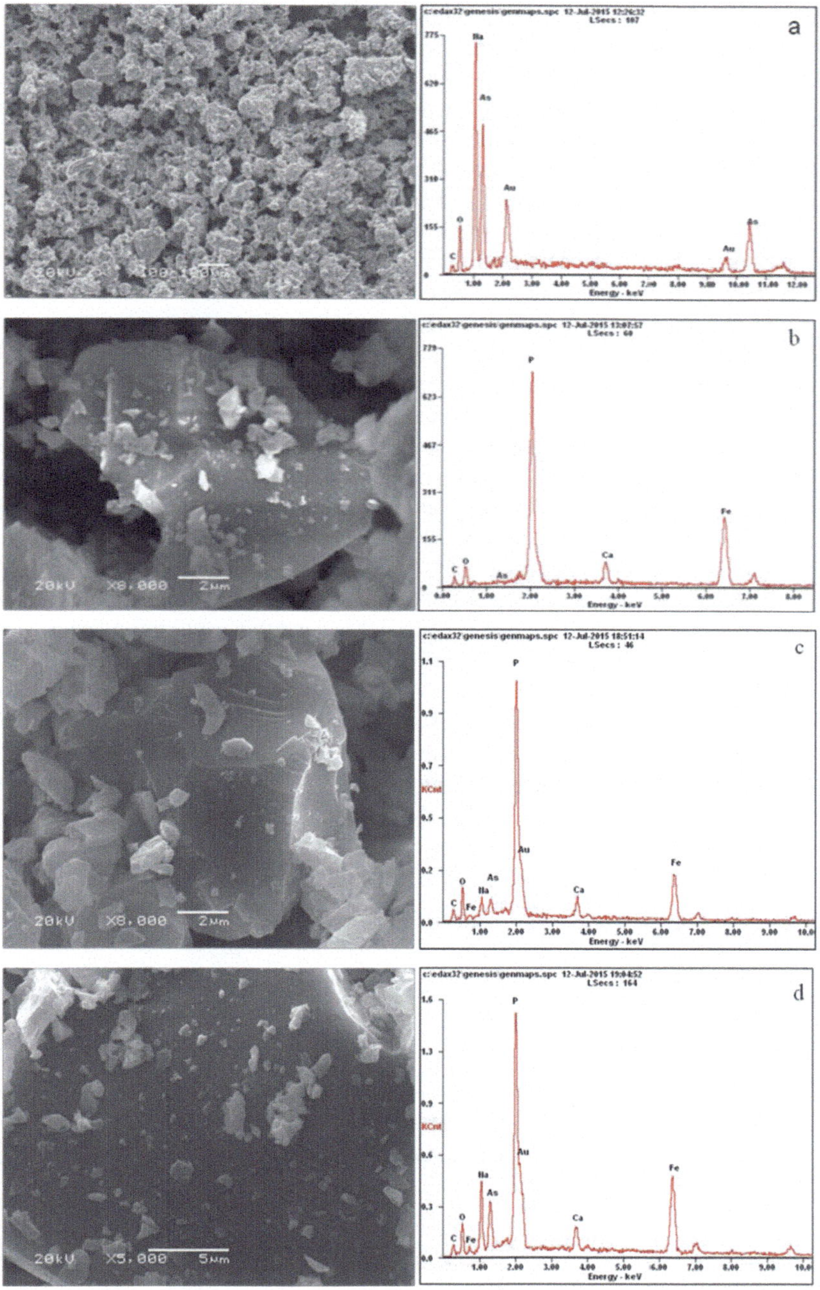


**Fig. 4.37** SEM images of ARR (**a**) and glasses for S1, S2 and S4. Reprinted from Ref. [102] Copyright 2017, with permission from the society of glass technology

participates in the glass network, resulting in a relatively chemical stable state for arsenic, which is consistent with previous study [106].

The ARR is white granular solids and the specimens containing different content ARR show black sheen in the surface, which becomes more noticeable with increase addition of ARR. The surface of the ARR particle is rough, with porous structure, resulting in large contact surface area with the environment (Fig. 4.37a). After being vitrified, the glasses show a regular and compact surface (Fig. 4.37b–d).

According to the EDX results, the least arsenic immobilization ratio is 97.37% (S5), and the highest arsenic immobilization ratio is over 99% (S2, S3, S6, and S7). High arsenic immobilization ratio has been achieved in the process of vitrification, indicating that the immobilization of ARR via vitrification is effective.

Raman spectra of the glasses with different ARR content are shown in Fig. 4.38. It is obvious that all glasses have similar spectra features. Low-scattering cross-section of borate units in the Raman spectra and their low content in the studied glasses allow us to neglect their effects on Raman spectra. All specimens except S2 and S4 show a weak broad band near 270 $cm^{-1}$ assigned to the bending $Q^0$ with Fe as modifier [107], which becomes distinct with more addition of ARR into the glass. A low shoulder appears at 342 $cm^{-1}$ except S2 can be likely related to the vibrations of $AsO_4$ polyhedral [108]. The special peak around 472 $cm^{-1}$ is due to the As–O–As stretch vibration [109] and appears to be merged with vibration of O–P–O bending vibrations of $Q^0$ units which exhibit a common meta centre about 472 $cm^{-1}$ [110]. Therefore, it is reasonable to speculate that the vibration around 472 $cm^{-1}$ is due to the existence of P–O–As bond, which is consistent with previous study [101]. The low shoulder around 586 $cm^{-1}$ is assigned to the overlapping vibrations involving iron oxygen polyhedral and $(P_2O_7)^{4-}$ groups which are characteristics for the structure dominated by $Q^1$ tetrahedrons [111]. Another band around 634 $cm^{-1}$ is ascribed to the asymmetric stretching of the bridging oxygen in metaphosphate group ($Q^2$ group) [112] and the intensity of the peak presents evident with more ARR added into the glass, indicating that the ARR

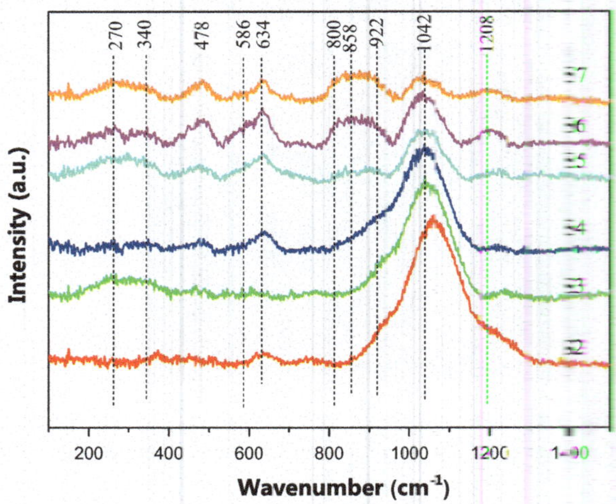


**Fig. 4.38** Raman spectra for glass containing different content of ARR. Reprinted from Ref. [102] Copyright 2017, with permission from the society of glass technology

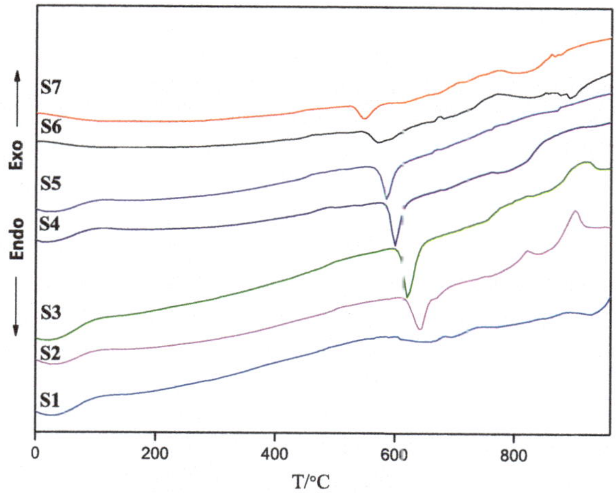


**Fig. 4.39** DTA parameters for glass containing different content of ARR. Reprinted from Ref. [102] Copyright 2017, with permission from the society of glass technology

immobilized into the glass could promote the relative content of $Q^2$ groups. Similar to the peak around 634 $cm^{-1}$, the peak at 855 $cm^{-1}$ corresponded to the asymmetric vibration of As–O–As [95] exhibits the same variation tendency and the peak becomes obvious with more ARR immobilized into the glass, especially for S6 and S7. Such change in the peak around 855 $cm^{-1}$ proves that arsenic immobilized into the glass may participate in the glass structure. The broad band around 1042 $cm^{-1}$ exhibiting decreased intensity is due to the symmetric stretching mode of O–P–O non-bridging bond in $Q^1$ groups [112], which indicates the decrease of $Q^1$ groups in the glass. Moreover, the new band appeared around 1208 $cm^{-1}$ is attributed to the asymmetric stretching motions of the O–P–O non-bridging oxygen in $Q^2$ groups [111] and the intensity of the new band increased with more addition of ARR into the glass, implying the increase of $Q^2$ groups in the glass.

Moreover, the peaks around 1042 and 1208 $cm^{-1}$ change with the content of ARR, which indicates the conversion of $Q^1$ into $Q^2$ units in the glass network and less depolymerization of phosphate chains. Therefore, it is reasonable to speculate that arsenic is inserted in the glass and participates in the glass structure, which results in greater capacity for glass modifier (Na) and promoting the conversion of $Q^1$ into $Q^2$ units in the glass.

The thermal stability of the glass containing different content ARR can be described by DTA in Fig. 4.39 (Tg for glass transition temperature, and Tc for the first crystallization temperature). Characteristic temperatures of the glasses are summarized in Table 4.15.

**Table 4.15** Characteristic temperatures (°C) for glass containing different content of ARR

| Specimen name | S1 | S2 | S3 | S4 | S5 | S6 | S7 |
|---|---|---|---|---|---|---|---|
| Tg | 596 | 611 | 600 | 585 | 559 | 545 | 521 |
| Tc | 736 | 815 | 806 | 793 | 773 | 765 | 780 |
| Tc − Tg | 140 | 204 | 206 | 208 | 214 | 220 | 259 |

The glass transition temperature (Tg) of all glasses are given in Table 4.15. All glasses exhibit a slight endothermic effect between 521 and 611 °C, which is ascribed to the glass transition temperature. The glass transition temperature of S1 corresponded to no ARR in the glass is 596 °C (Table 4.15). With the addition of ARR into the glass, the glass transition temperature presents a downward trend from 611 °C (S2) to 521 °C (S7). According to previous study [113], the glass transition temperature plays a vital role in understanding the physical properties of glass. The glass transition temperature depends on the density of covalent cross-linking and the number and strength of the cross-links between the cation and oxygen atoms [114]. Therefore, the decrease in Tg is due to the weaker strength of the cross-links between the cations and oxygen atoms in the phosphate glass with increase content of ARR.

In addition, the value of (Tc − Tg) could be used to describe the thermal stability of the glass with the addition of ARR. A higher value of (Tc − Tg) stands for a greater thermal stability in the glass. With the addition of ARR, the value of (Tc − Tg) varies from 140 °C (S1) to 259 °C (S7), indicating that the thermal stability of the glasses are improved [115].

After being vitrified, all specimens containing ARR formed x-amorphous glass and showed a regular and homogeneous surface with shiny, metallic luster on the external surface. Vitrification is an effective method to treat ARR. The arsenic existed in the glasses in the form of $AsO_4$ and connected with $PO_4$ tetrahedron via bridging oxygen. Besides, the thermal stability of the glasses with the addition of ARR was improved.

### 4.3.3 Sodium Arsenate Vitrification by Borosilicate Glass

It has been reported that arsenic can be chemically immobilized into cementitious environment of the solidified/stabilized matrices by three important immobilization mechanisms: sorption onto C–S–H surface [116], replacing $SO_4^{2-}$ of ettringite [117, 118], and reaction with cement components to form calcium–arsenic compounds, the insolubility phase [119, 120]. However, arsenic can be released from cement-solidified immobilization products due to the desorption onto C–S–H surface or the substitution of arsenate by sulfate in ettringite, which causes serious second pollution [7, 99, 121]. Therefore, new method, such as vitrification, must be brought forth to solve arsenic residue pollution.

Vitrification is a well-established technology that involves the conversion of the waste into a stable and homogeneous (silicate) glass through the thermal treatment

**Table 4.16** Average chemical compositions of the cullet (wt%)

| $SiO_2$ | $B_2O_3$ | CaO | $Na_2O$ | $Fe_2O_3$ | $K_2O$ | MgO | $Al_2O_3$ | ZrO | Other oxides |
|---|---|---|---|---|---|---|---|---|---|
| 72.95 | 14.85 | 0.35 | 5.54 | 0.23 | 0.048 | 0.025 | 2.161 | 0.079 | 3.766 |

of fusion [122]. After being vitrified, hazardous waste can be incorporated into the inorganic amorphous network and turn to stable [123]. However, there is a little studies concern about the arsenic residue treated by vitrification. In our present work [104] the glass modifier (antimony oxide) was introduced into iron phosphate glass to strengthen glass structure containing arsenic. However, the previously studies did not directly illustrate arsenic evolution process during the vitrification treatment.

In this section, the cullet was used as host to vitrify arsenate waste. The using cullet as the host can significantly reduce firing temperature by 200 °C, which could effectually avoid the arsenic volatilization and reduce the vitrification cost [124]; Additionally, the cullet can be collected from glass fragments and recycled the resources [105, 125, 126]; Above all, the cullet is mainly composed of glass network former oxide ($SiO_2$), together with small amounts of some modifier oxides ($K_2O$, $Na_2O$, CaO, MgO) and other ceramic oxides ($Al_2O_3$, $Fe_2O_3$), which are responsible for the physicochemical property of the final produced glass to vitrify arsenate [127, 128].

The cullet was collected from laboratory and was ground into powder. The compositions of the cullet (Table 4.16) are glass network formers ($SiO_2$, $Al_2O_3$, $B_2O_3$), glass modifiers (CaO, MgO) and glass intermediate ($Fe_2O_3$) [126]. The cullet, due to their oxide content, can provide by melting a useful preliminary batch

**Table 4.17** Content of cullet powder and sodium arsenate in mixture undergo vitrification (g)

| Specimen name | Cullet powder | $Na_3AsO_4·12H_2O$ |
|---|---|---|
| G0 | 35 | 0 |
| G1 | 35 | 6 |
| G2 | 35 | 12 |
| G3 | 35 | 24 |

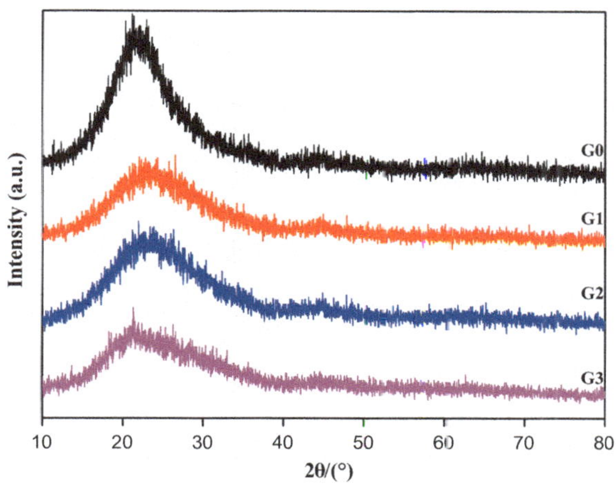


**Fig. 4.40** XRD patterns of the vitrification. Reprinted from Ref. [50] Copyright 2016, with permission from Elsevier

for the preparation of glasses to host sodium arsenate. Moreover, the use of cullet can lower the glass melting temperature.

Analytical grade $Na_3AsO_4 \cdot 12H_2O$ (purity >99%) was used as starting materials. The mixture of cullet and sodium arsenate was homogenized by grinding in a mortar with various mass ratio (Table 4.17). The sample were heated to 1200 °C (heating rate is 10 °C $min^{-1}$) in corundum crucible and melted at 1200 °C for 1 h. After that the vitrification products were then quenched to room temperature in the air without annealing to produce vitreous samples.

XRD has been used to confirm the glassy state of the as-quenched samples [129]. As shown in Fig. 4.40, a broad hump around $2\theta = 22°$ is observed with no indication of any diffraction peaks in all samples. No evidence of arsenic partitioning in any crystallized samples were observed, indicating great compatibility of sodium arsenate in the cullet.

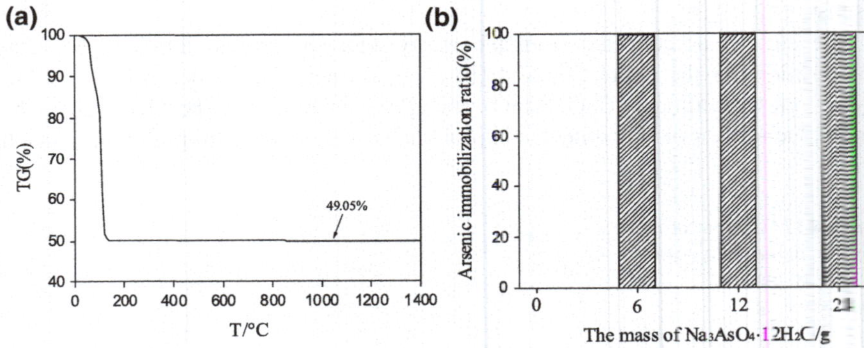


**Fig. 4.41** TG curves for sodium arsenate dodecahydrate (**a**) and arsenic immobilization ratio of the glasses (**b**). Reprinted from Ref. [50] Copyright 2016, with permission from Elsevier

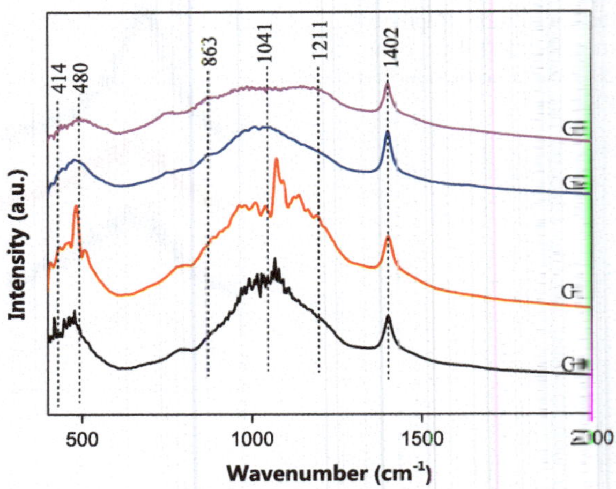


**Fig. 4.42** FTIR spectra of glasses for G0, G1, G2, G3. Reprinted from Ref. [50] Copyright 2016, with permission from Elsevier

The thermal gravimetric curve of $Na_3AsO_4{\cdot}12H_2O$ can be seen in Fig. 4.41a. There is a sharp mass reduction below 138 °C, which is due to the loss of crystal water in $Na_3AsO_4{\cdot}12H_2O$. The remaining mass ratio keeps constant at 49.05% (the mass ratio of $Na_3AsO_4/Na_3AsO_4{\cdot}12H_2O$ is 49.05%) up to 1400 °C. It proves that the arsenic in sodium arsenate is stable and little arsenic volatilizes in the process of vitrification, which consists with previous study [130].

The arsenic immobilization ratio in the glass (Fig. 4.41b) indicates that the arsenic immobilization ratio keeps constant over 99%, which means the arsenic rarely volatilized in the process of vitrification.

As for the structural groups involved in the glass network, FTIR spectra (Fig. 4.42) can be classified into two characteristic ranges: (1) The range from 1400 to 620 $cm^{-1}$ connected with the asymmetric stretching vibration of Si–O–Si, Si–O–Al and $Si–O^-$ bonds [129, 131], Si–O–Si symmetric stretching vibration of tetrahedral bridging oxygen [132]. (2) The range from 600 to 400 $cm^{-1}$ in which the bonds are mainly due to the bending vibration of Si–O–Si and Si–O–Al/Fe bridges [133].

Inspecting these spectra, the prominent bands are seen around 1402, 1211, 1041, 863, 480 and 414 $cm^{-1}$. The peak around 1402 $cm^{-1}$ corresponds to B–O

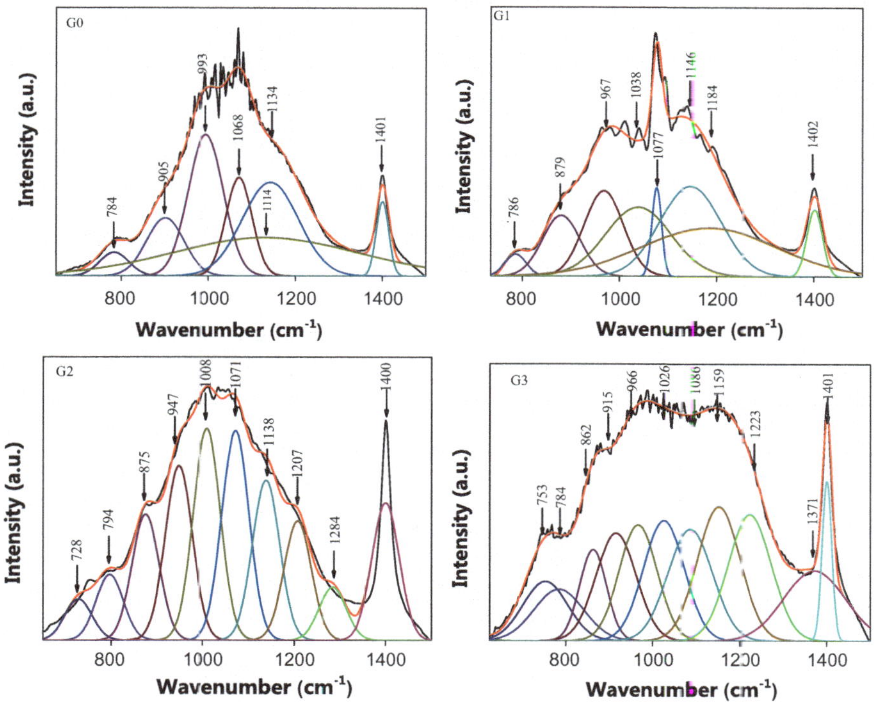


**Fig. 4.43** Deconvoluted FTIR spectra of glasses for G0, G1, G2, G3. Reprinted from Ref. [50] Copyright 2016, with permission from Elsevier

stretching vibration in Si–O–B bond [134], and the peak remains constant with more addition of $Na_3AsO_4{\cdot}12H_2O$ into the glass. The peak near 1211 cm$^{-1}$ relates to Si–O– in $Q^4$ structure [135], and the intensity of the peak becomes enhanced with increase of $Na_3AsO_4{\cdot}12H_2O$, which indicates that this structure has been strengthened. The peak at 1041 cm$^{-1}$ is due to the band of Si–O stretching vibration [136] and it becomes broad with more addition of $Na_3AsO_4{\cdot}12H_2O$. The peak near 863 cm$^{-1}$ is related to the Si–O–Si antisymmetric stretching vibrations of bridging oxygens between tetrahedra [137]. The two peaks located at 480 cm$^{-1}$ and 414 cm$^{-1}$ correspond to bending modes of O–Si–O [138] and modifier cations in their residing sites neighboring bridging and non-bridging types [137], respectively. The majority absorption bands located at high wavenumber (700–1500 cm$^{-1}$) are large with shoulders in the antisymmetric band, which are closely linked with $Q^0$, $Q^1$, $Q^2$, and $Q^3$ units in the glass. In order to verify the relative change of $Q^0$, $Q^1$, $Q^2$, and $Q^3$ units in the glass, a deconvolution of the experimental spectra is necessary.

The specific deconvoluted spectrum for G0 glass is given in Fig. 4.43. The small band near 784 cm$^{-1}$ is attributed to bending of bridged $SiO_4$ tetrahedral group [135]. The bands located between 900 and 1200 cm$^{-1}$ are related to the Si–O in $Q^n$ (n = 1, 2, 3, 4) structure. The band at 905 cm$^{-1}$ is due to the vibration of Si–O–[NB] in $Q^1$ structure [132]. The obvious peak around 993 cm$^{-1}$ is corresponded to $SiO_4$ tetrahedron with one non bridging oxygen atom [129]. The band near 1068 cm$^{-1}$ is due to Si–O–Si antisymmetric stretching vibrations of bridging oxygen within the tetrahedral [137]. Bands around 1114 and 1134 cm$^{-1}$ are corresponded to Si–O–[NB] in $Q^3$ structure [137]. The last peak around 1401 cm$^{-1}$ is due to B–O stretching vibration in Si–O–B bond [134]. Changes in the FTIR spectra of the glass with increasing sodium arsenate content unambiguously manifest the incorporation of arsenic and sodium into the silicate glass. Structural changes in the glass network are reflected especially in the region of 800–1200 cm$^{-1}$ of the FTIR spectra, whereas

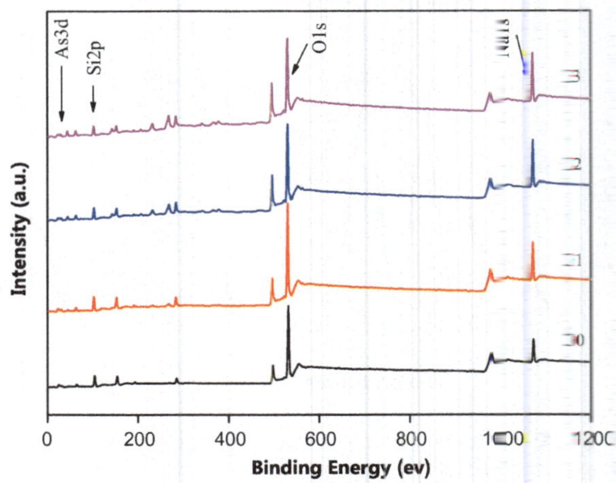


**Fig. 4.44** XPS survey scans of glasses for G0, G1, G2, G3. Reprinted from Ref. [50] Copyright 2016, with permission from Elsevier

vibrations of Si–O–Si in $Q^n$ structure can be used to identify the arsenic participating in the silicon glass structure [139].

The G3 has been used as the example to illustrate this transformation in the silicate glass. The specific peak related to the glass modifier of $Na^+$, $Ca^{2+}$ etc. around 414 $cm^{-1}$ in G0 has shifted to high-frequency region in G3 glass, which is resulted from more sodium added into the glass. $Na^+$ acts as glass modifier and provokes the disruption of continuity of the glassy network due to the breaking of some Si–O–Si bonds in the glass [140]. Other corresponded bands in G0 also shift to high-frequency region in G3, with the more addition of sodium arsenate into the glass, which means the $Q^n$ structure in silicate glass has been strengthened. Moreover, new bands have emerged around 753, 862, 1223 and 1371 $cm^{-1}$, which correspond to Si–O–Si symmetric stretching vibrations of bridging oxygens between tetrahedra, Si–O–Si antisymmetric stretching vibrations of bridging oxygens between tetrahedra [137], Si–O– in $Q^4$ structure [135], and B–O vibration [141], respectively. These new bands have directly proved that the silicate glass network structure has been strengthened by sodium arsenate. Though Na in the sodium arsenate can act as glass modifier and break the silicate glass network, arsenic which exists in the form of $AsO_4^{3-}$ tetrahedra in the arsenate [142] can form chemical bonds with silicon tetrahedra, and participate in the glass structure, which results in greater capacity for glass modifier.

Survey scans of glasses with different content of $Na_3AsO_4 \cdot 12H_2O$ are shown in Fig. 4.44. Silicon, oxygen, arsenic, and sodium have been detected in the glass, which indicates that sodium arsenate has been immobilized into the glass. Each spectrum includes the photoelectron and X-ray induced Auger peaks from the constituent elements of the glass [143]. To get a well understanding of Si 2p, O 1s, As 3d and Na 1s in the glass, the spectra have been deconvolved and the results are presented in the following discussion.

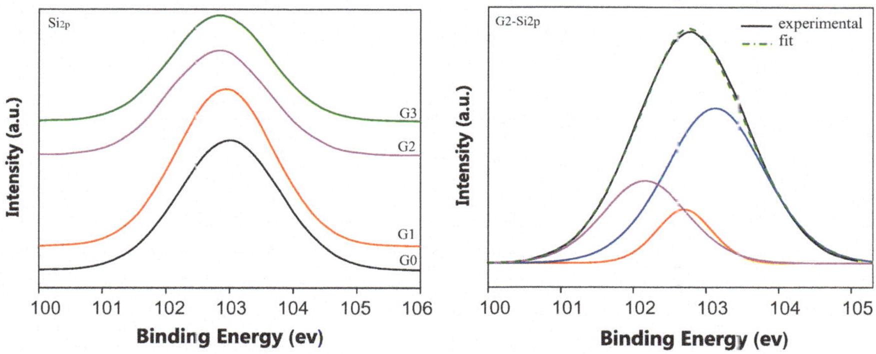


**Fig. 4.45** Si 2p spectra of glasses for G0, G1, G2, G3 (left) and a typical deconvoluted peak for G2 (right). Reprinted from Ref. [50] Copyright 2016, with permission from Elsevier

**Table 4.18** Peak positions, FWHM, and relative abundance from the curve fitting of Si 2p and O 1s

| Mass (g) of $Na_3AsO_4 \cdot 12H_2O$ | Si 2p | | | O 1s | | |
|---|---|---|---|---|---|---|
| | BE (ev) | FWHM (ev) | Percentage (%) | BE (ev) | FWHM (ev) | Percentage (%) |
| 0 | 103.11 | 1.64 | 82.53 | 532.12 | 1.67 | 54.21 |
| | 102.12 | 1.27 | 17.47 | 531.31 | 2.18 | 35.79 |
| 6 | 103.25 | 1.71 | 65.88 | 532.29 | 1.57 | 53.87 |
| | 102.90 | 0.83 | 8.93 | 531.95 | 1.80 | 3.28 |
| | 102.29 | 1.33 | 25.19 | 531.37 | 0.73 | 42.85 |
| 12 | 103.13 | 1.60 | 58.52 | 532.34 | 1.47 | 45.34 |
| | 102.71 | 0.90 | 11.83 | 531.70 | 0.83 | 11.39 |
| | 102.16 | 1.37 | 29.65 | 530.87 | 1.70 | 43.27 |
| 24 | 103.08 | 1.52 | 50.72 | 532.39 | 1.39 | 40.06 |
| | 102.78 | 1.09 | 15.22 | 531.83 | 1.07 | 15.26 |
| | 102.18 | 1.46 | 34.06 | 530.86 | 1.74 | 44.06 |

(a) Si 2p spectra

The evolution of binding energy and relative intensity of Si 2p line components with different sodium arsenate content are shown in Fig. 4.45. The shift to the low binding energy with more addition of sodium arsenate and the maximum shift is approximately 0.25 eV, indicating that the valence electron density around the Si nuclei increases with the sodium arsenate content increasing [144]. Moreover the broadening of the Si 2p peaks are caused by differences in the chemical environments changing the electron density at the silicon atom and hence induces small chemical shifts broadening the Si 2p spectra. The introduce of sodium arsenate, of which Na acts as glass modifier and Arsenic plays the role of glass former, into the cullet can result in the formation of Si–O(NBO)–Me (Me = Na, Ca, Fe, B etc.) and Si–O(BO)–As at the expense of Si–O(BO)–Si in the glass. A Gaussian (80%)–Lorentzian (20%) mixed function has been used to determine this change in Si 2p. The peak positions, FWHM and relative abundance are given in Table 4.18.

In G0 glass, there exist two types of Si. One is boned to BO which shows high binding energy and the other is boned to NBO, presenting low binding energy. The corresponding binding energy of the two types of Si are 103.11 eV, 102.12 eV, which accounts for 82.53%, 17.47% in the glass, respectively. Therefore, it can be deduced that the main structure of the glass in G0 is formed by Si–O–Si chains. However, with the addition of sodium arsenate, the type of Si bones to BO transition to Si–O(BO)–Si and Si–O(BO)–As, and this change makes arsenic immobilized in the glass.

In order to illustrate this change in the glass, the relative percentages of the three types of Si are used to explain the change of Si 2p. The binding energy of the three types of Si complies with the following order [144]: Si–O(BO)–Si, Si–O(BO)–As, Si–O(NBO)–Me. The relative percentages of Si–O(BO)–Si, Si–O(BO)–As, Si–O (NBO)–Me are 65.88%, 8.93%, 25.19% in G1 glass and 58.52%, 11.83%, 29.65%

in G2 glass, and 50.72%, 15.22%, 34.06% in G3 glass, respectively. The addition of sodium arsenate, of which the chemical formula can be seen as $Na_3O_{1.5}\cdot AsO_{2.5}\cdot 12H_2O$ (glass modifier is $Na_3O_{1.5}$ and glass former is $AsO_{2.5}$ in the glass), results in the relative increase of Si–O(BO)–As and Si–O(NBO)–Me at the decrease of Si–O(BO)–Si.

(b) O 1s spectra

These peaks of O 1s provide us with very useful information to identify the chemical bonds in the silicate glasses. The characteristic of O 1s is the shift to the high binding energy, with more addition of the sodium arsenate into the glass. The maximum shift is about 0.20 eV over the composition range studied, which means a decrease of valence electron around the O nuclei. According to the previous works [145], the low binding energy of O 1s line is interpreted to represent the non-bridging oxygen atoms (Si–O–Me); the high energy line corresponds to the bridging oxygen (Si–O–Si). The shape of O 1s peak is asymmetric, and it evidently contains three components in G1–G3 glasses. Then, it is possible to analyze the bonding states of the oxide ions by deconvoluting the O 1s spectrum. The peak positions, FWHM and relative abundance are also given in Table 4.18.

The curve fitting of a typical O 1s photoelectron spectrum for the G0 is shown in Fig. 4.46. The peak at 532.12 eV is ascribed to BO (Si–O(BO)–Si) [146], and the other peak located at 531.31 eV is due to NBO (Si–O(NBO)–Me) [147]. However, the peaks present significant changes when the sodium arsenate has been added. The new formation of Si–O(BO)–As results in the second BO (Si–O(BO)–As), of which the binding energy locates between Si–O(BO)–Si and Si–O(NBO)–Me.

The initial relative percentage of Si–O(BO)–Si, Si–O(NBO)–Me are 64.21%, 35.79% in G0 glass. With the addition of sodium arsenate, the relative contents of Si–O(BO)–Si, Si–O(BO)–As, Si–O(NBO)–Me are 53.87, 3.28, 42.85% in G1 and 45.34, 11.39, 43.27% in G2, and 40.06, 15.26, 44.06% in G3. It is obvious that the addition of sodium arsenate, which is the combination of glass former (As) and

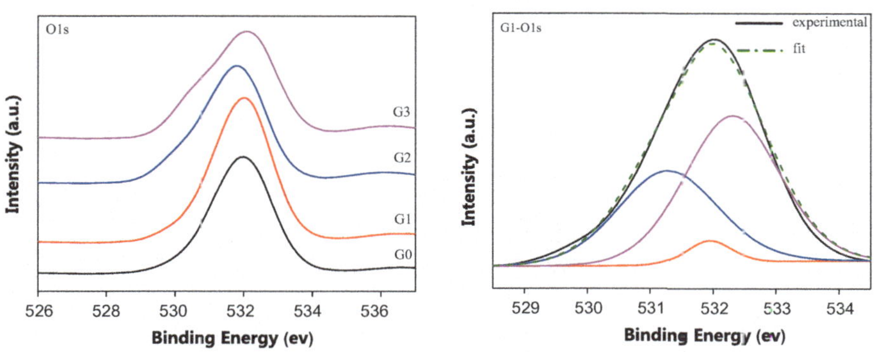


**Fig. 4.46** O 1s spectra of glasses for G0, G1, G2, G3 (left) and a typical deconvoluted peak for G1(right). Reprinted from Ref. [50] Copyright 2016, with permission from Elsevier

glass modifier (Na), results in the formation of Si–O(BO)–As and Si–O(NBO)–Me. Moreover, the arsenic exists in the form of $AsO_4$ tetrahedron in sodium arsenate and could form chemical bonds with $SiO_4$ tetrahedron via bridging oxygen, which makes arsenic forming Si–O(BO)–As in the glass. Moreover, this change is consistent with the change in the FTIR spectra that the intensity of the peaks located between 700 and 1500 $cm^{-1}$ has been enhanced, namely, the new formation of Si–O(BO)–As in the glass resulting in strengthening the $Q^n$ structure in silicate glass and making the glass have greater capacity for glass modifier.

(c) As 3d

The core level spectra of As 3d in the glass are shown in Fig. 4.47. There exists no peak in G0 due to no sodium arsenate has been introduced into the specimen. When arsenic has been added into the glass, a small peak has appeared. The intensity of As 3d increases with increase of sodium arsenate into the glass. A small shift to high binding energy has been observed in the Fig. 4.47, which means the decrease of electron density around the arsenic atom. When arsenic has been added into the glass, arsenic has embedded itself into the silicate glass and participated in the glass structure, resulting in the shift to high binding energy. The chemical bond of arsenic has changed from As–O–Me to As–O–Si in glass, and this trend is more obvious with more arsenic in the glass, which makes arsenic immobilized in the glass.

Arsenic can successfully dissolve into the glass in 1200 °C, with a high arsenic immobilization ratio. FTIR spectra show that silicate glass structure has been strengthened with the addition of sodium arsenate. Arsenic existing in the form of $AsO_4$ could form chemical bonds with $SiO_4$ via bridging oxygen, resulting in a great capacity for glass modifier. The XPS results prove that the addition of sodium arsenate leads to the new formation of Si–O(BO)–As, which keeps arsenic stable in the glass.

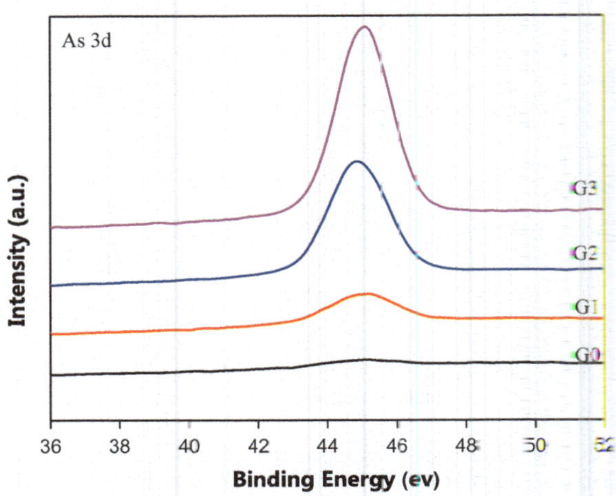


**Fig. 4.47** As 3d spectra of glasses for G0, G1, G2, G3. Reprinted from Ref. [50] Copyright 2016, with permission from Elsevier

### 4.3.4 *Vitrification of Sodium Arsenate by Copper Slag Based Glass*

Copper slag is a by-product obtained during the matte smelting and refining of copper [42], which is of great environment concern [148, 149]. Current options for management of copper slag include recycling, recovering of metal, production of value added products and disposal in slag dumps or stockpiles. New methods to reuse copper slag become very urgent. The main phases in copper slag are fayalite-$Fe_2SiO_4$, magnetite-$Fe_3O_4$, silica-$SiO_2$, and matte, which could be used as the raw material to form glass to host hazardous waste [150].

According to our previous works [104], arsenic trioxide could be incorporated into the phosphate based glass and lead to the conversion of $Q^1$–$Q^2$ in the glass structure. But little effort has been focused to use industrial waste to vitrify arsenic produced by industry, which is commonly in the form of arsenate, such as sodium arsenate, calcium arsenate and ferrous arsenate. In practical, the vitrification of sodium arsenate by industrial waste, such as copper slag, is of great practical significance to environment protection.

**Table 4.19** Average chemical compositions of the glass (g)[a]

| Specimen ID | $SiO_2$ | $B_2O_3$ | $Fe_2O_3$ | $Na_2O$ | $Na_3AsO_4$ | Arsenic mass fraction/% |
|---|---|---|---|---|---|---|
| G1 | 23.10 | 3.90 | 8.00 | 4.40 | 0.00 | 0 |
| G2 | 23.10 | 3.90 | 8.00 | 3.42 | 2.26 | 2 |
| G3 | 23.10 | 3.90 | 8.00 | 2.36 | 4.67 | 4 |
| G4 | 23.10 | 3.90 | 8.00 | 1.25 | 7.25 | 6 |
| G5 | 23.10 | 3.90 | 8.00 | 0.00 | 10.00 | 8 |

[a]All data were measured by electronic balance and the error range is ±0.01 g

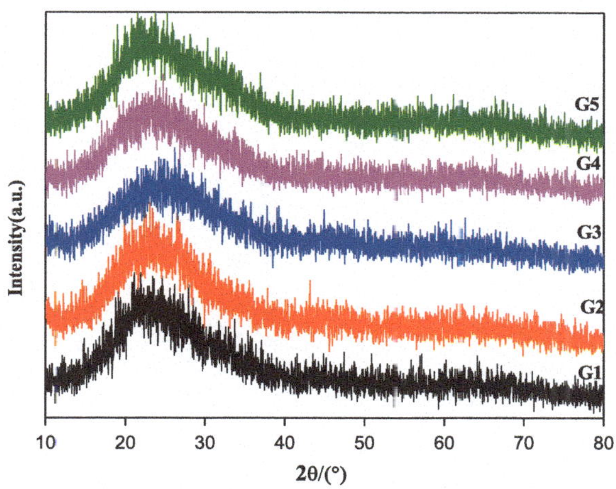


**Fig. 4.48** XRD patterns of the vitrification products. Reprinted from Ref. [101] Copyright 2017, with permission from Elsevier

Pure materials such as $Na_3AsO_4$, $Fe_2O_3$ and $SiO_2$, which are main components of arsenic-containing waste and copper slag, have been used to explore the arsenic vitrification mechanism and the chemical durability of the vitrification product. This provides a solution for the hazardous solid wastes generated by pyrometallurgy industry and minimizes the environmental impact caused by these industrial wastes [101].

The samples with different content of arsenic were prepared by using analytical grade >9% purity $SiO_2$, $H_3BO_3$, $Fe_2O_3$, $Na_2CO_3$ and $Na_3AsO_4$ as starting materials, and the chemical compositions of the $(100 - x)(58SiO_2–20Fe_2O_3–10B_2O_3–12Na_2O) - x$As ($x = 0, 2, 4, 6, 8$ wt%) specimens were presented in Table 4.19. In order to avoid the influence of sodium and to keep the arsenic mass fraction increase from 0 to 8%, different proportions of $Na_2CO_3$ and $Na_3AsO_4$ were added into the specimens.

The compositions containing up to 8% of arsenic form glasses and no crystalline phases are detected (Fig. 4.48). Moreover, no evidence of arsenic partitioning in any crystallized samples are observed, indicating great compatibility of arsenate in the glasses.

The overall FTIR spectra of $(100 - x)(58SiO_2–20Fe_2O_3–10B_2O_3–12Na_2O) - x$As ($x = 0, 2, 4, 6, 8\%$) glasses are shown in Fig. 4.49. Except for the weak band around 1627 cm$^{-1}$ which may be attributed to the B–O stretching vibration of $BO_3$ units [151], six prominent bands can be seen around 1395, 1196, 1045, 946, 785 and 448 cm$^{-1}$, respectively. The peak around 1395 cm$^{-1}$ corresponds to B–O stretching vibration in Si–O–B bond [134], which remains constant with more addition of arsenic into the glass. The low shoulders appear at 1196 and 946 cm$^{-1}$ are related to the vibrations of the Si–O tetrahedral groups in framework silicate [130], and Si–O stretching mode of non-bridging oxygen [152], respectively. Additionally, the intensity of the peaks becomes weak with the increase of arsenic, indicating that this structure has been weakened. Moreover, the strong and broad band at 1045 cm$^{-1}$ is related to the Si–O–Si asymmetric stretching [153], and the

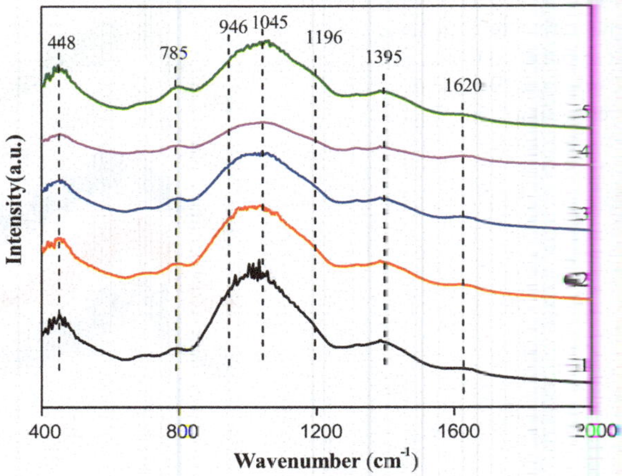


**Fig. 4.49** FTIR spectra of glasses for $(100 - x)(58SiO_2–20Fe_2O_3–10B_2O_3–12Na_2O) - x$As. Reprinted from Ref. [101] Copyright 2017, with permission from Elsevier

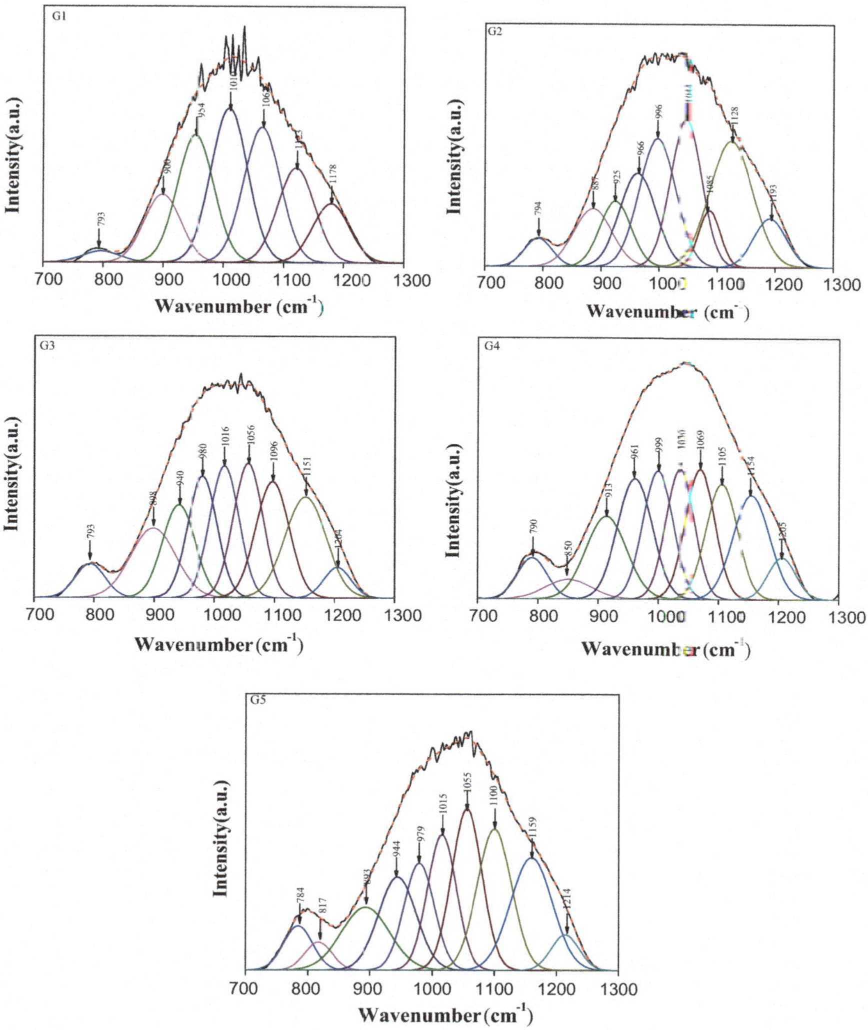


**Fig. 4.50** Deconvoluted FTIR spectra of glasses for (100 − *x*)($58SiO_2$–$20Fe_2O_3$–$10B_2O_3$–$12Na_2O$) − *x*As. Reprinted from Ref. [101] Copyright 2017, with permission from Elsevier

peak becomes broad with the increase content of arsenic. The small peaks around 785 and 448 $cm^{-1}$ are due to Si–O–Si symmetric stretching vibration and Si–O–Si bending vibration, of which the intensity increases with increasing arsenic into the glasses. Moreover, the broad absorption bands located at 700–1300 $cm^{-1}$ are closely linked with $Q^0$, $Q^1$, $Q^2$, and $Q^3$ units in the silicate glass. Therefore, the Gaussian peak fit method [154] has been chosen to verify the relative change of $Q^0$, $Q^1$, $Q^2$, and $Q^3$ units in the glass caused by the addition of arsenic and the deconvoluted spectra of the (100 − $x$)(58$SiO_2$–20$Fe_2O_3$–10$B_2O_3$–12$Na_2O$) − $x$As ($x$ = 0, 2, 4, 6, 8 wt%) glasses are presented in Fig. 4.50. The band near 793 $cm^{-1}$ is due to the vibration of bridged $SiO_4$ tetrahedral group [135], and the intensity of the peak becomes strong with more addition of arsenic into the glass, indicating the addition of arsenic can enhance this structure in the glasses. The distinct peak around 900 $cm^{-1}$ is assigned to Si–O–[NB] in $Q^1$ structure [132]. Other bands at 954 and 1062 $cm^{-1}$ are ascribed to Si–O–[NB] in $Q^2$ structure, and Si–O–Si antisymmetric stretching vibrations of bridging oxygen within the tetrahedral [137]. The peak at 1010 $cm^{-1}$ is due to Si–O–[NB] stretching vibration [155]. The high-frequency bands around 1123 and 1178 are related to Si–O–Si stretching vibration in $Q^4$ structure and Si–O–[NB] in $Q^3$ structure [135].

Changes in the FTIR spectra of (100 − $x$)(58$SiO_2$–20$Fe_2O_3$–10$B_2O_3$–12$Na_2O$) − $x$As ($x$ = 0, 2, 4, 6, 8 wt%) with increase content of arsenic unambiguously manifest the incorporation of arsenic into the structural network of the parent modified iron silicate glass. Structural changes in the silicate network are reflected especially in the high-frequency part of the FTIR spectra, whereas vibrations of structural units containing $AsO_4$ tetrahedral can be seen in the low-frequency region. The peak around 900 $cm^{-1}$ splits into two peaks with the addition of arsenic—the high-frequency peak near 925 $cm^{-1}$ corresponds to Si–O–[NB] in $Q^1$ structure [132] and the low-frequency peak near 887 $cm^{-1}$ is ascribed to the vibration of $AsO_4$ tetrahedral [156], of which the two peaks shift to low-frequency region with more addition of arsenic into the glasses. Previous XRD results have proved that no arsenic-containing crystal has been found, which means that arsenic participates in the glass structure in the form of $AsO_4$ tetrahedra. Besides, the peaks around 954, 1010, 1062, 1123 and 1178 $cm^{-1}$ shift to high-frequency region and these bands belong to the crosslinking (880–1105 $cm^{-1}$) of the glass network [157], which indicates that the silicate glass structure has been strengthened. Furthermore, a new peak has appeared around 1200 $cm^{-1}$ related to Si–O vibration in $Q^4$ structure

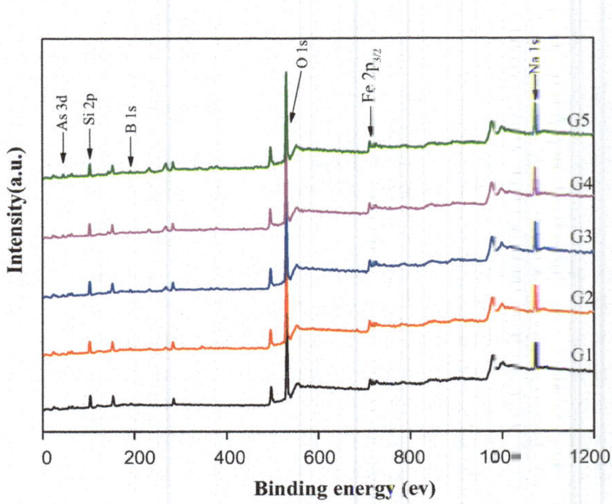


**Fig. 4.51** XPS survey scans of glasses for (100 − $x$)(58$SiO_2$–20$Fe_2O_3$–10$B_2O_3$–12$Na_2O$) − $x$As. Reprinted from Ref. [101] Copyright 2017, with permission from Elsevier

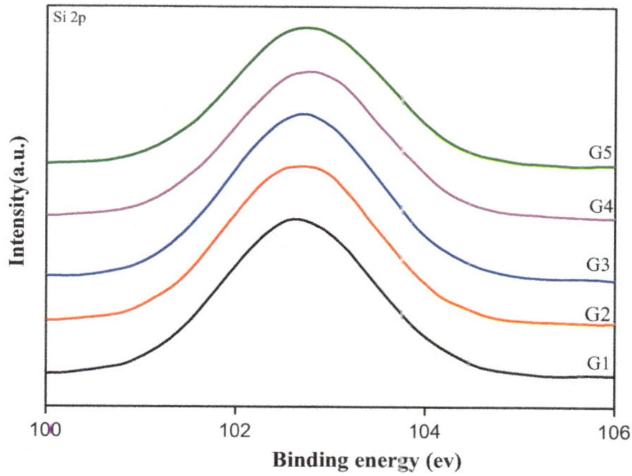


**Fig. 4.52** Si 2p spectra of glasses for $(100 - x)(58SiO_2–20Fe_2O_3–10B_2O_3–12Na_2O) - x$As. Reprinted from Ref. [101] Copyright 2017, with permission from Elsevier

[135], which proves that the arsenic in the glass could increase the crosslinking degree of the glass structure.

XPS was operated to study the detail structure information of $(100 - x)(58SiO_2–20Fe_2O_3–10B_2O_3–12Na_2O) - x$As ($x = 0, 2, 4, 6, 8\%$) glasses. Survey scans of glasses with different content of arsenic are shown in Fig. 4.51. Silicon, oxygen, arsenic, sodium, iron and boron have been detected in the glass, which indicates that arsenic has been immobilized into the glass. Each spectrum includes the photoelectron and X-ray induced Auger peaks from the constituent elements of the glass [143]. Based on previous FTIR results, the addition of arsenic into the glasses can influence the silicate glass structure.

(1) The XPS spectra of Si 2p

The Si 2p spectra containing different content of arsenic are shown in Fig. 4.52. The peaks shift to higher binding energies as more arsenic added into the glass. The binding energy of Si 2p in G1 is 102.60 eV, with the addition of arsenic into the glasses, and the binding energy increases to 102.78 eV in G5, of which the maximum shift is approximately 0.18 eV over the composition range studied. The increase in the electron density on the relevant atom screens the core electrons and hence decreases the measured binding energy [158]. Therefore, the shift to higher binding energy is assigned to the decrease of electron density around Si, which causes by arsenic in the glasses. According to Pauling' electronegativity theory [159], the ionicity of the chemical bond can be used to explain the binding energy change of Si 2p in the glasses. The ionicity of Si–O is more than that of As–O, the average electron density on $SiO_4$ tetrahedra will decrease with more arsenic in the glasses. Correspondingly, the decrease in electron density on Si causes the increase of the binding energy of Si 2p.

The silicate based glass is three-dimensional network composed of $SiO_4$ tetrahedra and the adjacent $SiO_4$ tetrahedra are connected by bridging oxygen, forming Si–O–Si and Si–O–Me bonds. The arsenic exists in the form of $AsO_4$ tetrahedral

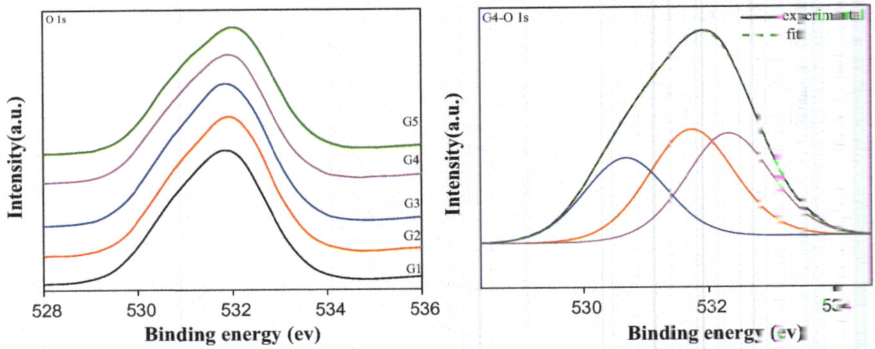


**Fig. 4.53** O 1s spectra of glasses for $(100 - x)(58SiO_2–20Fe_2O_3–10B_2O_3–12Na_2O) - xAs$ (left) and a typical deconvoluted peak for G4 (right). Reprinted from Ref. [101] Copyright 2017, with permission from Elsevier

and could substitute for $SiO_4$ in the Si–O–Si in the glasses. In addition, based on the fact that the maximum shift of binding energy of the Si 2p is 0.18 eV and little change appears in Si 2p peaks, it can be deduced that the $SiO_4$ tetrahedra based glass structure changes little with the addition of arsenic into the glass, which indicates that the arsenic participates in the glass structure in the form of $AsO_4$ tetrahedra and forms chemical bonds with $SiO_4$ tetrahedra in the form of Si–O–As [160]. Therefore, it is reasonable to attribute the increase binding energy of Si to the emergence of Si–O–As bond in the glasses.

**Table 4.20** Peak positions, FWHM, and relative abundance from the curve fitting of O 1s[a]

| Specimen ID | O 1s | | |
|---|---|---|---|
| | BE (eV) | FWHM (eV) | Percentage (%) |
| G1 | 530.72 | 1.64 | 25.08 |
| | 531.98 | 1.95 | 74.92 |
| G2 | 530.72 | 1.55 | 27.12 |
| | 531.86 | 1.85 | 43.48 |
| | 532.28 | 1.47 | 29.4 |
| G3 | 530.48 | 1.27 | 20.77 |
| | 531.66 | 1.59 | 47.05 |
| | 532.38 | 1.57 | 32.18 |
| G4 | 530.68 | 1.56 | 27.62 |
| | 531.73 | 1.61 | 37.31 |
| | 532.32 | 1.59 | 35.06 |
| G5 | 530.89 | 1.61 | 26.47 |
| | 531.78 | 1.92 | 35.48 |
| | 532.35 | 1.76 | 38.05 |

[a]The error between actual values and measured values is ±[illegible]%

(2) The XPS spectra of O 1s

Figure 4.53 shows the XPS O 1s spectra of the (100 − *x*)($58SiO_2$–$20Fe_2O_3$–$10B_2O_3$–$12Na_2O$) − *x*As (*x* = 0, 2, 4, 6, 8 wt%) glasses. The characteristic of O 1s is the shift to the high binding energy and the maximum shift is about 0.30 eV over the composition range studied, which means decrease of valence electron around the O nuclei. Since the oxide ions in the silicate glasses have different type of chemical bonding states, it is possible to analyze the bonding states of the oxide ions by deconvoluting the O 1s spectrum. The curve fitting of a typical O 1s photoelectron spectrum for the glass is shown in Fig. 4.53, and the peak positions, FWHM and relative abundance are also given in Table 4.20.

It is easy to assign the peak with higher binding energy to Si–O–Si (bridging oxygen) and to attribute the peak with lower binding energy to Si–O–Me (non-bridging oxygen) [50, 161]. Therefore, the peak at 531.98 eV is due to BO (Si–O(BO)–Si), and the other peak located at 531.31 eV is ascribed to NBO (Si–O (NBO)–Me), of which the corresponding relative contents are 74.92 and 25.08% in G1. With the addition of arsenic into the glasses, a new peak around 532 has emerged. Since the electronegativity of arsenic is bigger than silicate, the electron density around arsenic is more than that of silicate in As–O and Si–O. Moreover, arsenic is a strong network former [162], therefore, the new peak around 532 eV is attributed to Si–O(BO)–As. It is obvious that, the relative contents of Si–O(BO)–As increase from 29.4% (G2) to 38.05% (G5) and the relative contents of Si–O(NBO)–Me keep constant around 25%, and meanwhile the relative contents of Si–O(BO)–Si decrease from 74.92% (G1) to 35.48% (G5), proving that the arsenic form Si–O (BO)–As in the glasses. Moreover, this change is consistent with the change in the FTIR spectra that the intensity of the peaks located between 700 and 1500 $cm^{-1}$ has been enhanced, namely, the new formation of Si–O(BO)–As in the glass resulting in strengthening the $Q^n$ structure in silicate glass.

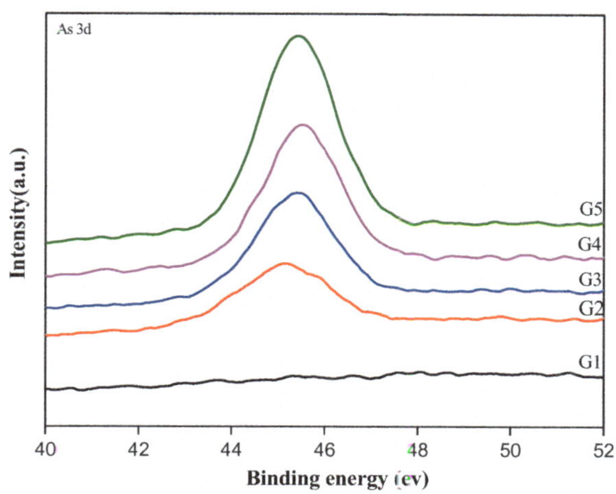


**Fig. 4.54** As 3d spectra of glasses for (100 − *x*)($58SiO_2$–$20Fe_2O_3$–$10B_2O_3$–$12Na_2O$) − *x*As. Reprinted from Ref. [101] Copyright 2017, with permission from Elsevier

(3) The XPS spectra of As 3d

Figure 4.54 shows the binding energy of As 3d as a function of arsenic content in the glasses. No peak concerning arsenic in G0 has been observed. A clear peak around 45 eV has appeared and the intensity of the peak increases with the addition of arsenic into the glasses. Additionally, the clear peak is due to As(V) in $AsO_4$ tetrahedra in the glasses [163]. With the increase of arsenic into the glasses, a small shift to lower binding energy has been observed in As 3d. A reasonable explanation is that the electronegativity of arsenic is more than silicate, therefore, the electron density of As is more than Si in Si–O–As, causing the binding energies of arsenic shift to lower energy. In addition, this change can also prove that arsenic has embedded itself into the silicate glass and participated in the glass structure in the form of Si–O–As.

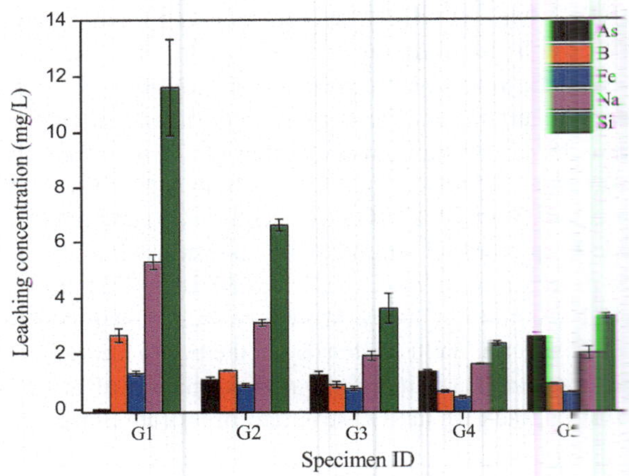


**Fig. 4.55** Leaching concentrations of As, B, Fe, Na, and Si of (100 − $x$)($58SiO_2$–$20Fe_2O_3$–$10B_2O_3$–$12Na_2O$) − $x$As glasses. Reprinted from Ref. [101] Copyright 2017, with permission from Elsevier

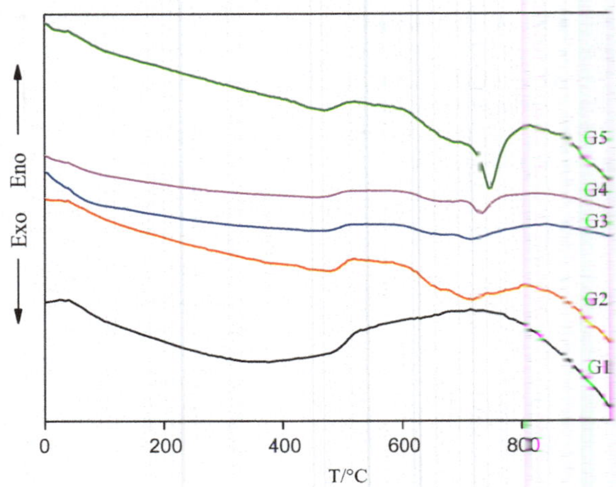


**Fig. 4.56** DSC curve for (100 − $x$)($58SiO_2$–$20Fe_2O_3$–$10B_2O_3$–$12Na_2O$) − $x$As glasses. Reprinted from Ref. [101] Copyright 2017, with permission from Elsevier

**Table 4.21** DTA parameters (°C) for $(100 - x)(58SiO_2–20Fe_2O_3–10B_2O_3–12Na_2O) - x$As glasses[a]

| Specimen ID | *Tg* | *Tc* | *ΔT* |
|---|---|---|---|
| G1 | 522 | – | – |
| G2 | 513 | 621 | 108 |
| G3 | 508 | 700 | 192 |
| G4 | 509 | 718 | 209 |
| G5 | 513 | 729 | 216 |

[a]The error is ±3 °C

TCLP test results, presented in Fig. 4.55, indicate that the leaching concentrations of As, B, Fe, Na, and Si are quite low. Moreover, the leaching concentrations present a decreasing trend with the addition of arsenic and the lowest leaching concentrations of As, B, Fe, Na, and Si are $1.22 \pm 0.14$ mg $L^{-1}$, $0.63 \pm 0.03$ mg $L^{-1}$, $0.42 \pm 0.06$ mg $L^{-1}$, $1.6 \pm 0.03$ mg $L^{-1}$, $2.33 \pm 0.08$ mg $L^{-1}$, respectively, in the glasses. In addition, the arsenic leaching concentrations in all glasses are lower than the USEPA' regulatory threshold (5 mg $L^{-1}$), which indicates that the addition of arsenic into the glasses can improve the chemical stability of the glasses. Importantly, the $SiO_2$ exists in the glasses with a net-like structure and the arsenic can inset itself in the structure and can form chemical bond with silicate, and other elements, such as B, Fe, Na, can also be immobilized in the structure, so it is difficult to extract.

The DSC (Differential Scanning Calorimeter) curve recorded for the $(100 - x)(58SiO_2–20Fe_2O_3–10B_2O_3–12Na_2O) - x$As ($x = 0, 2, 4, 6, 8$ wt%) glasses are depicted in Fig. 4.56. The glass transition temperature (*Tg*) and the onset crystallization temperature (*Tc*) are analyzed from the DSC curve and the results are showed in Table 4.21.

The *Tg* of the glasses vary from 522 °C (G1) to 513 °C (G5) and the small temperature fluctuations mean the arsenic influence little of the *Tg*. The *Tc* of the glass exhibits positive correlation with the increase of arsenic in the glasses, which indicates the arsenic can enhance the ability of glass to avoid crystallization. However, the *Tc* of the G1 cannot be determined due to the baseline shift is not clearly pronounced. The *ΔT* (*Tc* − *Tg*) of the glasses also presents an increasing trend with the increase of arsenic, and the *ΔT* increases from 108 °C in G2 to 216 °C in G5. Previous studies [164] has proved that the higher the value of *ΔT*, better is the thermal stability of the glass system against crystallization during heat treatment. Therefore, the DSC test can be exactly used to prove that the addition of arsenic into the glass can strengthen the glass structure and this change has been enforced with the increase of arsenic.

All samples can form glass and the arsenic is compatible with the silicate based vitreous framework. The arsenic exists in the form of $AsO_4$ tetrahedra in the copper slag based glasses and the glass structure has been strengthened by the increase addition of arsenic. The changes of the binding energies of Si 2p, O 1s, Fe 2p and As 3d have proved the formation of Si–O–As and Fe–O–Si/As bonds in the glass. The leaching concentrations of As, B, Fe, Na, and Si are under safety thresholds and increasing the addition of arsenic into the glasses can decrease the leaching concentrations, indicating that the arsenic can improve the chemical stability of the glass. The arsenic added into the glass can improve the $\Delta T$, which strengthens the thermal stability of the glass.

## References


1. Singh, T.S., Pant, K.K.: Solidification/stabilization of arsenic containing solid wastes using Portland cement, fly ash and polymeric materials. J. Hazard. Mater. **131**(1–3), 29–36 (2006)
2. Shaw, J.K., Fathordoobadi, S., Zelinski, B.J., Ela, W.P., Saez, A.E.: Stabilization of arsenic-bearing solid residuals in polymeric matrices. J. Hazard. Mater. **152**(3), 1115–1121 (2008)
3. Shi, M.Q., Liang, Y.J., Chai, L.Y., et al.: Raman and FTIR spectra of modified iron phosphate glasses containing arsenic. J. Mol. Struct. **1081**, 389–394 (2015)
4. Ke, Y., Chai, L.Y., Min, X.B., et al.: Sulfidation of heavy-metal-containing neutralization sludge using zinc leaching residue as the sulfur source for metal recovery and stabilization. Miner. Eng. **61**, 105–112 (2014)
5. Yang, Z.H., Liu, L., Chai, L.Y., et al.: Arsenic immobilization in the contaminated soil using poorly crystalline Fe-oxyhydroxy sulfate. Environ. Sci. Pollut. R. **22**(16), 12624–12632 (2015)
6. Paktunc, D., Bruggeman, K.: Solubility of nanocrystalline scorodite and amorphous ferric arsenate: implications for stabilization of arsenic in mine wastes. Appl. Geochem. **25**(5), 674–683 (2010)
7. Kundu, S., Gupta, A.K.: Immobilization and leaching characteristics of arsenic from cement and/or lime solidified/stabilized spent adsorbent containing arsenic. J. Hazard. Mater. **153**(1–2), 434–443 (2008)
8. Mendonca, A.A., Galvao, T.C.B., Lima, D.C., et al.: Stabilization of arsenic-bearing sludges using lime. J. Mater. Civil. Eng. **18**(2), 135–139 (2006)
9. Yoon, I.H., Moon, D.H., Kim, K.W., et al.: Mechanism for the stabilization/solidification of arsenic-contaminated soils with Portland cement and cement kiln dust. J. Environ. Manag. **91**(11), 2322–2328 (2010)
10. Peng, B., Lei, J., Min, X., et al.: Physicochemical properties of arsenic-bearing lime-ferrate sludge and its leaching behaviors. Trans. Nonferrous Met. Soc. China **27**, 1188–1198 (2017)
11. Bothe, J.V., Brown, P.W.: Arsenic immobilization by calcium arsenate formation. Environ. Sci. Technol. **33**(21), 3806–3811 (1999)
12. Donahue, R., Hendry, M.J.: Geochemistry of arsenic in uranium mine mill tailings, Saskatchewan, Canada. Appl. Geochem. **18**(11), 1733–1750 (2003)
13. Camacho, J., Wee, H.Y., Kramer, T.A., Autenrieth, R.: Arsenic stabilization on water treatment residuals by calcium addition. J. Hazard. Mater. **165**(1–3), 599–603 (2009)
14. Guo, X.J., Wang, K.P., He, M.C., et al.: Antimony smelting process generating solid wastes and dust: characterization and leaching behaviors. J. Environ. Sci. China. **26**(7), 1549–1556 (2014)
15. Baskan, M.B., Pala, A.: Determination of arsenic removal efficiency by ferric ions using response surface methodology. J. Hazard. Mater. **166**(2–3), 796–801 (2009)
16. Lei, J., Peng, B., Min, X., et al.: Modeling and optimization of lime-based stabilization in high alkaline arsenic-bearing sludges with a central composite design. J. Environ. Sci. Health **52**(5), 449–458 (2017)
17. Divsar, F., Habibzadeh, K., Shariati, S., Shahriarinour, M.: Aptamer conjugated silver nanoparticles for the colorimetric detection of arsenic ions using response surface methodology. Anal. Methods UK **7**(11), 4568–4576 (2015)

18. Kowalski, K.P., Søgaard, E.G.: Implementation of zero-valent iron (ZVI) into drinking water supply—role of the ZVI and biological processes. Chemosphere **117**, 108–114 (2014)
19. Wen, Z., Zhang, Y., Dai, C., et al.: Synthesis of ordered mesoporous iron manganese bimetal oxides for arsenic removal from aqueous solutions. Microporous Mesoporous Mater. **200**, 235–244 (2014)
20. An, B., Zhao, D.: Immobilization of As(III) in soil and groundwater using a new class of polysaccharide stabilized Fe-Mn oxide nanoparticles. J. Hazard. Mater. **211–212**, 332–341 (2012)
21. Liang, Y., Min, X., Chai, L., et al.: Stabilization of arsenic sludge with mechanochemically modified zero valent iron. Chemosphere **168**, 1142–1151 (2017)
22. Selena, M., Alessandro, C., Massimo, P., et al.: Remediation of heavy metals contaminated soils by ball milling. Chemosphere **67**, 631–639 (2007)
23. Kim, J.Y., Allen, P.D.: Stabilization of available arsenic in highly contaminated mine tailings using iron. Environ. Sci. Technol. **37**, 189–195 (2003)
24. Lagno, F., Rocha, S.D., Chryssoulis, S., et al.: Scorodite encapsulation by controlled deposition of aluminum phosphate coatings. J. Hazard. Mater. **181**, 526–534 (2010)
25. Yang, H., McCormick, P.G.: Combustion reaction of zinc oxide with magnesium during mechanical milling. J. Solid State Chem. **107**, 258–263 (1993)
26. Zhang, D., Richmond, J.: Microstructural evolution during combustion reaction between CuO and Al induced by high energy ball milling. J. Mater. Sci. **34**, 701–706 (1999)
27. Laszlo, T.: Self-sustaining reactions induced by ball milling. Prog. Mater. Sci. **47**, 355–414 (2002)
28. Chai, L., Liang, Y., Ke, Y., et al.: Mechano-chemical sulfidization of zinc oxide by grinding with sulfur and reductive additives. Trans. Nonferrous Met. Soc. China **23**, 1129–1138 (2013)
29. Li, M., Sun, C., Gau, S., et al.: Effects of wet ball milling on lead stabilization and particle size variation in municipal solid waste incinerator fly ash. J. Hazard. Mater. **174**, 586–591 (2010)
30. Grosvenor, A.P., Kobe, B.A., Biesinger, M.C., et al.: Investigation of multiplet splitting of Fe 2p XPS spectra and bonding in iron compounds. Surf. Interface Anal. **36**, 1564–1574 (2004)
31. Sun, F., Kwadwo, A.O., Chen, Y., et al.: Reduction of As(V) to As(III) by commercial ZVI or As(0) with acid-treated ZVI. J. Hazard. Mater. **196**, 311–317 (2011)
32. Cyril, W.C., Naoto, M., Yoshinaga, N.: Ferrihydrite deposits in paddy races, Aso-Dani. Clay Sci. **8**, 9–15 (1990)
33. Song, J., Jia, S., Yu, B., et al.: Formation of iron (hydr)oxides during the abiotic oxidation of Fe(II) in the presence of arsenate. J. Hazard. Mater. **294**, 70–79 (2015)
34. Himemstra, T., Riemsdijk, W.H.V.: Surface structural adsorption modeling of competitive binding of oxyanions by metal(hydr)oxides. J. Colloid Interface Sci. **210**, 182–193 (1999)
35. Jia, Y., Xu, L., Fang, Z., et al.: Observation of surface precipitation of arsenate on ferrihydrite. Environ. Sci. Technol. **40**, 3248–3253 (2006)
36. Cui, H., Li, Q., Gao, S., et al.: Strong adsorption of arsenic species by amorphous zirconium oxide nanoparticles. J. Ind. Eng. Chem. **18**, 1418–1427 (2012)
37. Cornell, R.M., Giovanoli, R.: Effect of manganese on the transformation of ferrihydrite into goethite and jacobsite in alkaline media. Clays Clay Miner. **35**, 11–20 (1987)
38. Ouvrard, S., Donato, P.H., Simonnot, M.O.: Natural manganese oxide: combined analytical approach for solid characterization and arsenic retention. Geochim. Cosmochim. Acta **69**, 2715–2724 (2005)
39. Rauret, G., Nchez, J., Sahuquillo, A., et al.: Improvement of the BCR three step sequential extraction procedure prior to the certification of new sediment and soil reference materials. J. Environ. Monit. **1**, 57–61 (1999)
40. Xie, X., Min, X., Chai, L., et al.: Quantitative evaluation of environmental risks of flotation tailings from hydrothermal sulfidation–flotation process. Environ. Sci. Pollut. Res. **20**, 6050–6058 (2013)

41. Ke, Y., Shen, C., Min, X.-B., et al.: Separation of Cu and As in Cu-As-containing filter cakes by $Cu^{2+}$-assisted acid leaching. Hydrometallurgy **172**, 45–50 (2017)
42. Shi, C., Meyer, C., Behnood, A.: Utilization of copper slag in cement and concrete. Resour. Conserv. Recycl. **52**, 1115–1120 (2008)
43. Jing, C., Korfiatis, G.P., Meng, X.: Immobilization mechanisms of arsenate in iron hydroxide sludge stabilized with cement. Environ. Sci. Technol. **37**, 5050–5056 (2003)
44. Chai, L., Yue, M., Yang, J., et al.: Formation of tooeleite and the role of direct removal of As (III) from high-arsenic acid wastewater. J. Hazard. Mater. **320**, 620–627 (2016)
45. Choi, W.H., Lee, S.R., Park, J.Y.: Cement based solidification/stabilization of arsenic-contaminated mine tailings. Waste Manag. **29**, 1766–1771 (2009)
46. Li, Y.C., Min, X.B., Chai, L.Y., et al.: Co-treatment of gypsum sludge and Pb/Zn smelting slag for the solidification of sludge containing arsenic and heavy metals. J. Environ. Manag. **181**, 756–761 (2016)
47. Min, X.-B., Liao, Y.-P., Chai, L.-Y., et al.: Removal and stabilization of arsenic from anode slime by forming crystal scorodite. Trans. Nonferrous Met. Soc. **25**, 1298–1306 (2015)
48. Min, X., Li, Y., Ke, Y., et al.: Fe-$FeS_2$ adsorbent prepared with iron powder and pyrite by facile ball milling and its application for arsenic removal. Water Sci. Technol. **76**(1), 192–200 (2017)
49. Peng, B., Song, T., Wang, T., et al.: Facile synthesis of $Fe_3O_4@Cu(OH)_2$ composites and their arsenic adsorption application. Chem. Eng. J. **299**, 15–22 (2016)
50. Zhao, Z., Song, Y., Min, X., et al.: XPS and FTIR studies of sodium arsenate vitrification by cullet. J. Non-Cryst. Solids **452**, 238–244 (2016)
51. Bose, P., Sharma, A.: Role of iron in controlling speciation and mobilization of arsenic in subsurface environment. Water Res. 4916–4926 (2002)
52. Desogus, P., Manca, P.P., Orrù, G., et al.: Stabilization–solidification treatment of mine tailings using Portland cement, potassium dihydrogen phosphate and ferric chloride hexahydrate. Miner. Eng. **45**, 47–54 (2013)
53. Jaarsveld, J.G.S.V., Deventer, J.S.J.V., Lorenzeni, L.: The potential use of geopolymeric materials to immobilise toxic metals Part I. Theory and applications. Miner. Eng. **7**, 659–669 (1997)
54. Jaarsveld, J.G.S.V., Deventer, J.S.J.V., Lorenzeni, L.: The potential use of geopolymeric materials to immobilise toxic metals Part II. Material and leaching characteristics. Miner. Eng. **1**, 75–91 (1999)
55. Jang, A., Kim, I.S.: Solidification and stabilization of Pb, Zn, Cd and Cu in tailing wastes using cement and fly ash. Miner. Eng. **14**, 1659–1662 (2000)
56. Singh, T.S., Pant, K.K.: Solidification/stabilization of arsenic containing solid wastes using portland cement, fly ash and polymeric materials. J. Hazard. Mater. **131**, 29–36 (2006)
57. Qian, G., Cao, Y., Chui, P., et al.: Utilization of MSWI fly ash for stabilization/solidification of industrial waste sludge. J. Hazard. Mater. **129**, 274–281 (2006)
58. Liu, D.-G., Min, X.-B., Ke, Y., et al.: Co-treatment of flotation waste, neutralization sludge, and arsenic-containing gypsum sludge from copper smelting: solidification/stabilization of arsenic and heavy metals with minimal cement clinker. Environ. Sci. Pollut. Res. **25**, 7600–7607 (2018)
59. Kumar, S., Kumar, R., Bandopadhyay, A., et al.: Mechanical activation of granulated blast furnace slag and its effect on the properties and structure of portland slag cement. Cem. Concr. Compos. **30**, 679–685 (2008)
60. Horpibulsuk, S., Miura, N., Nagaraj, T.: Assessment of strength development in cement-admixed high water content clays with Abrams' law as a basis. Géotechnique **53**, 439–444 (2003)
61. Seco, J.I., Fernández-Pereira, C., Vale, J.: A study of the leachate toxicity of metal-containing solid wastes using Daphnia magna. Ecotoxicol. Environ. Saf. **56**, 339–350 (2003)
62. Coussy, S., Paktunc, D., Rose, J., et al.: Arsenic speciation in cemented paste backfills and synthetic calcium–silicate–hydrates. Miner. Eng. **39**, 51–61 (2012)

63. Phenrat, T., Marhaba, T.F., Rachakornkij, M.: A SEM and X-ray study for investigation of solidified/stabilized arsenic-iron hydroxide sludge. J. Hazard. Mater. **118**, 185–195 (2005)
64. Stronach, S., Walker, N., Macphee, D., et al.: Reactions between cement and As (III) oxide: the system $CaO·SiO_2·As_2O_3·H_2O$ at 25 °C. Waste Manag. **17**, 9–13 (1997)
65. Kumarathasan, P., McCarthy, G.J., Hassett, D.J., et al.: Oxyanion substituted ettringites: synthesis and characterization; and their potential role in immobilization of As, B, Cr, Se and V. MRS Online Proc. Library Arch. **178**, 83 (1989)
66. Vandecasteele, C., Dutré, V., Geysen, D., et al.: Solidification/stabilisation of arsenic bearing fly ash from the metallurgical industry. Immobilisation mechanism of arsenic. Waste Manag. **22**, 143–146 (2002)
67. Dutré, V., Vandecasteele, C.: Solidification/stabilisation of hazardous arsenic containing waste from a copper refining process. J. Hazard. Mater. **40**, 55–68 (1995)
68. Qiao, X.C., Poon, C.S., Cheeseman, C.R.: Investigation into the stabilization/solidification performance of Portland cement through cement clinker phases. J. Hazard. Mater. **139**, 238–243 (2007)
69. Li, Y.-C., Min, X.-B., Chai, L.-Y., et al.: Co-treatment of gypsum sludge and Pb/Zn smelting slag for the solidification of sludge containing arsenic and heavy metals. J. Environ. Manag. **181**, 756–761 (2016)
70. Boldyrev, V., Pavlov, S., Goldberg, E.: Interrelation between fine grinding and mechanical activation. Comminution **44**(95), 181–185 (1996)
71. Wei, B., Zhang, Y., Bao, S.: Preparation of geopolymers from vanadium tailings by mechanical activation. Constr. Build. Mater. **145**(Supplement C), 236–242 (2017)
72. Sulaymon, A.H., Faisal, A.A., Khaliefa, Q.M.: Cement kiln dust (CKD)-filter sand permeable reactive barrier for the removal of Cu(II) and Zn(II) from simulated acidic groundwater. J. Hazard. Mater. **297**, 160–72 (2015)
73. Doudart de la Grée, G.C.H., Yu, Q.L., Brouwers, H.J.H.: Assessing the effect of $CaSO_4$ content on the hydration kinetics, microstructure and mechanical properties of cements containing sugars. Constr. Build. Mater. **143**, 48–60 (2017)
74. Kang, S.-P., Kwon, S.-J.: Effects of red mud and alkali-activated slag cement on efflorescence in cement mortar. Constr. Build. Mater. **133**, 459–467 (2017)
75. Li, Y.-C., Min, X.-B., Ke, Y., et al.: Utilization of red mud and Pb/Zn smelter waste for the synthesis of a red mud-based cementitious material. J. Hazard. Mater. **344**, 343–349 (2018)
76. Minard, H., Garrault, S., Regnaud, L., et al.: Mechanisms and parameters controlling the tricalcium aluminate reactivity in the presence of gypsum. Cem. Concr. Res. **37**(10), 1418–1426 (2007)
77. Pontikes, Y., Angelopoulos, G.N.: Bauxite residue in cement and cementitious applications: current status and a possible way forward. Resour. Conserv. Recycl. **73**, 53–63 (2013)
78. Coussy, S., Paktunc, D., Rose, J., et al.: Arsenic speciation in cemented paste backfills and synthetic calcium–silicate–hydrates. Miner. Eng. **39**, 51–61 (2012)
79. Moon, D.H., Dermatas, D.: Arsenic and lead release from fly ash stabilized/solidified soils under modified semi-dynamic leaching conditions. J. Hazard. Mater. **141**, 388–394 (2007)
80. Phair, J.W., Van Deventer, J.S.J.: Effect of the silicate activator pH on the microstructural characteristics of waste-based geopolymers. Int. J. Miner. Process. **66**, 121–143 (2002)
81. Liu, X., Zhang, N.: Utilization of red mud in cement production: a review. Waste Manag. Res. **29**, 1053–1063 (2011)
82. Singh, M., Upadhayay, S.N., Prasad, P.M.: Preperation of iron rich cements using red mud. Cem. Concr. Res. **27**, 1037–1046 (1997)
83. Zhang, M., Yang, C., Zhao, M., et al.: Immobilization potential of Cr(VI) in sodium hydroxide activated slag pastes. J. Hazard. Mater. **321**, 281–289 (2017)
84. Andini, S., Cioffi, R., Colangelo, F., et al.: Coal fly ash as raw material for the manufacture of geopolymer-based products. Waste Manag. **28**, 416–423 (2008)
85. Goetz-Neunhoeffer, F., Neubauer, J., Schwesig, P.: Mineralogical characteristics of Ettringites synthesized from solutions and suspensions. Cem. Concr. Res. **36**, 65–70 (2006)

86. Bhatnagar, A., Minocha, A.K.: Utilization of industrial waste for cadmium removal from water and immobilization in cement. Chem. Eng. J. **150**, 145–151 (2009)
87. Choi, W.H., Lee, S.R., Park, J.Y.: Cement based solidification/stabilization of arsenic-contaminated mine tailings. Waste Manag. **29**(5), 1766–1771 (2009)
88. Miller, J., Akhter, H., Cartledge, F.K., et al.: Treatment of arsenic-contaminated soils. II: Treatability study and remediation. J. Environ. Eng. **126**(11), 1004–1012 (2000)
89. Kuo, Y.M., Wang, J.W., Chao, H.R., et al.: Effect of cooling rate and basicity during vitrification of fly ash: Part 2. On the chemical stability and acid resistance of slags. J. Hazard. Mater. **152**(2), 554–562 (2008)
90. Joseph, K., Kutty, K.G., Chandramohan, P., et al.: Studies on the synthesis and characterization of cesium-containing iron phosphate glasses. J. Nucl. Mater. **384**(3), 262–267 (2009)
91. Reis, S.T., Karabulut, M., Day, D.E.: Structural features and properties of lead-iron-phosphate nuclear wasteforms. J. Nucl. Mater. **304**(2–3), 87–95 (2002)
92. Chakraborty, S., Arora, A.K.: Temperature evolution of Raman spectrum of iron phosphate glass. Vib. Spectrosc. **61**, 99–104 (2012)
93. Bingham, P., Hand, R., Forder, S.: Doping of iron phosphate glasses with $Al_2O_3$, $SiO_2$ or $B_2O_3$ for improved thermal stability. Mater. Res. Bull. **41**(9), 1622–1630 (2006)
94. Shi, M., Liang, Y., Chai, L., et al.: Raman and FTIR spectra of modified iron phosphate glasses containing arsenic. J. Mol. Struct. **1081**, 389–394 (2015)
95. Krishna, S.B.M., Babu, A.R., Rajya Sree, C., et al.: Influence of molybdenum ions on the structure of $ZnO–As_2O_3–Sb_2O_3$ glass system by means of spectroscopic and dielectric studies. J. Non-Cryst. Solids **356**, 1754–1761 (2010)
96. Leist, M., Casey, R.J., Caridi, D.: The management of arsenic wastes: problems and prospects. J. Hazard. Mater. **76**, 125–138 (2000)
97. Tang, Y., Chan, S.-W., Shih, K.: Copper stabilization in beneficial use of waterworks sludge and copper-laden electroplating sludge for ceramic materials. Waste Manag. **34**, 1085–1091 (2014)
98. Li, H., Yang, X., Xu, W., et al.: Application of dry composite electroplating sludge into preparation of cement-based decorative mortar as green pigment. J. Clean. Prod. **66**, 101–106 (2014)
99. Huang, R., Huang, K.-L., Lin, Z.-Y., et al.: Recovery of valuable metals from electroplating sludge with reducing additives via vitrification. J. Environ. Manag. **129**, 586–592 (2013)
100. Chen, Y.-L., Shih, P.-H., Chiang, L.-C., et al.: The influence of heavy metals on the polymorphs of dicalcium silicate in the belite-rich clinkers produced from electroplating sludge. J. Hazard. Mater. **170**, 443–448 (2009)
101. wen Zhao, Z., yuan Chai, L., Peng, B., et al.: Arsenic vitrification by copper slag based glass: mechanism and stability studies. J. Non-Cryst. Solids **466**, 21–28 (2017)
102. Zhao, Z., Chai, L., Liang, Y., et al.: The vitrification of arsenic-rich residue using iron phosphate glass. Phys. Chem. Glasses Eur. J. Glass Sci. Technol. Part B **58**(3), 109–114 (2017)
103. Reis, S., Karabulut, M., Day, D.: Chemical durability and structure of zinc–iron phosphate glasses. J. Non-Cryst. Solids **292**, 150–157 (2001)
104. Zhao, Z., Liang, Y., Min, X., et al.: The effects of antimony oxide on the structure of iron phosphate glass for the immobilisation of arsenic. Glass Technol. Eur. J. Glass Sci. Technol. Part A **56**, 196–202 (2015)
105. Karamberi, A., Orkopoulos, K., Moutsatsou, A.: Synthesis of glass-ceramics using glass cullet and vitrified industrial by-products. J. Eur. Ceram. Soc. **27**, 629–636 (2007)
106. Moustakas, K., Mavropoulos, A., Katsou, E., et al.: Leaching properties of slag generated by a gasification/vitrification unit: the role of pH, particle size, contact time and cooling method used. J. Hazard. Mater. **207**, 44–50 (2012)

107. Moguš-Milanković, A., Šantić, A., Reis, S.T., et al.: Studies of lead–iron phosphate glasses by Raman, Mössbauer and impedance spectroscopy. J. Non-Cryst. Solids **351**, 3246–3258 (2005)
108. Rosli, A.N., Zabidi, N.A., Kassim, H.A., et al.: Ab initio calculation of vibrational frequencies of AsO glass. J. Non-Cryst. Solids **356**, 428–433 (2010)
109. Gilliam, S.J., Merrow, C.N., Kirkby, S.J., et al.: Raman spectroscopy of arsenolite: crystalline cubic $As_4O_6$. J. Solid State Chem. **173**, 54–58 (2003)
110. Zhang, L., Brow, R.K., Schlesinger, M.E., et al.: Glass formation from iron-rich phosphate melts. J. Non-Cryst. Solids **356**, 1252–1257 (2010)
111. Lai, Y.M., Liang, X.F., Yang, S.Y., et al.: Raman spectra study of iron phosphate glasses with sodium sulfate. J. Mol. Struct. **10**(13), 134–137 (2012)
112. Silva, A., Correia, R., Oliveira, J., et al.: Structural characterization of $TiO_2$–$P_2O_5$–CaO glasses by spectroscopy. J. Eur. Ceram. Soc. **30**, 1253–1258 (2010)
113. Couchman, P., Karasz, F.: A classical thermodynamic discussion of the effect of composition on glass-transition temperatures. Macromolecules **11**, 117–119 (1978)
114. Gayathri Devi, A.V., Rajendran, V., Rajendran, N.: Structure, solubility and bioactivity in $TiO_2$-doped phosphate-based bioglasses and glass–ceramics. Mater. Chem. Phys. **124**, 312–318 (2010)
115. Qian, B., Liang, X., Wang, C., et al.: Structure and properties of calcium iron phosphate glasses. J. Nucl. Mater. **443**, 140–144 (2013)
116. Glasser, F.: Chemistry of cement-solidified waste forms. Chem. Microstruct. Solidified Waste Forms 1–39 (1993)
117. Phenrat, T., Marhaba, T.F., Rachakornkij, M.: A SEM and X-ray study for investigation of solidified/stabilized arsenic–iron hydroxide sludge. J. Hazard. Mater. **118**, 185–195 (2005)
118. Myneni, S.C., Traina, S.J., Logan, T.J., et al.: Oxyanion behavior in alkaline environments: sorption and desorption of arsenate in ettringite. Environ. Sci. Technol. **31**, 1761–1768 (1997)
119. Yoon, I.-H., Moon, D.H., Kim, K.-W., et al.: Mechanism for the stabilization/solidification of arsenic-contaminated soils with Portland cement and cement kiln dust. J. Environ. Manag. **91**, 2322–2328 (2010)
120. Moon, D.H., Dermatas, D.: Arsenic and lead release from fly ash stabilized/solidified soils under modified semi-dynamic leaching conditions. J. Hazard. Mater. **141**, 388–394 (2007)
121. Sullivan, C., Tyrer, M., Cheeseman, C.R., et al.: Disposal of water treatment wastes containing arsenic—a review. Sci. Total Environ. **408**, 1770–1778 (2010)
122. Colombo, P., Brusatin, G., Bernardo, E., et al.: Inertization and reuse of waste materials by vitrification and fabrication of glass-based products. Curr. Opin. Solid State Mater. Sci. **7**, 225–239 (2003)
123. Park, Y.J., Heo, J.: Vitrification of fly ash from municipal solid waste incinerator. J. Hazard. Mater. **91**, 83–93 (2002)
124. El-Shimy, Y.N., Amin, S.K., El-Sherbiny, S.A., et al.: The use of cullet in the manufacture of vitrified clay pipes. Constr. Build. Mater. **73**, 452–457 (2014)
125. Federico, L., Chidiac, S.: Waste glass as a supplementary cementitious material in concrete–critical review of treatment methods. Cem. Concr. Compos. **31**, 606–610 (2009)
126. Bernardo, E., Doyle, J., Hampshire, S.: Sintered feldspar glass–ceramics and glass–ceramic matrix composites. Ceram. Int. **34**, 2037–2042 (2008)
127. Dalby, K.N., Nesbitt, H.W., Zakaznova-Herzog, V.P., et al.: Resolution of bridging oxygen signals from O 1s spectra of silicate glasses using XPS: implications for O and Si speciation. Geochim. Cosmochim. Acta **71**, 4297–4313 (2007)
128. Ribeiro, A.S.M., Monteiro, R.C.C., Davim, E.J.R., et al.: Ash from a pulp mill boiler—characterisation and vitrification. J. Hazard. Mater. **179**, 303–308 (2010)
129. Hassaan, M., Saudi, H., Saad, H.M., et al.: Structural study of glass and glass ceramics prepared with Egyptian Basalt. Silicon **7**, 383–391 (2015)
130. Contreras, M.L., Arostegui, J.M., Armesto, L.: Arsenic interactions during co-combustion processes based on thermodynamic equilibrium calculations. Fuel **88**, 539–546 (2009)

131. Sitarz, M., Mozgawa, W., Handke, M.: Rings in the structure of silicate glasses. J. Mol. Struct. **511**, 281–285 (1999)
132. Merzbacher, C.I., White, W.B.: The structure of alkaline earth aluminosilicate glasses as determined by vibrational spectroscopy. J. Non-Cryst. Solids **130**, 18–34 (1991)
133. Lubas, M., Sitarz, M., Fojud, Z., et al.: Structure of multicomponent $SiO_2$–$Al_2O_3$–$Fe_2O_3$–CaO–MgO glasses for the preparation of fibrous insulating materials. J. Mol. Struct. **744**, 615–619 (2005)
134. Villegas, M., Navarro, J.F.: Characterization of $B_2O_3$-$SiO_2$ glasses prepared via sol-gel. J. Mater. Sci. **23**, 2464–2478 (1988)
135. MacDonald, S.A., Schardt, C.R., Masiello, D.J., et al.: Dispersion analysis of FTIR reflection measurements in silicate glasses. J. Non-Cryst. Solids **275**, 72–82 (2000)
136. De Ferri, L., Bersani, D., Lorenzi, A., et al.: Structural and vibrational characterization of medieval like glass samples. J. Non-Cryst. Solids **358**, 814–819 (2012)
137. Ibrahim, M.M., Fanny, M.A., Hassaan, M., et al.: Optical, FTIR and DC conductivity of soda lime silicate glass containing cement dust and transition metal ions. Silicon **8**(3), 443–453 (2016)
138. ElBatal, F., Selim, M., Marzouk, S., et al.: UV-vis absorption of the transition metal-doped $SiO_2$–$B_2O_3$–$Na_2O$ glasses. Phys. B **398**, 126–134 (2007)
139. Akatov, A., Nikonov, B., Omel'yanenko, B., et al.: Structure of borosilicate glassy materials with high concentrations of sodium, iron, and aluminum oxides. Glass Phys. Chem. **35**, 245–259 (2009)
140. Serra, J., Gonzalez, P., Liste, S., et al.: Influence of the non-bridging oxygen groups on the bioactivity of silicate glasses. J. Mater. Sci. Mater. Med. **13**, 1221–1225 (2002)
141. Brotikovskii, O., Pen, K., Cherntsov, S.: IR spectroscopic investigation of the formation and properties of lead-silicate glass films on silicon surfaces. J. Appl. Spectrosc. **32**, 365–369 (1980)
142. zur Loye, K.D., Latshaw, A.M., Smith, M.D., et al.: Synthesis and crystal structure of sodium arsenate oxyhydroxide: $Na_4(AsO_4)OH$. J. Chem. Crystallogr. **45**, 20–25 (2015)
143. Mekki, A., Khattak, G., Wenger, L.: Structure and magnetic properties of lead vanadate glasses. J. Non-Cryst. Solids **330**, 156–167 (2003)
144. Sawyer, R., Nesbitt, H.W., Secco, R.A.: High resolution X-ray Photoelectron Spectroscopy (XPS) study of $K_2O$–$SiO_2$ glasses: Evidence for three types of O and at least two types of Si. J. Non-Cryst. Solids **358**, 290–302 (2012)
145. Gresch, R., Müller-Warmuth, W., Dutz, H.: X-ray photoelectron spectroscopy of sodium phosphate glasses. J. Non-Cryst. Solids **34**, 127–136 (1979)
146. Minami, T., Hayashi, A., Tatsumisago, M.: Preparation and characterization of lithium ion-conducting oxysulfide glasses. Solid State Ion. **136**, 1015–1023 (2000)
147. Flambard, A., Videau, J.-J., Delevoye, L., et al.: Structure and nonlinear optical properties of sodium–niobium phosphate glasses. J. Non-Cryst. Solids **354**, 3540–3547 (2008)
148. Fu, Z., Wu, F., Chen, L., et al.: Copper and zinc, but not other priority toxic metals, pose risks to native aquatic species in a large urban lake in Eastern China. Environ. Pollut. (2016)
149. Guo, X., Song, Y.: Substance flow analysis of copper in China. Resour. Conserv. Recycl. **52**(6), 874–882 (2008)
150. Kavouras, P., Komninou, P., Chrissafis, K., et al.: Microstructural changes of processed vitrified solid waste products. J. Eur. Ceram. Soc. **23**(8), 1305–1311 (2003)
151. El-Damrawi, G., El-Egili, K.: Characterization of novel $CeO_2$–$B_2O_3$ glasses, structure and properties. Phys. B **299**(1), 180–186 (2001)
152. El-Batal, F.H., Khalil, E.M., Hamdy, Y.M., et al.: FTIR spectral analysis of corrosion mechanisms in soda lime silica glasses doped with transition metal oxides. Silicon **2**(1), 41–47 (2010)
153. Wang, M., Mei, L.I., Cheng, J., et al.: Free volume and structure of $Gd_2O_3$ and $Y_2O_3$ co-doped silicate glasses. J. Non-Cryst. Solids **379**, 145–149 (2013)
154. Lu, M., Wang, F., Chen, K., et al.: The crystallization and structure features of barium-iron phosphate glasses. Spectrochim. Acta Part A Mol. Biomol. Spectrosc. **148**, 1–6 (2015)

155. Husung, R.D., Doremus, R.H.: The infrared transmission spectra of four silicate glasses before and after exposure to water. J. Mater. Res. **5**(10), 2209–2217 (1990)
156. Steger, E.: Spektroskopische Untersuchungen zum Bindungszustand in Phosphorsäurederivaten Amidoderivate. Zeitschrift Für Elektrochemie Berichte Der Bunsengesellschaft Für Physikalische Chemie **61**(61), 1004–1007 (2015)
157. Mansour, E.: Semi-quantitative analysis for FTIR spectra of $Al_2O_3$-PbO-$B_2O_3$-$SiO_2$ glasses. J. Non-Cryst. Solids **358**(3), 454–460 (2012)
158. Mekki, A., Holland, D., Mcconville, C.F., et al.: An XPS study of iron sodium silicate glass surfaces. J. Non-Cryst. Solids **208**(208), 267–276 (1996)
159. Pauling, L.: The nature of the chemical bond. IV. The energy of single bonds and the relative electronegativity of atoms. J. Am. Chem. Soc. **54**(9), 3570–3582 (1931)
160. Wang, P.W., Zhang, L.: Structural role of lead in lead silicate glasses derived from XPS spectra. J. Non-Cryst. Solids **194**(1–2), 129–134 (1996)
161. Serra, J., González, P., Liste, S., et al.: FTIR and XPS studies of bioactive silica based glasses. J. Non-Cryst. Solids **332**(1), 20–27 (2003)
162. Raghavaiah, B.V., Laxmikanth, C., Veeraiah, N.: Spectroscopic studies of titanium ions in PbO–$Sb_2O_3$–$As_2O_3$ glass system. Opt. Commun. **235**(4–6), 341–349 (2004)
163. Lee, K., Zimmerman, J.D., Xiao, X., et al.: Reuse of GaAs substrates for epitaxial lift-off by employing protection layers. J. Appl. Phys. **111**(3), 84–327 (2012)
164. Imran, M.M.A., Saxena, N.S., Bhandari, D., et al.: Transition phenomena, crystallization kinetics and enthalpy released in binary $Se_{100-x}In_x$ ($x = 2{,}4$ and 10) semiconducting glasses. Phys. Status Solidi **181**(2), 357–368 (2000)

# Chapter 5
# Separation and Recovery of Valuable Metals from Arsenic-Bearing Materials

**Li-Yuan Chai, Xiao-Bo Min, Wei-Chun Yang, Yong-Ming Chen, Yong Ke and Wei-Zhi Zeng**

Arsenic usually coexists with congeners such as antimony and bismuth in arsenic-bearing materials. Due to their similar properties, the separation of antimony and bismuth with arsenic is difficult during dearsenification process, which makes it hard to selectively separate arsenic in polymetallic materials.

Most lead ores have arsenic levels of 0.1–5.0% because of the similar physical–chemical properties of Pb and As. In the lead smelting process, most of the arsenic is extracted to the crude lead and is then finally gathered in the anode slime with a content of between 1 and 20%, the anode slime also contains silver, gold, and other valuable metals. Arsenic removal is a necessary process for recovery of the valuable metals. Hydrometallurgical methods, including acid, chloride, and alkaline leaching display the drawbacks of equipment corrosion, poor selectivity, and arsenic dispersion. Alkaline leaching process is a simple and reliable process in which arsenic can be dissolved to sodium arsenate in solution with a concentration of 1–50 g $L^{-1}$.

L.-Y. Chai · X.-B. Min · W.-C. Yang (✉) · Y.-M. Chen · Y. Ke · W.-Z. Zeng
School of Metallurgy and Environment, Central South University, Changsha, Hunan, China
e-mail: yang220222000@yahoo.com.cn

L.-Y. Chai
e-mail: lychai@csu.edu.cn

X.-B. Min
e-mail: mxb@csu.edu.cn

Y.-M. Chen
e-mail: chenyming_307@sina.com

Y. Ke
e-mail: keyong@csu.edu.cn

W.-Z. Zeng
e-mail: weizhizeng@foxmail.com

L.-Y. Chai
Chinese National Engineering Research Center for Control and Treatment of Heavy Metal Pollution (CNERC-CTHMP), Changsha, Hunan, China

 
L.-Y. Chai (ed.), *Arsenic Pollution Control in Nonferrous Metallurgy*,
https://doi.org/10.1007/978-981-13-6721-2_5

However, the disposal of As-rich solution remains a challenge. Problems related to arsenic pollution have attracted wide attention [1].

Cu–As-containing filter cakes from the copper smelting industry is a typical As-containing waste, with an annual production of approximately 250,000 tons in China. Usually, the filter cake from copper smelting has a low pH value and high contents of Cu, As and S. Due to the stringent environmental regulations, declining storage space and the need to recover valuable metals, the separation of metals from arsenic in filter cakes is currently of significant interest for both environment protection and resource conservation.

## 5.1 Recovery of Valuable Resources from Arsenic-Bearing Anode Slime

The clean utilization technology of lead anode slime combines selective dearsenification of arsenic-bearing materials and cascade recovery of valuable metals, leading to secure treatment and clean utilization of arsenic-containing acidic wastes from nonferrous metallurgical processes. First, with the screening and application of highly selective arsenic-seizing reagent, the dearsenification of polymetallic complex materials can be strengthened, while the dissolution of valuable metals (such as lead, antimony, bismuth, etc.) is suppressed. As a result, source trapping of arsenic can be achieved, thus avoiding its dispersion in intermediate materials and by-products while increasing recovery rates of valuable metals. Then arsenic in the concentrated solution is precipitated in the form of stable scorodite phase, resulting in harmless treatment of arsenic. In addition, efficient separation of bismuth and copper with low melting temperature metals such as lead and antimony in dearsenification residue can be achieved by electro-leaching in chloride system. Bismuth and copper can be further separated and recovered by taking advantage of their difference in hydrolysis pH, while lead, silver, and antimony in high lead-containing materials can be recovered and utilized through low-temperature oxygen-enriched bath smelting.

The chemical compositions of typical lead anode slime with the main components of Pb, As, Sb, and Bi, sampled from a company specializing in the production of lead and silver, in Chenzhou city, China are listed in Table 5.1.

**Table 5.1** The chemical compositions of lead anode slime

| No. | Pb (%) | As (%) | Sb (%) | Bi (%) | Cu (%) | Fe (%) | Au ($g\ t^{-1}$) | Ag (%) |
|---|---|---|---|---|---|---|---|---|
| Sample A | 14.51 | 4.21 | 42.44 | 7.97 | 0.96 | – | 36 | 4.95 |
| Sample B | 14.73 | 5.65 | 43.14 | 4.68 | – | 3.04 | 38 | 3.19 |
| Sample C | 16.74 | 7.29 | 47.16 | 8.69 | 3.36 | – | 36 | 6.08 |
| Sample D | 17.39 | 7.81 | 45.43 | 8.32 | 3.71 | – | 44 | 6.30 |

### 5.1.1 Atmospheric Normal Pressure Alkaline Leaching Pretreatment

Effect of dearsenification reagent dosage on the leaching rates of As, Sb, and Pb is shown in Fig. 5.1a (air flow rate of 160 L $h^{-1}$, temperature of 80 °C, liquid-to-solid ratio of 5:1, original anode slime size, and leaching time of 2 h). In order to decrease the reagent consumption and improve the arsenic leaching efficiency, the optimal dearsenification reagent dosage was determined to be 1.4 times of theoretical dosage.

Effect of air flow rate on the leaching rates of As, Sb, and Pb is shown in Fig. 5.1b ($W_{dearsenification\_reagent}/W_{Theory}$ = 1.4, temperature of 80 °C, liquid-to-solid ratio of 5:1, original anode slime size, and leaching time of 2 h). The arsenic leaching rate increases with the increase in air flow rate. Considering the arsenic leaching rate and power consumption, the optimal air flow rate was determined to be 100 L $h^{-1}$.

Effect of liquid-to-solid ratio on the leaching rates of As, Sb, and Pb is shown in Fig. 5.1c ($W_{dearsenification\_reagent}/W_{Theory}$ = 1.4, air flow rate of 100 L $h^{-1}$, temperature of 80 °C, original anode slime size, and leaching time of 2 h). Taking the arsenic leaching rate into account, the utilization rate of equipment and alkaline solution concentration, the optimal liquid-to-solid ratio was determined to be 5:1.

Temperature (Fig. 5.2a, $W_{dearsenification\_reagent}/W_{Theory}$ = 1.4, air flow rate of 100 L $h^{-1}$, liquid-to-solid ratio of 5:1, original anode slime size, and leaching time of 2 h): the higher the temperature results in the faster the leaching reaction rate, and the greater the solubility of sodium arsenate. Based on the energy consumption and arsenic leaching rate, the optimal leaching temperature was confirmed at 80 °C.

Particle size (Fig. 5.2b, $W_{dearsenification\_reagent}/W_{Theory}$ = 1.4, air flow rate of 100 L $h^{-1}$, liquid-to-solid ratio of 5:1, temperature of 80 °C, and leaching time of 2 h): the smaller the particle size of anode slime leads to the larger the interface area of leaching reaction, and the shorter the reactant diffusion route, which is beneficial to accelerating the leaching reaction. Considering both grinding cost and arsenic leaching rate, the optimal particle size of anode slime was determined to be 0.15 and 0.178 mm.

Leaching time (Fig. 5.3c, $W_{dearsenification\_reagent}/W_{Theory}$ = 1.4, air flow rate of 100 L $h^{-1}$, liquid-to-solid ratio of 5:1, temperature of 80 °C, and anode slime size of 0.15–0.178 mm): the alkaline leaching dearsenification reaction is very fast. Aim to improve work efficiency, reduce production cost, and increase arsenic leaching rate, the optimal leaching time was determined to be 2 h.

### 5.1.2 Oxygen Pressure Alkaline Leaching Pretreatment

Dearsenification reagent dosage (Fig. 5.3a, $PO_2$ = 1.0 MPa, L/S = 6:1, T = 160 °C, and t = 2 h): considering the arsenic removal rate, lead and antimony loss ratio, and

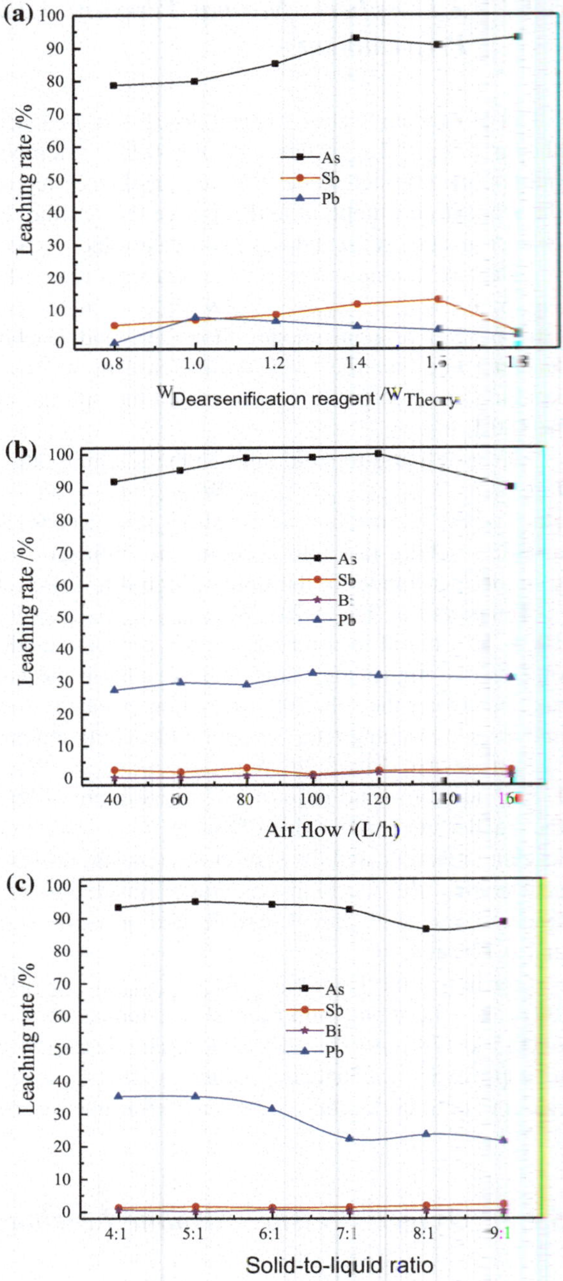


**Fig. 5.1** The effects of dearsenification reagent dosage (**a**), air flow rate (**b**), and liquid-to-solid ration (**c**) on the leaching rates of As, Sb, and Pb

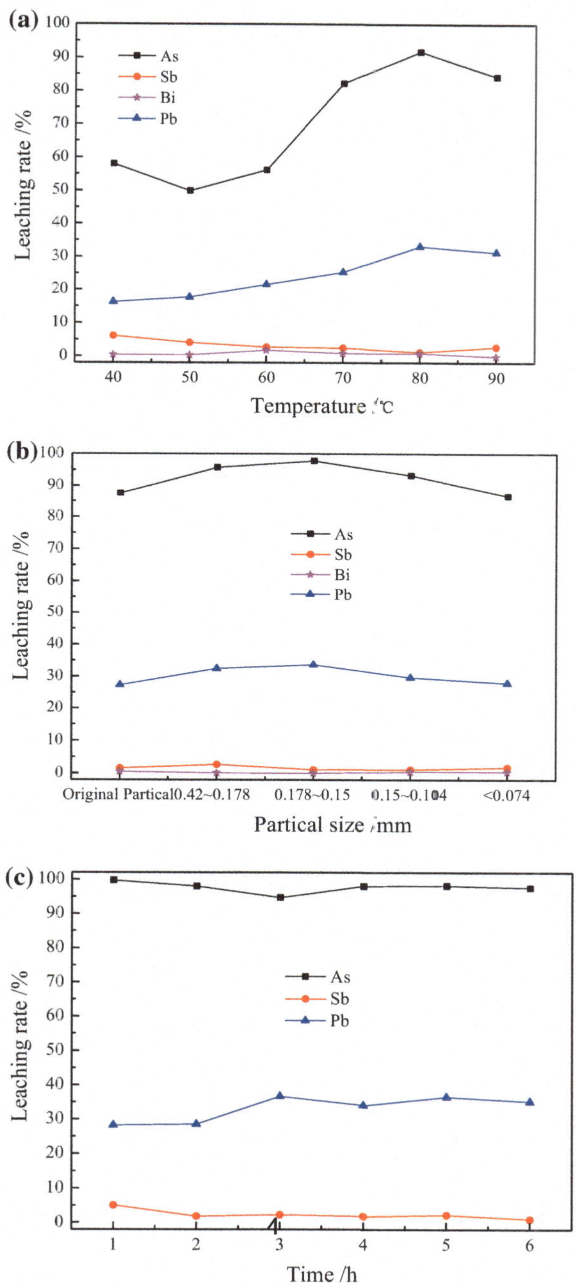


**Fig. 5.2** Effects of temperature (**a**), anode slime size (**b**), and leaching time (**c**) on the leaching rates of As, Sb, Bi, Pb

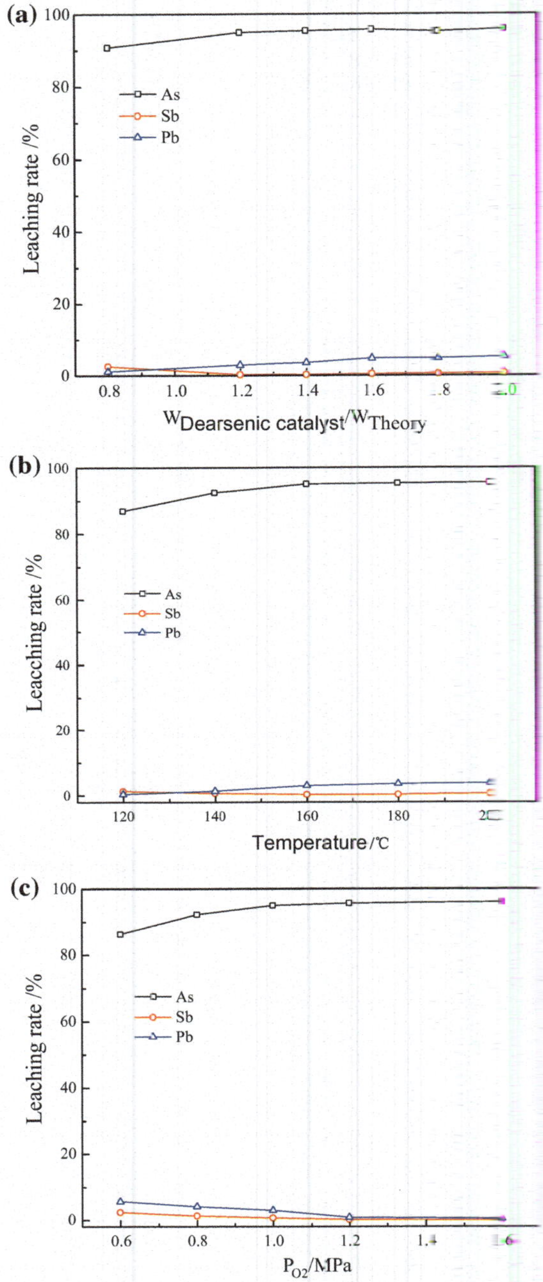


**Fig. 5.3** The effects of dearsenification reagent dosage (**a**), oxygen partial pressure (**b**), and temperature (**c**) on the leaching rates of As, Sb, and Pb

reagent consumption, the optimal dearsenification reagent dosage was determined to be 1.2 times of the theoretical amount.

Temperature (Fig. 5.3b, $W/W_{Theory} = 1.2$, $PO_2 = 1.0$ MPa, L/S = 6:1, and t = 2 h): in order to improve arsenic removal rate and reduce the loss ratio of lead and antimony, the optimal oxygen pressure alkaline leaching temperature was determined to be 160 °C.

Oxygen partial pressure (Fig. 5.3c, $W/W_{Theory} = 1.2$, L/S = 6:1, T = 180 °C, and t = 2.5 h): larger oxygen partial pressure is beneficial to increasing arsenic removal rate and reducing lead and antimony loss. Taking arsenic removal rate and the loss ratio of lead and antimony into overall consideration, the optimal oxygen partial pressure was determined to be 1–1.2 Mpa.

### *5.1.3 Oxidative Lead Precipitation of Arsenic Removal Solution*

When pretreated by alkaline leaching at atmospheric pressure, sufficient alkali concentration must be guaranteed to achieve an ideal arsenic removal rate. Nevertheless, lead originated from raw materials also dissolves in large quantities simultaneously. Thus, lead in the alkaline leaching solution needs to be removed and recovered.

Oxidative lead precipitation of alkaline leaching solution, a process that directly adds oxidizing agent into high pH alkaline solution, can oxidize $PbO_2^{2-}$ in solution to high valence $PbO_2$ precipitate. After liquid–solid separation, the solution is returned to alkaline leaching procedure for recycle use.

Initial pH (Fig. 5.4a, oxidative excess coefficient of 1.8, temperature of 30 °C, and duration of 1 h): system initial pH has profound effects on oxidative lead precipitation of arsenic removal solution. Considering lead precipitation rate and the possibility of reusing yield solution, the optimal pH for oxidative lead precipitation was set at 14.

Oxidative agent amount (Fig. 5.4b, initial pH of 14, temperature of 30 °C, and duration time of 1 h): taking lead precipitation rate and oxidizing agent consumption into account, the best oxidative excess coefficient was chosen at 1.8.

Temperature (Fig. 5.4c, oxidative excess coefficient of 1.8, initial pH of 14, and duration of 1 h): Considering aspects such as lead precipitation rate and energy consumption, the optimal reacting temperature was determined to be 30 °C.

Duration time (Fig. 5.4d, oxidative excess coefficient of 1.8, initial pH of 14, and temperature of 30 °C): with overall consideration of lead precipitation rate and productivity, the optimal oxidative lead precipitation duration was determined to be 60–90 min.

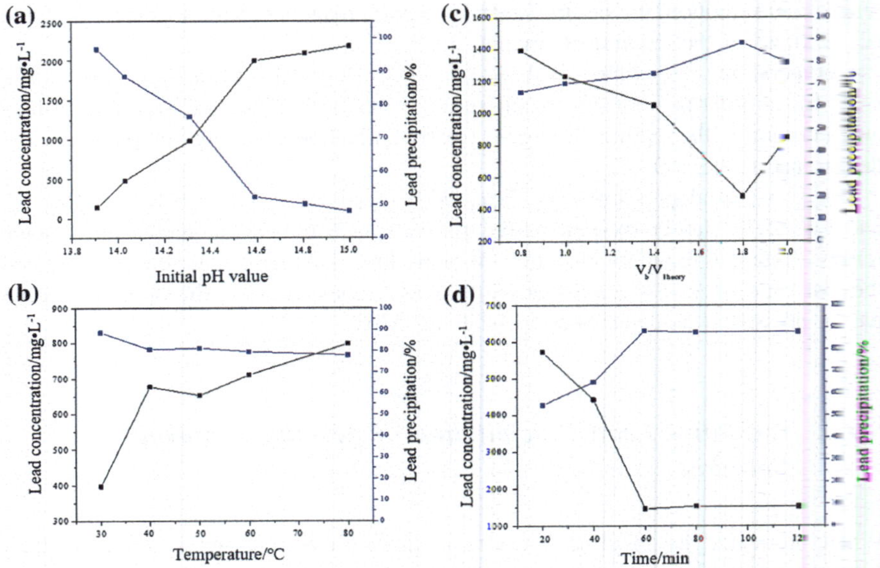

**Fig. 5.4** The effects of system initial pH (**a**), oxidative excess coefficient (**b**), temperature (**c**) and reaction time (**d**) on oxidative lead precipitation process

## *5.1.4 Process Effect*

(1) Atmospheric normal pressure alkaline leaching pretreatment

The optimal conditions of selectively dearsenification process are determined as follows: $W_{reagent}/W_{thoery} = 1.4$ (W—dosage of dearsenification reagent), air flow of 100 L $h^{-1}$, the liquid-to-solid ratio of 5:1, the temperature of 80 °C, the granularity of anode slime of 0.15–0.178 mm, and the retention time of 2 h. The chemical components of leachate and residue in five comprehensive expanded experiments under the above conditions are shown in Table 5.2 and Table 5.3, respectively. Arsenic can be removed efficiently in the selective leaching dearsenification process with the arsenic content in the leaching residue less than 1%. Nevertheless, due to apparent difference of arsenic valence in the collected anode slimes from different batches, the results of five expanded experiments fluctuate obviously. The arsenic

**Table 5.2** Chemical composition of leachate in comprehensive expanded experiments (g $L^{-1}$)

| No. | Pb | As | Sb | Bi | Cu |
|---|---|---|---|---|---|
| ZS-1 | 5.79 | 4.67 | 0.70 | 0.26 | 0.3[illegible] |
| ZS-2 | 4.52 | 3.74 | 0.70 | 0.14 | 0.4[illegible] |
| ZS-3 | 6.25 | 3.98 | 0.53 | 0.19 | 0.6[illegible] |
| ZS-4 | 5.12 | 4.49 | 0.69 | 0.17 | 0.5[illegible] |
| ZS-5 | 5.01 | 3.97 | 1.12 | 0.24 | 0.2[illegible] |

Table 5.3 Chemical composition of slag in comprehensive expanded experiments (%)

| No. | Pb | As | Sb | Bi | Cu | Au/(g $t^{-1}$) | Ag (g $t^{-1}$) |
|---|---|---|---|---|---|---|---|
| ZS-1 | 9.18 | 0.85 | 32.95 | 9.27 | 0.75 | 33 | 5.99 |
| ZS-2 | 9.28 | 0.89 | 32.94 | 9.82 | 0.67 | 42 | 5.94 |
| ZS-3 | 8.85 | 0.25 | 34.41 | 8.93 | 0.66 | 39 | 6.01 |
| ZS-4 | 9.64 | 0.84 | 31.37 | 8.86 | 0.61 | 24 | 6.30 |
| ZS-5 | 9.10 | 0.54 | 36.39 | 8.66 | 0.58 | 48 | 6.79 |

Table 5.4 Leaching rate of valuable metals in comprehensive expanded experiments (%)

| No. | Pb | As | Sb | Bi | Cu |
|---|---|---|---|---|---|
| ZS-1 | 19.01 | 67.16 | 0.78 | 1.03 | 15.16 |
| ZS-2 | 17.96 | 65.39 | 0.95 | 0.64 | 24.37 |
| ZS-3 | 21.64 | 86.16 | 0.60 | 0.82 | 28.42 |
| ZS-4 | 18.02 | 68.86 | 0.90 | 0.79 | 28.58 |
| ZS-5 | 20.26 | 77.24 | 1.40 | 1.26 | 38.30 |
| Average | 19.38 | 72.96 | 0.93 | 0.91 | 26.97 |

contents of leaching slag vary from 0.2 to 0.9%. The highest arsenic removal rate reaches 86.16%, while the lowest arsenic removal rate is 65.39%, and the average arsenic removal rate is 72.96%. In addition, poor arsenic in raw material results in the arsenic concentration in leachate being only 4 g $L^{-1}$, which is harmful to subsequent crystallization process. Thus, it is necessary to leach raw material cyclically to improve the arsenic concentration in leachate. Table 5.4 illustrates that in alkaline leaching dearsenification process, most of antimony, bismuth, silver, and gold remained in leaching slag, while their distribution rates in leachate are less than 1%. Meanwhile, 20% of lead in raw material dissolves into leachate, in which lead concentration is 9 g $L^{-1}$ approximately; thus, it is ought to set up a leachate lead removal process.

(2) Oxygen pressure alkaline leaching pretreatment

Figure 5.5 shows the experimental apparatus for tertiary oxygen pressure alkali leaching pretreatment, which uses large amount of arsenic and lead as raw materials.

The process parameters are listed in Table 5.5.

The volume and chemical composition of leachate in comprehensive expanded experiments are listed in Table 5.6. Table 5.7 shows the mass and chemical composition of the leaching slag, and Table 5.8 shows the calculated metal leaching rate in oxygen pressure alkali leaching pretreatment.

Under the condition that $W_{actual}/W_{thoery}$ = 1.8 (W—dosage of dearsenification reagent), the liquid-to-solid ratio of 10:1, the oxygen partial pressure of 1–1.2 MPa, the temperature of 180–200 °C, the retention time of 3 h, arsenic removal rate is 95.52%, and arsenic content in leaching slag is only 0.49%, while lead is hardly leached. When $W_{actual}/W_{thoery}$ declines to 1.5, arsenic removal rate remains

**Fig. 5.5** Oxygen pressure alkaline leaching dearsenification pilot-test workshop and autoclave

97.72%, and arsenic content in leaching slag is only 0.29%. When $W_{actual}/W_{theory}$ declines to 1.2, at the same time the liquid-to-solid ratio changes to 7:1 and the oxygen partial pressure changes to 0.8 MPa, arsenic removal rate declines to 91.47%, and arsenic content in leaching slag increase to 2.21%. In conclusion, proper alkali concentration and high oxygen partial pressure are the requirements for removing arsenic deeply and reducing lead loss.

During the pretreatment of oxygen pressure alkali leaching, bismuth, copper, gold, silver, and tellurium remain in the leaching slag, while selenium is completely dissolved; thus, tellurium and selenium can be effectively separated.

(3) Oxidative lead precipitation of arsenic removal solution

The optimal technological conditions are that oxidant excess coefficient of 1.8, initial pH of 14.0, temperature of 30 °C, and retention time of 1–1.5 h. The simulated solution, with 4.0634 (g $L^{-1}$) lead and 0.1936 g $L^{-1}$ arsenic, consumes 800 mL alkaline leaching solution at one time under the above conditions. The results of comprehensive expanded test of arsenic alkaline leachate oxidative lead precipitation are listed in Table 5.9.

The oxidative lead precipitation in comprehensive expanded tests performs well, and lead precipitation rate is higher than that of conditional experiments. The lead precipitation rate of three lead precipitation comprehensive expanded experiments were 88.98%, 89.21%, and 88.46% respectively, and the average is 88.88%. The XRD pattern of dried lead precipitation slag is shown in Fig. 5.6. It demonstrates that lead precipitation slag contains lead oxide such as $PbO_2$, $Pb_3O_4$, PbO, and sulfate such as $PbSO_4$ and $PbSO_3$.

High-arsenic materials such as lead anode slime were treated in processes as follows: "intensified arsenic removal and trap with selective arsenic removal reagent", "efficient grouping extraction by chloride system potential controlled leaching", and "low-temperature enriched-oxygen bath smelting recovering silver, lead, and antimony". After these processes, the arsenic removal rate can reach over 96%, primary arsenic precipitation rate can reach over 80%, the recovery rate of

**Table 5.5** Process parameters of oxygen pressure alkaline leaching pretreatment comprehensive expanded experiments

| No. | Feeding amount/kg | Dosage of arsenic removal reagent/kg | Liquid-to-solid ratio | Temperature/°C | Time/h | Oxygen partial pressure/MPa | Consumption of oxygen/kg |
|---|---|---|---|---|---|---|---|
| YGS-1 | 60 | 53.60 | 10:1 | 180–200 | 3 | 1–1.2 | 10.25 |
| YGS-2 | 62 | 45.00 | 10:1 | 180–200 | 3 | 1–1.2 | 8.00 |
| YGS-3 | 104 | 60.50 | 7:1 | 180–200 | 3 | 0.8 | 9.65 |

**Table 5.6** Volume and chemical composition of leachate in oxygen pressure alkaline leaching pretreatment comprehensive expanded experiments

| No. | Volume/L | Chemical composition/% | | | | | | | | |
|---|---|---|---|---|---|---|---|---|---|---|
| | | Pb | As | Sb | Bi | Cu | Au | Ag | Se | Te |
| YGS-1 | 620 | 0.015 | 10.46 | 0.108 | Trace | Trace | Trace | Trace | 0.041[illegible] | Trace |
| YGS-2 | 620 | Trace | 10.98 | 0.007 | Trace | Trace | Trace | Trace | 0.0433 | Trace |
| YGS-3 | 750 | 0.04 | 14.25 | 0.066 | Trace | Trace | Trace | Trace | 0.0583 | Trace |

**Table 5.7** Mass and chemical composition of leaching slag in oxygen pressure alkaline leaching pretreatment comprehensive expanded experiments

| No. | Mass/g | Chemical composition/% | | | | | | | | |
|---|---|---|---|---|---|---|---|---|---|---|
| | | Pb | As | Sb | Bi | Cu | Au/($g\ t^{-1}$) | Ag | Se | Te |
| YGS-1 | 54762.20 | 13.35 | 0.49 | 30.87 | 19.46 | 2.27 | 82 | 6.6486 | 0.0028 | 0.3[illegible]7 |
| YGS-2 | 60774.37 | 12.48 | 0.29 | 29.39 | 18.00 | 2.11 | 76 | 6.1316 | Trace | 0.29[illegible]6 |
| YGS-3 | 103452.53 | 12.64 | 0.95 | 27.97 | 17.86 | 2.05 | 76 | 5.9212 | Trace | 0.29[illegible]9 |

**Table 5.8** Computational metal leaching rate in oxygen pressure alkaline leaching pretreatment comprehensive expanded experiments/%

| No. | Pb | As | Sb | Se | Te |
|---|---|---|---|---|---|
| YGS-1 | 0.09 | 95.52 | 0.45 | 94.39 | Trace |
| YGS-2 | Trace | 97.72 | 0.03 | 100 | Trace |
| YGS-3 | 2.21 | 91.47 | 0.18 | 100 | Trace |

**Table 5.9** Results of arsenic alkaline leachate oxidative lead precipitation comprehensive expanded experiments

| No. | Leachate volume and compositions | | | | Lead precipitation rate/% | Slag mass |
|---|---|---|---|---|---|---|
| | V/mL | Pb/($g\ L^{-1}$) | As/($g\ L^{-1}$) | pH | | W/g |
| ZH-N-1 | 978 | 0.3663 | 0.1659 | 14.02 | 88.98 | 6.1[illegible]04 |
| ZH-N-2 | 982 | 0.3572 | 0.1601 | 14.03 | 89.21 | 6.0[illegible]75 |
| ZH-N-3 | 980 | 0.3828 | 0.1646 | 14.04 | 88.46 | 5.7[illegible]74 |
| Average | 980 | 0.3686 | 0.1635 | 14.03 | 88.88 | 6.0[illegible]84 |

silver and gold can be increased from 98.5% to above 99%, the recovery rate of bismuth can be increased from below 30% to above 95%, and the transfer rate of antimony to antimony oxide dust can reach over 95%. The grade of antimony oxide is more than 75%. Additionally, a 5000 t $a^{-1}$ high-arsenic polymetallic materials treatment production line has been established. This process has a potential application for dearsenification from anode slime in the nonferrous smelters.

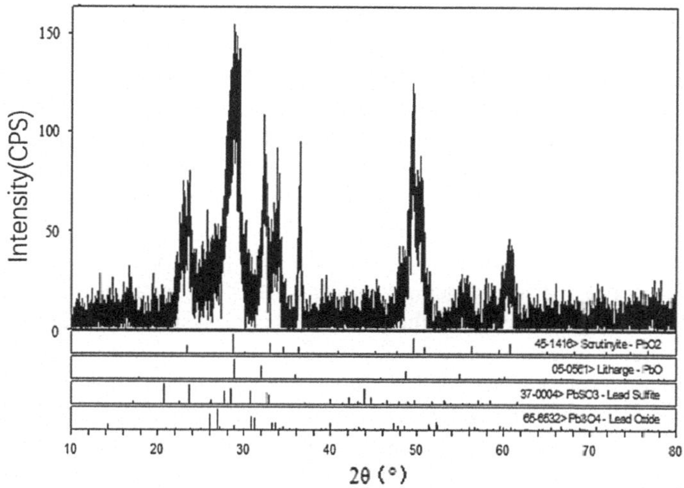


**Fig. 5.6** XRD pattern of lead precipitation slag

## 5.2 Separation and Recovery of Copper from Arsenic in Arsenic-Bearing Materials

Until now, the separation methods employed in As-bearing wastes mainly involved roasting and leaching [2–5]. Clearly, roasting As-bearing wastes is impossible or even useless because of its low separation efficiency, serious secondary pollution, and expensive cleaning facilities [6, 7]. Thus, leaching is the most suitable alternative for separating valuable metals from As-bearing wastes [8, 9]. In this regard, although the alkali extraction method can effectively separate copper and arsenic [10–12], a large amount of alkali is consumed in this process due to the acidity of the Cu–As-containing filter cakes. In acid leaching, both valuable metals and arsenic dissolve into solution [13]. And then the arsenic-containing leachate generated should be treated with calcium salt or ferric salt to form calcium arsenate or ferric arsenate to recover the valuable metals [14, 15].

Metathesis reactions between sulfides and copper salt with the formation of a CuS precipitate were observed in the Nikkelverk process for separating nickel and copper [16] and in the study of Lundström et al. [17]. Inspired by these findings, we proposed a $Cu^{2+}$-assisted acid leaching method for the separation of Cu and As in Cu–As-containing filter cakes. The reaction between $As_2S_3$ and $Cu^{2+}$ are shown in Eq. (5.1).

$$3Cu^{2+} + As_2S_3 + 6H_2O = 3CuS + 2H_3AsO_3 + 6H^+ \tag{5.1}$$

According to Eq. (5.1), insoluble CuS and soluble $H_3AsO_3$ are generated as copper displaces arsenic in form of $As_2S_3$. Thus, the separation of copper and arsenic can be achieved while the CuS-bearing residue can be sent back to copper smelting.

Characteristic of arsenic speciation and investigation of arsenic mobility in the solid phase–water system for a number of secondary arsenic minerals is critical. The ability of secondary arsenic minerals to immobilize arsenic and control its dissolved concentrations depends on the solubility of these phases, which is highly variable.

This chapter presents the separation of Cu and As in Cu–As-containing filter cakes by $Cu^{2+}$-assisted acid leaching process [18]. We analyzed the Eh-pH diagrams and leaching chemistry of the Cu–As–S–$H_2O$ system first, and then investigated the effects of pH, dosage of $Cu^{2+}$, temperature, liquid-to-solid (L/S) ratio, and leaching time on the leaching procedure. The optimal leaching conditions were achieved when the original pH value was used with the $CuCl_2{\cdot}2H_2O$ dosage of 99 mmol $L^{-1}$, the leaching temperature of 85 °C, the L/S ratio of 10:1 and the leaching time of 6 h. The leaching efficiencies were 97.4% for As and 0.1% for Cu, with final As and Cu concentrations of 17.99 g $L^{-1}$ and 0.02 g $L^{-1}$ in the leaching solution, respectively. The Cu content in the leach residue increased to 33.05% with CuS as the major Cu phase. Notably, only 0.93% of the arsenic remained in the leach residue. The results confirm this process is a viable method for the effective separation of Cu and As in Cu–As-containing filter cakes.

### 5.2.1 Thermodynamic Considerations

(1) Eh-pH diagrams of the Cu–As–S–$H_2O$ system

The Eh-pH diagrams of the Cu–As–S–$H_2O$ system at 25 and 85 °C are presented in Fig. 5.7. The area surrounded by the blue dashed lines is the region possessing a high prevalence of $H_3AsO_3$. The region surrounded by the red dashed lines is the stability area of CuS, whereas the area surrounded by the green dashed lines is the region with a high prevalence of $Cu_2S$. Overall, it should be straightforward to separate copper and arsenic under alkaline conditions. However, as an acidic type of residue, the filter cake requires large amounts of alkali when it is disposed via alkaline leaching. Thus, it is more cost-effective to separate Cu and As in the filter cake under acidic conditions. The major area of interest for Cu and As separation under acidic conditions is that marked with the shadow in Fig. 5.7b, which encompasses the pH value ranges from −1 to 8.724 and the oxidation-reduction potential ranges from −0.404 to 0.486. Adjusting the redox potential and the solution pH value can achieve separation.

(2) Leaching chemistry

The filter cake contains a high amount of S, As, and Cu. Only 10.5% As in the filter cake exists as sulfides, while 79.9% of As is oxides, which is attributed to the

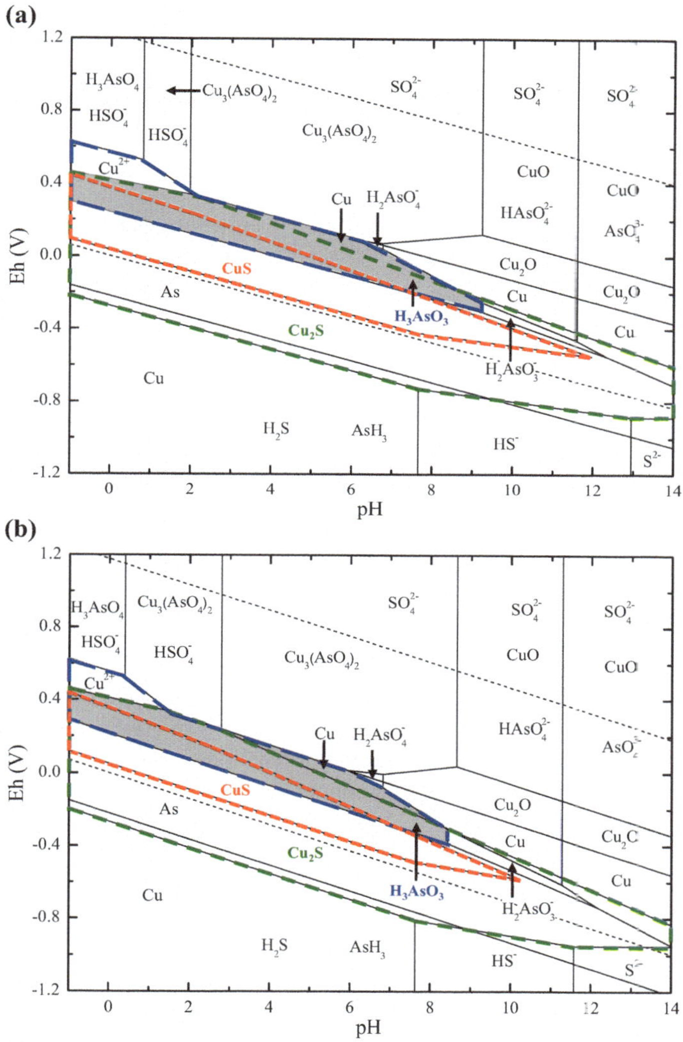


**Fig. 5.7** Eh-pH diagrams of Cu–As–S–$H_2O$ system at **a** 25 °C and **b** 85 °C, drawn using the HSC Chemistry Software Version 6.0 (Outokumpu Research Oy, Finland). The ion concentration of the system was set at 0.5 mol $L^{-1}$. Reprinted from Ref. [18] Copyright 2017, with permission from Elsevier

oxidation of arsenic sulfide during the storage period or drying process. Cu in the filter cake exists primarily as sulfide (55.7%) and soluble copper (42.3%). The XRD analysis shows $S_8$ is the only observable mineral phase in this filter cake. Arsenic in

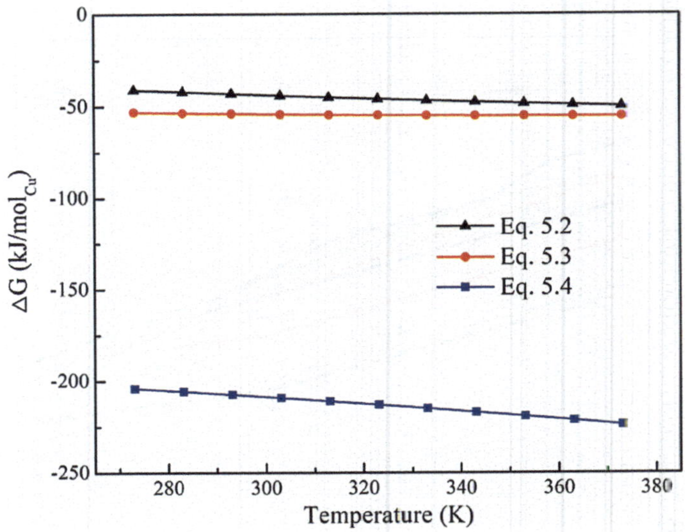


**Fig. 5.8** Standard Gibbs free energy changes versus temperature for Eqs. 5.2–5.4. Thermodynamic data were obtained from HSC Software Version 6.0 (Outokumpu Research Oy, Finland). Reprinted from Ref. [18] Copyright 2017, with permission from Elsevier

its oxide state is soluble in acidic media, whereas $As_2S_3$ and CuS are insoluble. To separate Cu and As, it is vital to enhance the leaching of $As_2S_3$ and to precipitate soluble copper as CuS. To this end, the major reactions involved in the leaching of $As_2S_3$ and the precipitation of soluble copper are

$$3Cu^{2+} + As_2S_3 + 6H_2O = 3CuS + 2H_3AsO_3 + 6H^+ \quad (5.2)$$

$$3Cu^{2+} + 4S + 4H_2O = 3CuS + SO_4{}^{2-} + 8H^+ \quad (5.3)$$

$$Cu^{2+} + S^{2-} = CuS \quad (5.4)$$

On the basis of thermodynamic calculations, the changes of the standard Gibbs free energy (ΔG) for Eqs. 5.2–5.4 over the temperature range of 273–373 K are shown in Fig. 5.8. The negative ΔG for reactions Eqs. 5.2–5.4 indicates that the precipitation or transformation of soluble Cu is thermodynamically feasible. Because the ΔG of Eqs. 5.3 and 5.4 are more negative than that of Eq. 5.2, the reactions given by them have priority over Eq. 5.2. Thus, the reaction depicted in Eq. 5.2 hardly occurs when the amount of soluble copper is insufficient and implies that the occurrence of this reaction can be controlled by the addition of $Cu^{2+}$.

### 5.2.2 *Effect of Various Parameters on the Leaching Efficiency of Copper and Arsenic*

(1) pH value

The pH value has a relatively small effect on the leaching of Cu and As. The leaching efficiencies of Cu and As vary slightly as the pH value increases (Fig. 5.9). For a pH value of 1, the leaching efficiencies of As and Cu are 97.1% and 0.1%, respectively, yielding As and Cu concentrations in the filtrate of 17.68 g $L^{-1}$ and 0.02 g $L^{-1}$, respectively. The leaching efficiency and As concentration decrease slowly as the pH value rises and then increase when the pH exceeds 7. In contrast, the leaching behavior of Cu is completely opposite to that of As. Thus, the separation of Cu and As in filter cakes can be realized under both basic condition and acidic conditions. Because the original pH value of As–Cu-containing filter cakes is 1.5, the separation of Cu and As is suggested to be applicable without pH value adjustment to reduce costs.

(2) Dosage of $CuCl_2{\cdot}2H_2O$

Based on the leaching analysis, the addition of $Cu^{2+}$ is one of the most significant factors in the process. Hence, the dosage of $CuCl_2{\cdot}2H_2O$ was investigated (Fig. 5.10). Without the addition of extra $Cu^{2+}$, the leaching efficiencies of As and Cu are 89.1% and 0.2%, respectively. Because $As_2S_3$ accounts for 10.5% of the total As content, it can be inferred that $As_2S_3$ in the filter cake is scarcely leached

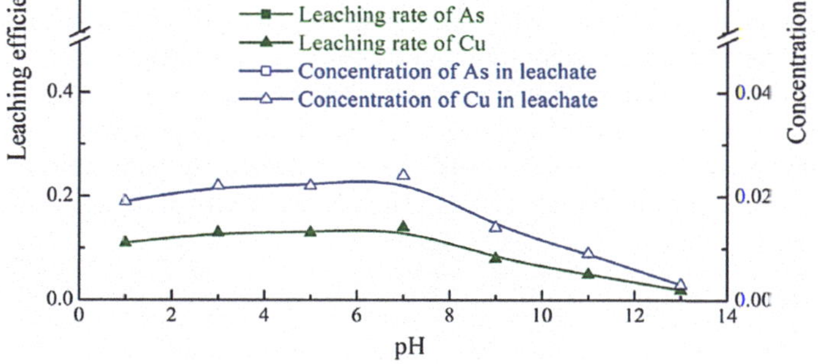


**Fig. 5.9** Effect of pH on the filter cake leaching ($CuCl_2{\cdot}2H_2O$ dosage of 99 mmol $L^{-1}$, leaching temperature of 85 °C, liquid-to-solid ratio of 10:1, and leaching time of 6 h). Reprinted from Ref. [18] Copyright 2017, with permission from Elsevier

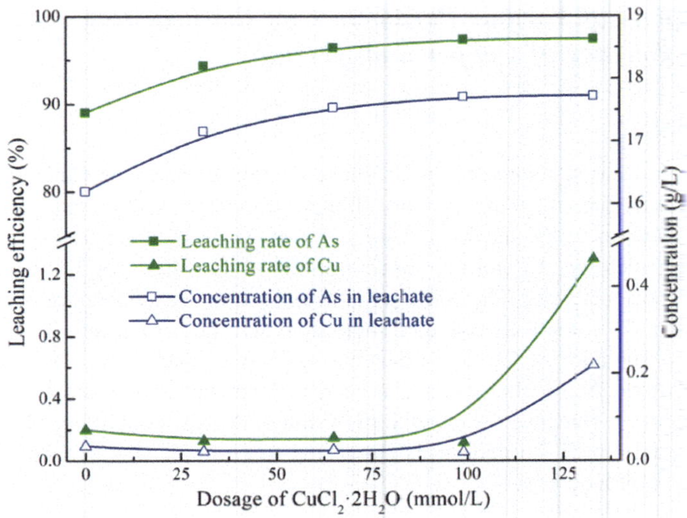


**Fig. 5.10** Effect of $CuCl_2 \cdot 2H_2O$ dosage on the filter cake leaching (leaching temperature of 85 °C, liquid-to-solid ratio of 10:1, and leaching time of 6 h). Reprinted from Ref. [18] Copyright 2017, with permission from Elsevier

when no extra $Cu^{2+}$ is added. The leaching efficiency of As increases gradually as the $Cu^{2+}$ concentration increases. When the dosage of $CuCl_2 \cdot 2H_2O$ increases from 99 to 133 mmol $L^{-1}$, the leaching efficiency of Cu improves from 0.1 to 1.3%, indicating an excess of $Cu^{2+}$. Hence, the optimal $CuCl_2 \cdot 2H_2O$ dosage is determined to be 99 mmol $L^{-1}$.

(3) Leaching temperature

Temperature affects both the prevalence of leaching and the solubility of $As_2O_3$ in acidic systems. At a temperature of 55 °C, the leaching efficiencies of As and Cu are 73.2% and 13.5%, respectively (Fig. 5.11). When the temperature increases, the leaching efficiency of As increases and the leaching efficiency of Cu decreases. This tendency of As and Cu concentrations in leachate is consistent with the results of leaching efficiency. The leaching efficiency of As increases to 97.4% and the leaching efficiency of Cu decreases to 0.2% when the temperature is 85 °C. The leaching efficiencies of As and Cu were barely affected by temperature when the temperature varied from 85 to 95 °C. Hence, the optimal temperature for the separation of Cu and As is chosen to be 85 °C.

(4) Liquid-to-solid ratio

The effect of the L/S ratio on the filter cake leaching is shown in Fig. 5.12. As and Cu concentrations are 17.02 g $L^{-1}$ and 0.16 g $L^{-1}$, respectively, with a low L/S ratio of 6:1. In the L/S ratio range from 6:1 to 10:1, the leaching efficiency of As increases, whereas that of Cu decreases as the L/S ratio increases. When the L/S

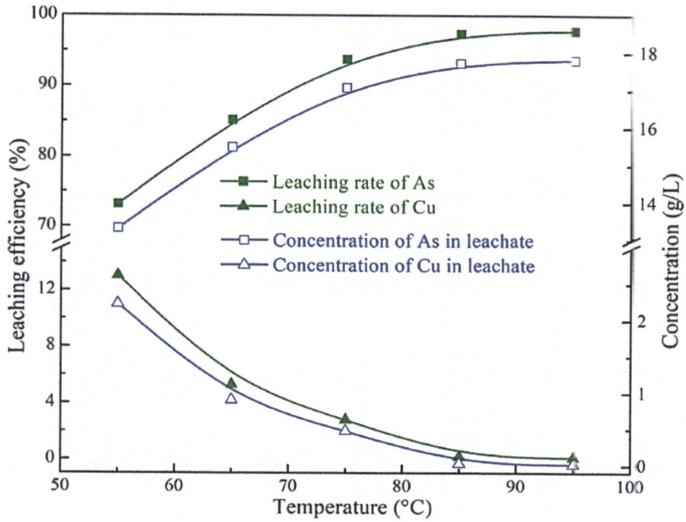


**Fig. 5.11** Effect of temperature on the filter cake leaching (liquid-to-solid ratio of 7:10 and leaching time of 6 h). Reprinted from Ref. [18] Copyright 2017, with permission from Elsevier

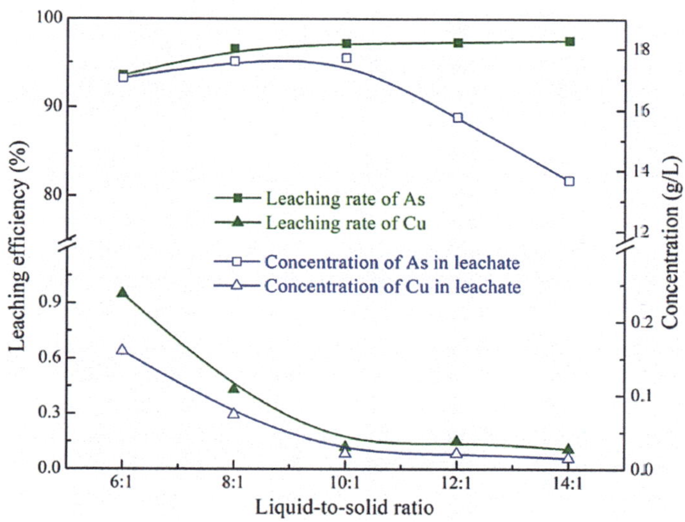


**Fig. 5.12** Effect of L/S on the filter cake leaching (leaching time of 6 h). Reprinted from Ref. [18] Copyright 2017, with permission from Elsevier

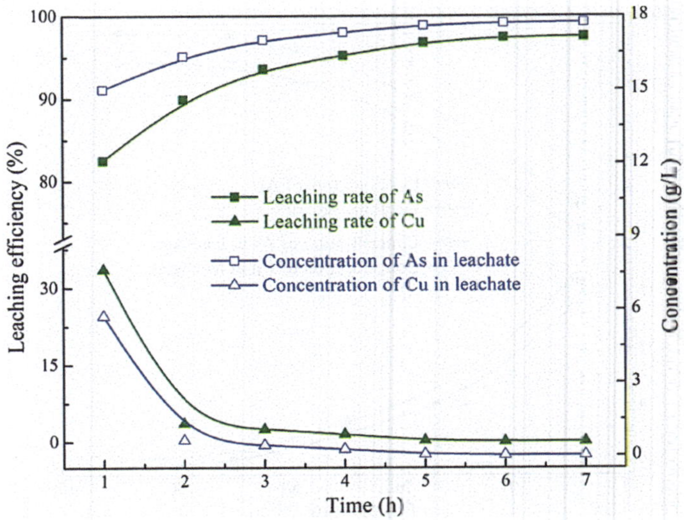


**Fig. 5.13** Effect of leaching time on the filter cake leaching. Reprinted from Ref. [18] Copyright 2017, with permission from Elsevier

ratio is 10:1, the leaching efficiencies of As and Cu are 97.2% and 0.1%, respectively. The leaching efficiencies of As and Cu barely change when L/S ratio is greater than 10:1. When L/S ratio exceeds 10:1, however, the concentration of As decreases drastically because of the dilution effect. Accordingly, the optimum L/S ratio is determined to be 10:1.

(5) Leaching time

The leaching efficiencies and concentrations of As and Cu were investigated with the leaching time varying from 1 to 7 h (Fig. 5.13). The leaching efficiencies of As and Cu are 82.6% and 33.6%, respectively, when the leaching time is 1 h. The leaching efficiency of As increases with prolonged leaching time; however, the leaching efficiency of Cu decreases with the time prolonging, which is attributed to the precipitation of $Cu^{2+}$ ions. At a leaching time of 6 h, the leaching efficiency of As increases to 97.4% and the leaching efficiency of Cu decreases to 0.2%. Moreover, the leaching solution becomes colorless and transparent. Increasing the leaching time further changes the leaching efficiency of neither Cu nor As, indicating that the leaching time of 6 h is sufficient for optimal separation.

### 5.2.3 *Characterization of the Leach Residue*

After leaching, only 0.93% of As remains in the leach residue, and the content of each As phase in the residue is reduced significantly compared with the original

**Table 5.10** Comparison of the phase contents of As and Cu in the original filter cake and leach residue

| As phase | Total As | Oxide | Sulfide | Arsenate | Others |
|---|---|---|---|---|---|
| Content in original filter cake (%) | 18.2 | 14.54 | 1.92 | 0.75 | 0.99 |
| Content in leach residue (%) | 0.93 | 0.22 | 0.21 | 0.23 | 0.27 |
| Cu phase | Total Cu | Sulfide | Soluble copper | Oxide | Others |
| Content in original filter cake (%) | 17.1 | 9.53 | 7.23 | 0.16 | 0.18 |
| Content in leach residue (%) | 33.05 | 32.38 | 0.52 | 0.09 | 0.06 |

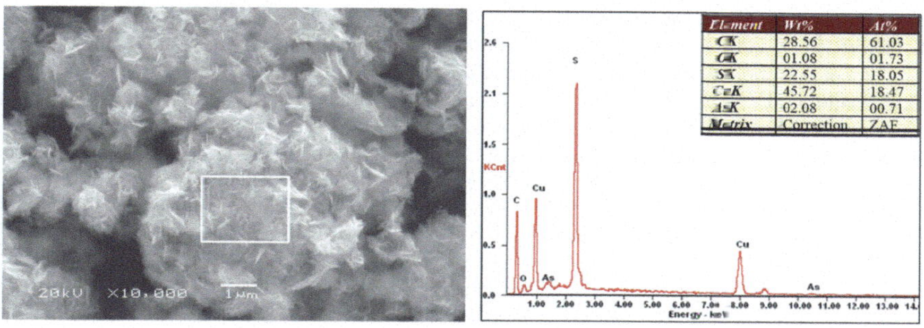


**Fig. 5.14** SEM and corresponding EDS images of the leach residues. Reprinted from Ref. [18] Copyright 2017, with permission from Elsevier

filter cake (Table 5.10). Cu content in the leach residue reaches 33.05% and Cu content in form of sulfide accounts for 32.38%, indicating that most of the Cu in the leach residue exists as CuS. The leach residue is composed of many flocculent amorphous particles (Fig. 5.14). The molar ratio of S to Cu is close to 1, which further demonstrates the conversion of Cu to CuS and proves the successful separation of Cu and As.

The $Cu^{2+}$-assisted acid leaching method developed herein successfully achieves the separation of Cu and As in a Cu–As-containing filter cake. The optimal leaching conditions include maintaining the original pH without adjustment with a $CuCl_2 \cdot 2H_2O$ dosage of 99 mmol $L^{-1}$ at a leaching temperature of 85 °C for 6 h with a L/S ratio of 10:1. Under these conditions, the leaching efficiencies of As and Cu are 97.4% and 0.1%, respectively. After leaching, Cu content in the leach residue is 33.05% and As content reduces to 0.93%. Because Cu in the leach residue mainly exists in the form of CuS, the leach residue can be returned to copper smelting for Cu recycling.

## References


1. Chen, Y., Liu, N., Ye, L., Xiong, S., Yang, S.: A cleaning process for the removal and stabilisation of arsenic from arsenic-rich lead anode slime. J. Clean. Prod. **176**, 26–35 (2018)
2. Li, Y., Liu, Z., Li, Q., et al.: Removal of arsenic from arsenate complex contained in secondary zinc oxide. Hydrometallurgy **109**(3–4), 237–244 (2011)
3. Riveros, G., Utigard, T.A.: Disposal of arsenic in copper discharge slags. J. Hazard. Mater. **77** (1–3), 241–252 (2000)
4. Shibayama, A., Takasaki, Y., William, T., et al.: Treatment of smelting residue for arsenic removal and recovery of copper using pyro-hydrometallurgical process. J. Hazard. Mater. **181** (1–3), 1016–1023 (2010)
5. Xu, Z.F., Li, Q., Nie, H.P.: Pressure leaching technique of smelter dust with high-copper and high-arsenic. Trans. Nonferrous Met. Soc. China **20**, s176–s181 (2010)
6. Dutré, V., Vandecasteele, C.: Immobilization mechanism of arsenic in waste solidified using cement and lime. Environ. Sci. Technol. **32**(18), 2782–2787 (1998)
7. Mihajlovic, I., Strbac, N., Zivkovic, Z., et al.: A potential method for arsenic removal from copper concentrates. Miner. Eng. **20**(1), 26–33 (2007)
8. Min, X.-B., Liao, Y.-P., Chai, L.-Y., et al.: Removal and stabilization of arsenic from anode slime by forming crystal scorodite. Trans. Nonferrous Met. Soc. China **25**, 1298–1306 (2015)
9. Tongamp, W., Takasaki, Y., Shibayama, A.: Arsenic removal from copper ores and concentrates through alkaline leaching in NaHS media. Hydrometallurgy **98**(3–4), 213–218 (2009)
10. Li, D., Guo, X., Xu, Z., et al.: Leaching behavior of metals from copper anode slime using an alkali fusion-leaching process. Hydrometallurgy **157**, 9–12 (2015)
11. Liu, W., Yang, T., Zhang, D., et al.: Pretreatment of copper anode slime with alkaline pressure oxidative leaching. Int. J. Miner. Process. **128**, 48–54 (2014)
12. Tongamp, W., Takasaki, Y., Shibayama, A.: Precipitation of arsenic as $Na_3AsS_4$ from $Cu_3AsS_4$-NaHS-NaOH leach solutions. Hydrometallurgy **105**(1–2), 42–46 (2010)
13. Padilla, R., Rodríguez, G., Ruiz, M.C.: Copper and arsenic dissolution from chalcopyrite-enargite concentrate by sulfidation and pressure leaching in $H_2SO_4$-$O_2$. Hydrometallurgy **100**(3–4), 152–156 (2010)
14. Leist, M., Casey, R.J., Caridi, D.: The management of arsenic wastes: problems and prospects. J. Hazard. Mater. **76**, 125–138 (2000)
15. Li, Y., Liu, Z., Li, Q., et al.: Removal of arsenic from Waelz zinc oxide using a mixed NaOH-$Na_2S$ leach. Hydrometallurgy **108**(3–4), 165–170 (2011)
16. Crundwell, F.K., Moats, M., Ramachandran, V.: Chapter 23—Hydrometallurgical Production of High-Purity Nickel and Cobalt, Extractive Metallurgy of Nickel, Cobalt and Platinum Group Metals, pp. 281–299. Elsevier, Oxford, (2011)
17. Lundström, M., Liipo, J., Taskinen, P., et al.: Copper precipitation during leaching of various copper sulfide concentrates with cupric chloride in acidic solutions. Hydrometallurgy **166**, 136–142 (2016)
18. Ke, Y., Shen, C., Min, X.-B., et al.: Separation of Cu and As in Cu-As-containing filter cakes by $Cu^{2+}$-assisted acid leaching. Hydrometallurgy **172**, 45–50 (2017)

The manufacturer's authorised representative in the EU is Springer Nature Customer Service Centre GmbH, Europaplatz 3, 69115 Heidelberg Germany. If you have any concerns regarding our products, please contact ProductSafety@springernature.com

Printed and bound by CPI Group (UK) Ltd, Croydon, CR0 4YY
24/07/2026
02175903-0003